Advances in Experimental Medicine and Biology

Volume 936

Advances in Experimental Medicine and Biology presents multidisciplinary and dynamic findings in the broad fields of experimental medicine and biology. The wide variety in topics it presents offers readers multiple perspectives on a variety of disciplines including neuroscience, microbiology, immunology, biochemistry, biomedical engineering and cancer research. Advances in Experimental Medicine and Biology has been publishing exceptional works in the field for over 30 years and is indexed in Medline, Scopus, EMBASE, BIOSIS, Biological Abstracts, CSA, Biological Sciences and Living Resources (ASFA-1), and Biological Sciences. The series also provides scientists with up to date information on emerging topics and techniques.

More information about this series at http://www.springer.com/series/5584

Katarzyna A. Rejniak
Editor

Systems Biology of Tumor Microenvironment

Quantitative Modeling and Simulations

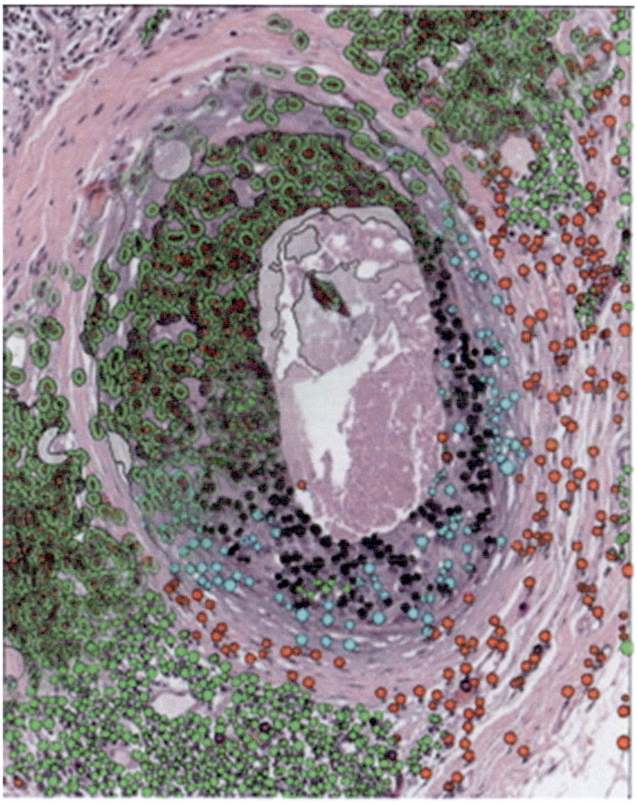

Springer

Editor
Katarzyna A. Rejniak
Integrated Mathematical Oncology Department
H. Lee Moffitt Cancer Center and Research Institute
Tampa, FL, USA

ISSN 0065-2598 ISSN 2214-8019 (electronic)
Advances in Experimental Medicine and Biology
ISBN 978-3-319-42021-9 ISBN 978-3-319-42023-3 (eBook)
DOI 10.1007/978-3-319-42023-3

Library of Congress Control Number: 2016955061

Printed on acid-free paper

This Springer imprint is published by Springer Nature
The registered company is Springer International Publishing AG Switzerland

Foreword

Despite recent advances in the development of new targeted anticancer therapies, further efforts are necessary to account for the elusive behavior of cancer cells involving tumor heterogeneity and its associated stroma of the tumor microenvironment, which are providing continuous challenges for the design of new effective anti-tumor therapies.

A new approach to understanding cancer biology and designing more effective therapies is mathematical modeling. Mathematical models are highly adaptable tools to deconvolute the complex, multidimensional datasets and make them amenable to analysis from different angles. The results from such studies may be instrumental in making this step forward.

There is a multitude of tumor and microenvironment-associated signaling molecules, which include numerous cytokines, growth factors, hormones, proteolytic enzymes such as metalloproteinases and metabolic components produced both by the tumor cells and the tumor-associated stroma [10, 18]. These components interact with the tumor cells and the stromal cells, thereby affecting tumor cell migration and invasion into nearby tissue or leading to the metastatic tumor spread into blood as circulating tumor cells, and into distant organs.

The process of invasion and migration through the extracellular matrix (ECM) is aided by numerous ECM structural components that include various fibrous elements such as collagens, fibronectin, laminin and many others. In addition, the abundance of space-filling components including glycosaminoglycans (GAGs) and attached proteoglycans (PGs) [19, 20] provide a rich microenvironment for the tumor cells to migrate through the extracellular matrix. In fact, it was shown that these structural ECM components and their increased rigidity actually promote migration of tumor cells such as glioma [12].

In addition to the ECM signaling molecules and the structural ECM elements, the tumor microenvironment also contains stromal cells such as fibroblasts and endothelial cells as well as pericytes of the angiogenic tumor vasculature. Also, cells of immune system such as mononuclear cells; monocytes and their derivative macrophages; granulocytes including neutrophils, eosinophils, basophils and mast cells, and also B and T lymphocytes are found in the tumor microenvironment. These cells can interact to the certain degree with tumor antigens and secrete various signaling molecules. It is now known that presence of these cells in the tumor vicinity would indicate an "inflamed" status of the tumor expressing Programmed Death Ligand-1 (PD-L1) and resulting in a better patient prognosis compared to "non-inflamed" tumors" [5].

The issue of quiescent tumor cell populations, often termed cancer stem cells, provides yet another challenge for designing new and effective therapeutic approaches. These quiescent cell populations frequently require a specific microenvironment, a perivascular and hypoxic niche to keep their "stemness" with few antigenic markers. As a consequence, these cells are difficult to target by any therapeutic approaches. Furthermore, variations in stem cells' behavior due to heterogeneity of the tumor microenvironment may contribute to the genetic heterogeneity of the tumor [6].

Based on the complexity of the tumor microenvironment, therapeutic agents targeting tumors must overcome a variety of hurdles like capture by multiple ECM components, leaky blood vessels within tumors, the tumor intestinal fluid pressure caused by accumulation of inflammatory components and a hypoxic environment. The role of hypoxia in the tumor microenvironment and its contribution to immune resistance and immune suppression is already well documented [9]. In addition, any targeting therapeutics would have to reach the target at a sufficient therapeutic concentration to have a therapeutic effect.

One of the examples that can be used to portray the tumor microenvironment complexity and its significance for a therapeutics delivery- is glioma, with glioblastoma (GBM) being the most advanced subtype. It is a primary brain tumor with highly invasive characteristics and short (6 months to 2 years) patient survival time (reviewed by [19]). The main ECM components of glioma, which invades brain parenchyma just within centimeters from a lesion [2] are the GAG hyaluronan (HA) and PGs such as chondroitin sulfate proteoglycans CSPGs) and heparan sulfate proteoglycans (HSPGs). All of these molecules play important roles in cell signaling and migration [16]. In addition, HA was recognized as main ECM component that forms a microenvironment in which stem cells can undergo self-renewal [4].

The HA receptor CD44 adhesion molecule, is highly expressed on the leading edge of glioma at the interface with the normal brain tissue signifying the importance of these ECM molecules in glioma invasion [13, 15, 17, 18, 20, 21]. It was found recently that HA and its CD44 receptor may play an important role in the "stemness" and survival of cancer stem cells [4]. In addition, CSPG proteoglycan known as neuroglial protein-2 (NG2) was recognized as a cell biomarker for oligodendrocyte progenitor cells and found in gliomas [11, 14], therefore emphasizing the importance of this ECM molecule.

The HSPGs components of the glioma microenvironment are also part of blood vessels and serve as a location for growth factor and cytokine storage, therefore contributing to the creation of a niche in which glioma stem cells can receive signals from its microenvironment [3]. The blood vessels and myelinated nerve fibers which serve as the infiltrative path for disseminating glioma cells have higher rigidity and together with the increased rigidity of ECM contribute to glioma migration [8, 12].

Recent therapeutic approaches to glioma and other tumors already take into account the importance of cancer stem cells and their niches [7]. In addition, detection of circulating tumor cells in blood of cancer patients, including glioma patients are viewed as "Liquid biopsies" that have the high clinical importance in tumor diagnosis and follow up [1].

Overall, the complexity of the tumor and tumor microenvironment and their multiple interactive processes could only be better understood and targeted when new analytical methods such as mathematical modeling could be applied to understand this highly complex system. This could aid in the development of new therapeutic strategies that can account for and possibly unravel some of the complex and elusive behavior of cancer.

Tampa, FL, USA Marzenna Wiranowska
April 2016

References

1. Adamczyk LA, Williams H, Frankow A, Ellis HP, Haynes HR, Perks C, Holly JM, Kurian KM (2015) Current understanding of circulating tumor cells – potential value in malignancies of the central nervous system. Front Neurol 6:174
2. Bolteus AJ, Berens ME, Pilkington GJ (2001) Migration and invasion in brain neoplasms. Curr Neurol Neurosci Rep 1(3):225–232
3. Brightman MW, Kaya M (2000) Permeable endothelium and the interstitial space of brain. Cell Mol Neurobiol 20(2):111–130
4. Chanmee T, Ontong P, Kimata K, Itano N (2015) Key roles of Hyaluronan and its CD44 receptor in the stemness and survival of cancer stem cells. Front Oncol 5:180
5. Chen L, Han X (2015) Anti-PD-1/PD-L1 therapy of human cancer: past, present, and future. J Clin Invest 125(9):3384–3391
6. Fuchs E (2016) Epithelial skin biology: three decades of developmental biology, a hundred questions answered and a thousand new ones to address. Curr Top Dev Biol 116:357–374
7. Lathia JD, Mack SC, Mulkearns-Hubert EE, Valentim CL, Rich JN (2015) Cancer stem cells in glioblastoma. Genes Dev 29(12):1203–1217
8. Lefranc F, Brotchi J, Kiss R (2005) Possible future issues in the treatment of glioblastomas: special emphasis on cell migration and the resistance of migrating glioblastoma cells to apoptosis. J Clin Oncol 23(10):2411–2422
9. Noman MZ, Hasmim M, Messai Y, Terry S, Kieda C, Janji B, Chouaib S (2015) Hypoxia: a key player in antitumor immune response. A review in the theme: cellular responses to Hypoxia. Am J Physiol Cell Physiol 309(9):C569–579
10. Rojiani MV, Wiranowska M, Rojiani AM (2011) Matrix metalloproteinases and their inhibitors friend or foe in tumor microenvironment In: Siemann DW (ed). Wiley
11. Stallcup WB, Huang FJ (2008) A role for the NG2 proteoglycan in glioma progression. Cell Adh Migr 2(3):192–201
12. Ulrich TA, de Juan Pardo EM, Kumar S (2009) The mechanical rigidity of the extracellular matrix regulates the structure, motility, and proliferation of glioma cells. Cancer Res 69(10):4167–4174
13. Wiranowska M, Ladd S, Moscinski LC, Hill B, Haller E, Mikecz K, Plaas A (2010) Modulation of hyaluronan production by CD44 positive glioma cells. Int J Cancer 127:532–542
14. Wiranowska M, Ladd S, Smith SR, Gottschall PE (2006) CD44 adhesion molecule and neuro-glial proteoglycan NG2 as invasive markers of glioma. Brain Cell Biol 35(2–3):159–172
15. Wiranowska M, Naidu AK (1994) Interferon effect on glycosaminoglycans in mouse glioma in vitro. J Neurooncol 18(1):9–17

16. Wiranowska M, Plaas A (2008) Cytokines and extracellular matrix remodeling in the central nervous system. In: Berczi I, Szentivanyi A (eds) Neuroimmune biology: cytokines and the brain. Elsevier B.V. Science
17. Wiranowska M, Rojiani AM, Gottschall PE, Moscinski LC, Johnson J, Saporta S (2000) CD44 expression and MMP-2 secretion by mouse glioma cells: effect of interferon and anti-CD44 antibody. Anticancer Res 20(6B):4301–4306
18. Wiranowska M, Rojiani AM, Rojiani MV (2015) Matrix metalloproteinases-modulating the tumor microenvironment. J Carcinog Mutagen 6:3
19. Wiranowska M, Rojiani MV (2011) Extracellular matrix microenvironment in glioma progression. In: Ghosh A (ed) Glioma/book 1-exploring its biology and practical relevance. InTech Open Access Publisher
20. Wiranowska M, Rojiani MV (2013) Glioma extracellular matrix molecules as therapeutic targets In: Wiranowska M, Vrionis FD (eds) Gliomas: symptoms, diagnosis and treatment options. Nova Science Publishers, Inc., New York
21. Wiranowska M, Tresser N, Saporta S (1998) The effect of interferon and anti-CD44 antibody on mouse glioma invasiveness in vitro. Anticancer Res 18(5A):3331–3338

Preface

The complexity and heterogeneity of tumor microenvironment, as well as its dynamic interactions with tumor cells are a very attractive topic for mathematical modeling. Several quite diverse modeling approaches have been developed over the last couple of years to address the role of the microenvironment in tumor initiation, progression and its response to treatments. In order to provide the readers both biologically- and mathematically-oriented with the recent achievements in this area, I invited several researchers to share their mathematical and computational models of tumor microenvironment and their perspectives on the future of this field.

Both normal and tumor cells are embedded into a complex and dynamically changing environment. That environment can regulate the behavior of individual cells and modulate homeostatic balance of the whole tissue. The complexity of tumor microenvironment arises from multiple players that interact with one another. Various types of cells reside in or migrate through the tumor stroma, including endothelial cells and pericytes forming the capillaries; immune cells, such as T cells, B cells, or macrophages; as well as adipocytes, fibroblast and other stromal cells. The extracellular matrix proteins (collagen, elastin, fibronectin, laminin) form fibril meshes defining their orientation, stiffness and overall physical characteristics. The interstitial fluid that penetrates space between the cells and the fibers allows for diffusion of numerous chemical factors (nutrients, oxygen, glucose, growth factors, chemokines, matrix metalloproteinases) and enable their transport to all stromal components.

At all stages of tumor development from initiation to growth and invasion, to metastasis, the tumor cells are subjected to cues and interactions from the surrounding microenvironment, and also modulate the environment in their vicinity. Additionally, when a given treatment (i.e., surgery, chemo-, radio-, immune- hormone or combination therapy) is applied, the tumor and its microenvironment may undergo significant alterations. As a result, the microenvironmental selection forces and tumor physico-chemical landscape may shift.

Due to the complexity, heterogeneity, and dynamic changes that take place in the tumor microenvironment, it is difficult to investigate experimentally, in a precise and quantitative way, all potential interactions between the tumor and its surrounding stroma. Thus, laboratory experiments are designed to address these issues at different scales of complexity. For example, genetic modifications, protein interactions, signaling pathways, cellular phenotypic functions or whole organism studies. However, integrating the results obtained from such studies into a common coherent description is as complex as the disease itself. Mathematical models and computational techniques provide researchers with invaluable tools for integration of this knowledge into organizing principles. Systems biology approaches, in turn, provide a way to discover emergent properties of cells, tissues or organs functioning as one system. Therefore, *in silico* models grounded in cancer biology and driven by experimental or clinical data can integrate knowledge across different biological scales, combine tools from various scientific fields, provide rigorous quantitative analyses, and produce testable hypotheses.

This book presents state-of-the-art mathematical and computational models addressing a broad range of tumor-microenvironment interactions from tumor initiation, to invasion to metastatic spread. A special consideration is also given to modeling tumor microenvironment under chemotactic treatment. Mathematically, these models embrace the continuous, agent-based and hybrid models, solid mechanics and fluid dynamics frameworks, optimal control theory, Monte-Carlo and final element methods.

With the advent of imaging technologies, researchers can now observe how tumor and stromal cells interact with one another, what factors they secrete and sense, and how they remodel the extracellular matrix (Fig. 1). This data can be utilized for mathematical model calibration and validation. In the first chapter of this book, Lloyd and colleagues discuss various image analysis tools for high resolution images of tumor tissue in order to identify and quantify the components of tumor microenvironment and their relationship with the tumor. This chapter provides an overview of currently available cell-scale imaging resources.

Since tumor vasculature is distorted and often not fully functional, the natural metabolic tissue milieu is also altered. In the second chapter, Martínez-González and co-authors address the importance of the hypoxic environment for glioma progression. The authors discuss the role of low oxygenation in the formation of cellular pseudopalisades both in vivo and in microfluidic devices. They also propose how such environmental conditions can be used for therapeutic interventions against brain tumors.

The third chapter, by Welter & Rieger, presents a discussion about the differences between normal and tumor vasculature; the various processes involved in tumor vasculature formation, including angiogenesis, and vessel cooption, remodeling and regression. The authors also show simulation results indicating the consequences of irregular tumor vasculature for metabolites and drug distribution into the tumor tissue.

The environment surrounding the tumors serves not only as a nutrient supply, but also provides a structural support for the tumor cells. It has been shown experimentally that tumor cells respond differently to extracellular matrices (ECM) of different composition, as well as to the local flow of the interstitial fluid.

In the fourth chapter, He and co-authors discuss the biophysical and biomechanical properties of the ECM as a crucial component in tumor cell invasion. The authors present a review of computational methods that address tumor cell-ECM interactions and ECM remodeling during the tumor

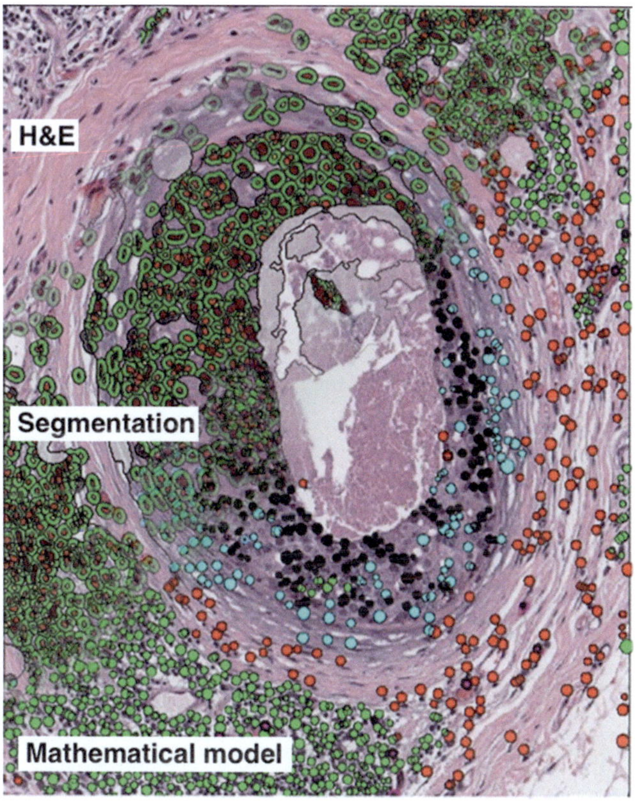

Fig. 1 A composition of three research images: a histology image in the background showing a section of breast tissue with a ductal carcinoma in situ stained with the H&E, an image of cell segmentation (green cytoplasm enclosing the red nuclei; along the diagonal) used for quantification of individual cell features, and an image of a computational agent-based model representing the same tumor tissue (black, cyan, green and red dots below the diagonal). The histology data and segmentation was provided by Mark Lloyd; the computational model by Katarzyna Rejniak; the graphics designed by Kamil Rejniak

cell migration, which is a first step in a metastatic cascade that leads to tumor spread (metastasis) to the distant organs.

In the fifth chapter, Rejniak discusses the microenvironment that the tumor cells encounter upon entering the blood or lymph circulation system. While this is a crucial step in the tumor metastatic cascade, only a small fraction of cells is able to withstand hemodynamic forces and overcome effects of blood shear in circulation. The author presents a fluid-structure interaction model to address the mechanical aspects of circulating tumor cells that allow them to survive in the intravascular fluid microenvironment.

In the sixth chapter, Lolas and co-authors discuss lymphangiogenesis, a process of the formation of new lymph vessels, and provide evidence both computational and experimental of tumor cell migration and dissemination through the lymphatic network.

The final step in the metastatic cascade is the colonization of the secondary site by tumor cells. In this new environment, tumor cells need to adapt to the local conditions or remodel the surroundings, creating metastatic niches. In the seventh chapter Kianercy & Pienta discuss a bone microenvironment as a common destination for metastasis of various types of cancer. The authors present a model of the construction of a metastatic niche in the bone, as well as complex interactions between tumor cells, and the bone niche components that are often referred to as a vicious cycle of bone metastasis.

In the eighth chapter, Pérez-Velázquez and co-authors discuss the role of microenvironmental niches and sanctuaries in the emergence of acquired drug resistance in tumor micrometastases. The authors use a model of the heterogeneous tumor microenvironment in which a small tumor cluster is exposed to a DNA damaging drug, and demonstrate that tumor cells can develop anti-drug resistance when they reside for a prolonged time in the hypoxic niches or in the pharmacological sanctuaries.

Anti-cancer treatments are usually designed to kill tumor cells or to suppress tumor cell growth. However, adding this new component to tumor milieu may have a profound effect on the tumor microenvironment itself, leading to modifications in its physico-chemical composition and to shifts in microenvironmental selection forces. In the ninth chapter, Curtis & Frieboes discuss microenvironmental barriers to a successful delivery of nanotherapeutic drugs. The authors analyze which parameters are crucial for interactions between nanodrugs and the tumor microenvironment, and provide suggestions for better trans-disciplinary collaborative efforts to overcome these microenvironmental obstructions to nanodrug transport.

In the tenth chapter, Gevertz discusses two classes of vascular-targeting drugs and explores clinically-relevant questions related to these drugs using a multi-scale hybrid model. The author addresses a question of drug redundancy and how to design optimal scheduling protocols for combination therapies.

In the eleventh chapter, Ledzewicz & Schaettler review various types of clinically relevant drug administration protocols and analyze the corresponding mathematical optimal controls designed for optimization of chemotherapeutic treatments. They also suggest treatment schedules that would be the most effective when the microenvironmental conditions are taken into account.

In the final chapter, Macklin and co-authors also present a review of mathematical models that incorporate various components of the tumor microenvironment. The authors discuss the key features that mathematical and computational models of cancer should account for in order to recapitulate biological complexity in enough detail to provide a platform for generating experimentally testable hypotheses.

I would like to thank all contributors of this book for enthusiastically supporting the concept and form of this publication, and all co-authors for their hard work and dedication. I would like to thank Marzenna Wiranowska for writing a foreword.

Tampa, FL, USA Katarzyna A. Rejniak
May 2016

Contents

Contributors

Jose M. Ayuso Aragón Institute of Engineering Research (I3A), University of Zaragoza, Zaragoza, Spain

George C. Bourantas Faculty of Science, Technology and Communication, University of Luxembourg, Luxembourg City, Luxembourg

Marilyn M. Bui Analytic Microscopy Core, Department of Anatomic Pathology, H. Lee Moffitt Cancer Center and Research Institute, Tampa, FL, USA

Gabriel F. Calvo Mathematical Oncology Laboratory (MôLAB), University of Castilla-La Mancha, Castilla-La Mancha, Spain

Louis T. Curtis Department of Bioengineering, University of Louisville, Louisville, KY, USA

Luis J. Fernández Aragón Institute of Engineering Research (I3A), University of Zaragoza, Zaragoza, Spain

Hermann B. Frieboes University of Louisville, Louisville, KY, USA

Samuel H. Friedman Lawrence J. Ellison Institute for Transformative Medicine, University of Southern California, Los Angeles, CA, USA

Ahmadreza Ghaffarizadeh Lawrence J. Ellison Institute for Transformative Medicine, University of Southern California, Los Angeles, CA, USA

Jana L. Gevertz Department of Mathematics and Statistics, The College of New Jersey, Ewing, NJ, USA

Xiuxin He Department of Mathematics and Statistics, Georgia State University, Atlanta, GA, USA

Lasse Jensen Department of Microbiology, Tumor and Cell Biology (MTC), C1, Karolinska Institute, Stockholm, Sweden

Department of Medical and Health Sciences, Linköping University, Linköping, Sweden

Yi Jiang Department of Mathematics and Statistics, Georgia State University, Atlanta, GA, USA

Joseph O. Johnson Analytic Microscopy Core, H. Lee Moffitt Cancer Center and Research Institute, Tampa, FL, USA

Edmond Jonckheere Department of Electrical Engineering, University of Southern California, Los Angeles, CA, USA

Edwin F. Juarez Lawrence J. Ellison Institute for Transformative Medicine, University of Southern California, Los Angeles, CA, USA

Department of Electrical Engineering, University of Southern California, Los Angeles, CA, USA

Aleksandra Karolak Integrated Mathematical Oncology, H. Lee Moffitt Cancer Center & Research Institute, Tampa, FL, USA

Agnieszka Kasprzak Analytic Microscopy Core, H. Lee Moffitt Cancer Center and Research Institute, Tampa, FL, USA

Ardeshir Kianercy Department of Molecular Physiology and Biophysics, Vanderbilt University, Nashville, TN, USA

Byoungkoo Lee Department of Mathematics and Statistics, Georgia State University, Atlanta, GA, USA

Urszula Ledzewicz Department of Mathematics and Statistics, Southern Illinois University Edwardsville, Edwardsville, IL, USA

Institute of Mathematics, Lodz University of Technology, Lodz, Poland

Mark C. Lloyd Analytic Microscopy Core, H. Lee Moffitt Cancer Center and Research Institute, Tampa, FL, USA

Department of Biological Sciences, University of Chicago Illinois, Chicago, IL, USA

Georgios Lolas Oncology Unit, 3rd Department of Internal Medicine, Sotiria General Hospital, Athens School of Medicine, Athens, Greece

Paul Macklin Lawrence J. Ellison Institute for Transformative Medicine, University of Southern California, Los Angeles, CA, USA

Alicia Martínez-González Mathematical Oncology Laboratory (MôLAB), University of Castilla-La Mancha, Castilla-La Mancha, Spain

Shannon M. Mumenthaler Lawrence J. Ellison Institute for Transformative Medicine, University of Southern California, Los Angeles, CA, USA

Ignacio Ochoa Aragón Institute of Engineering Research (I3A), University of Zaragoza, Zaragoza, Spain

Víctor M. Pérez-García Mathematical Oncology Laboratory (MôLAB), University of Castilla-La Mancha, Castilla-La Mancha, Spain

Judith Pérez-Velázquez Mathematical Modeling of Biological Systems, Centre for Mathematical Science, Technical University of Munich, Munich, Germany

Kenneth J. Pienta Brady Urological Institute, Johns Hopkins Hospital, Baltimore, MD, USA

Katarzyna A. Rejniak Integrated Mathematical Oncology Department, Center of Excellence in Cancer Imaging and Technology, H. Lee Moffitt Cancer Center & Research Institute, Tampa, FL, USA

Department of Oncologic Sciences, College of Medicine, University of South Florida, Tampa, FL, USA

Heiko Rieger Theoretical Physics, Saarland University, Saarbrücken, Germany

Heinz Schaettler Department of Electrical and Systems Engineering, Washington University, St. Louis, MO, USA

Jessica L. Sparks Department of Chemical, Paper, and Biomedical Engineering, Miami University, Oxford, OH, USA

Konstantinos Syrigos Oncology Unit, 3rd Department of Internal Medicine, Sotiria General Hospital, Athens School of Medicine, Athens, Greece

Vasiliki Tsikourkitoudi Oncology Unit, 3rd Department of Internal Medicine, Sotiria General Hospital, Athens School of Medicine, Athens, Greece

Michael Welter Theoretical Physics, Saarland University, Saarbrücken, Germany

Marzenna Wiranowska Department of Pathology and Cell Biology, College of Medicine University of South Florida, Tampa, FL, USA

Image Analysis of the Tumor Microenvironment

1

Mark C. Lloyd, Joseph O. Johnson, Agnieszka Kasprzak, and Marilyn M. Bui

Abstract

In the field of pathology it is clear that molecular genomics and digital imaging represent two promising future directions, and both are as relevant to the tumor microenvironment as they are to the tumor itself (Beck AH et al. Sci Transl Med 3(108):108ra113–08ra113, 2011). Digital imaging, or whole slide imaging (WSI), of glass histology slides facilitates a number of value-added competencies which were not previously possible with the traditional analog review of these slides under a microscope by a pathologist. As an important tool for investigational research, digital pathology can leverage the quantification and reproducibility offered by image analysis to add value to the pathology field. This chapter will focus on the application of image analysis to investigate the tumor microenvironment and how quantitative investigation can provide deeper insight into our understanding of the tumor to tumor microenvironment relationship.

Keywords

Image analysis • Tumor microenvironment • Whole slide imaging • Quantifiable pathology

M.C. Lloyd (✉)
Analytic Microscopy Core, H. Lee Moffitt Cancer Center and Research Institute, 12902 Magnolia Dr., Tampa, FL 33612, USA

Department of Biological Sciences, University of Chicago Illinois, 845 W. Taylor St., Chicago, IL 60607, USA
e-mail: Mlloyd8@uic.edu

J.O. Johnson • A. Kasprzak
Analytic Microscopy Core, H. Lee Moffitt Cancer Center and Research Institute, 12902 Magnolia Dr., Tampa, FL 33612, USA

M.M. Bui
Analytic Microscopy Core, H. Lee Moffitt Cancer Center and Research Institute, 12902 Magnolia Dr., Tampa, FL 33612, USA

Analytic Microscopy Core, Department of Anatomic Pathology, H. Lee Moffitt Cancer Center and Research Institute, 12902 Magnolia Dr., Tampa, FL 33612, USA
e-mail: Marilyn.Bui@moffitt.org

K.A. Rejniak (eds.), *Systems Biology of Tumor Microenvironment*, Advances in Experimental Medicine and Biology 936, DOI 10.1007/978-3-319-42023-3_1

1.1 Introduction

Before the use of image analysis investigators have been able to observe changes in the tumor microenvironment and infer how these changes may effect tumor growth, progression or the effects of specific therapies [9, 20]. The tumor microenvironment describes the non-neoplastic cells and stroma present in the tumor. These include fibroblasts, blood vessels and the immune cells [30, 41, 42]. Instances specific to breast cancer, for example, may include the number of inflammatory cells, the thickness of the basement member of ducts, the 'reactivity' of stroma or other observations [11, 22]. Unfortunately these associations are difficult to repeat reliably and therefore, have never largely been made part of a clinical standard for staging, grading or otherwise evaluating cancers. This is due, in part, to inter- and intra-observer variability [13, 19, 36].

Image analysis is a tool which may be used to extract meaningful information from a digital image [40]. Given the fact that image analysis is processed and reported by a computer, it is typically highly reproducible and objective. The results are no longer qualitative records observed by a human investigator but rather a quantified mathematical value which can be mined. Examples may include the number of objects, values or intensity of colors, or even patterns including the distribution of objects like vessels or ectopic lymph nodes [29]. These are relatively simple examples and far more multifaceted analysis are possible with modern image analysis tools including content based image retrieval, pattern recognition, computer learning and deep learning, to name a few. In fact, pathologists are often the users who train the algorithms regarding what and how to identify specific regions of interest. It is critical that pathologists are involved in this process to ensure accurate identification of each region and for the quality control of what the algorithm is classifying.

The use of image analysis technologies for a more standardized and repeatable measurement of biological processes in tissue samples has become increasingly popular [37]. Computers are consistent and dependable for quantification of samples, for example, algorithms may be used for counting, searching large numbers of records or areas [14, 27]. Examples include FDA approved algorithms for counting positively stained immunohistochemical slides (e.g. ER, PR or HER2) [23] or searching for rare events such as cytology abnormalities (e.g. Pap smear testing) [8].

While image analysis using these computer algorithms can help us quantify some aspects of a histological section they are not capable of performing the detailed and intricate diagnosis as it is rendered by a pathologist. Thus the benefit of a computer algorithm is best used in conjunction with a pathologist and this relationship can help us better understand new aspects of oncology and pathology which remain currently unknown. A prime example of leveraging image analysis for pathology diagnosis may in fact be the investigation of the tumor microenvironment and its role in understanding and treating cancer [4].

This chapter will focus on three examples of imaging modalities for interrogating the tumor microenvironment and will use specific use cases.

1.2 Imaging the Tumor Microenvironment

Image analysis may be used on various types of images to analyze the involvement of the tumor microenvironment in a specific disease. In fact, the types of images which may be acquired and studied are quite vast, including radiology, endoscopy and other imaging modalities [28, 31]. For the purposes of this chapter the authors have chosen to exclusively discuss histological sections of pathology samples. This allows the authors to use 'real-world' examples which will then result in quantifiable to design in vitro, intravital, mathematical modeling experiments [2].

Additionally, the authors have chosen to focus on the most ubiquitous and available image acquisition modalities for the investigation of histological samples including (1) brightfield microscopy including hematoxylin and eosin, as well as immunohistochemically stained samples

[14, 23]; (2) fluorescent microscopy stained samples [26]; and (3) Second Harmonic Generation [1]. Many additional image acquisition methods exist however the authors chose the rather diverse set of three methods listed above in order to demonstrate broad differences in the images available for image analysis of the tumor microenvironment as well as the types of information which may be gleaned regarding how the tumor microenvironment may be studied.

1.3 Brightfield Microscopy to Evaluate the Tumor Microenvironment

Hematoxylin and eosin stained samples are the standard for initial formulation of a pathology diagnosis. In the workflow of a pathologist these samples are the first slides reviewed before requesting additional studies, including ImmunoHistoChemistry (IHC) or fluorescence stained samples (i.e. fluorescent in situ hybridization). These type of slide preparations include the tumor as well as the tumor microenvironment. Both enable a pathologist to render an accurate diagnosis. These samples may be imaged with whole slide scanning technologies or digital cameras to create samples available for image analysis [39]. Once the brightfield microscopy images of tissue samples are acquired, many different types of investigations regarding the tumor microenvironment are plausible.

One method of the quantitative evaluation of the tumor microenvironment may include disease progression [5]. For example of the progression of DCIS to invasive breast cancer is well studied and described at a very high level. However, a deeper interrogation of the tumor microenvironment and tumor to stroma interface may provide insight into the ways in which malignant but not yet invasive disease may first break through the basement membrane [35]. This raises a number of very important questions related to the tumor microenvironment which can be derived by image analysis. For example, the thickness of the basement membrane (BM) can be measured quantitatively by using image analysis tools to segment the BM and measure the shortest distance across the membrane [44].

Many clinical samples that are prepared for brightfield imaging are also used to identify the tumor areas for molecular analysis (i.e. QRT-PCR or Next Generation Sequencing) (see Table 1.1 and Fig. 1.1). In this example the tumor and tumor microenvironment were segmented by image analysis and the area of only the tumor component was measured. The same sample was evaluated for protein content by tandem mass spectrometry and showed a much higher correlation coefficient (0.81) than the area of the entire section (not shown).

Unfortunately, it is difficult to isolate the tumor from its microenvironment which often extends into the tumor region like fingers or rivers. The ability to accurately and reliably quantify the area of tumor and/or the area of the microenvironment, as shown in Table 1.1 and Fig. 1.1, may prove to be important for accurately investigating the tumor microenvironment in pathology.

However, investigating more than the amount of each tissue type of the tumor and microenvironment in a histological sample can be accomplished using image analysis. A number of aspects of the tumor microenvironment on a slide may be quantified by measuring the distribution

Table 1.1 This table is the competent breakdown of area of each segment of both the tumor and microenvironment for a subset of specimens. The segmentation was determined by a pathologist trained image analysis algorithm

Epithelial tumour	Stroma	Inflammation	Necrosis	Normal	Adenoma
33.52	30.56	13.24	0.00	0.00	0.00
76.85	21.24	0.00	1.14	0.00	0.00
54.60	25.21	8.42	2.76	0.00	0.00
8.66	26.70	4.13	0.00	54.84	5.67

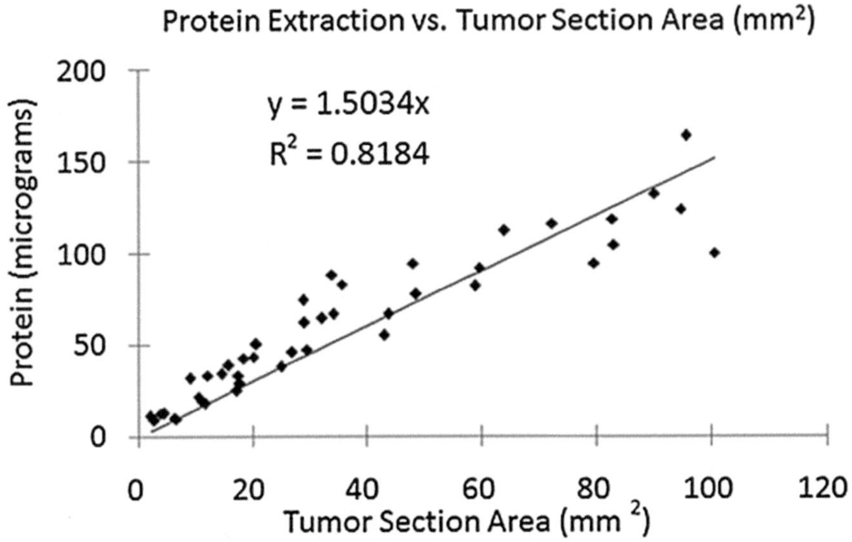

Fig. 1.1 This figure is the complete sample set of colon cancer cases plotting the tumor section area (as determined in the Table 1.1) against the micrograms of protein extracted by MS/MS. This shows a correlation between the target protein and the actual tumor content of the sample and can be compared to the same protein measurement for a whole specimen (not shown)

of areas in the tumor and microenvironment (e.g. heterogeneity) as depicted in Fig. 1.2 [17].

Additionally, we can use image analysis to measure other important aspects of the tumor microenvironment. Examples include, the use of pattern recognition image analysis to quantify tumor necrosis in tumor sections [24] and the combined use of digital pathology, image analysis and immunoscore metrics to measure the immunological response to a disease (see Fig. 1.3) [3, 29].

Finally, brightfield images provide features within the H&E such as distribution of histocytes or arrangement of stromal fibers, as well as IHC stained features including the number of inflammatory cells (i.e. CD8), number of vessels (i.e. CD31 or CD34), metabolic features such as hypoxia stained by HIF1a, or glucose use or transport by GLUT1 staining, . Together, these brightfield stain applications when coupled with image analysis provide insight into the evaluation of the tumor microenvironment and may in turn be extremely valuable for diagnosis, prognosis and prediction of therapy [15, 18, 38].

1.4 Fluorescent Microscopy to Evaluate the Tumor Microenvironment

In addition to the standard diagnostic brightfield stains, samples may be stained with fluorescent markers that identify specific tumor characteristics. This type sample prep requires different instruments to excite and capture the emission of fluorescence markers such as fluorescent, confocal or super-resolution microscopes. While these instruments represent broad categories of technologies, they each provide unique data to interrogate the tumor microenvironment. Key benefits of fluorescent microcopy include label specificity and the ability to multiplex multiple labels [6].

For example, while it has been shown that vascular measurements of the tumor microenvironment correlate with estrogen receptor status [24], the ability to specifically label vessels while minimizing non-specific binding remains a significant challenge using brightfield microscopy alone. Furthermore, the ability to use spectral

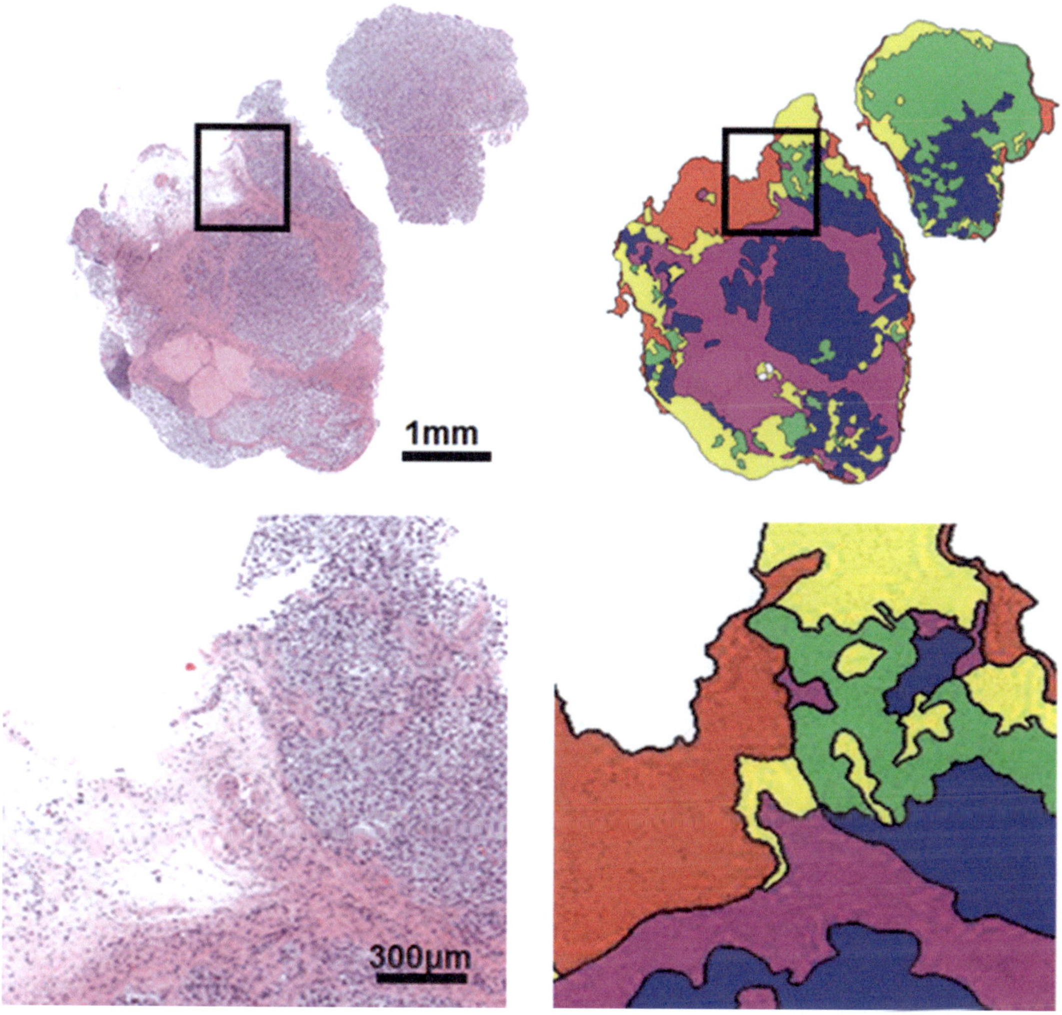

Fig. 1.2 Demonstrates at low (*top*) and high (*bottom*) magnification of automated segmentation of tumor and tumor microenvironment regions which can be quantitatively evaluated for distribution or other features (e.g. Ripley's K or Moran's I measurements). The masks overlay areas of necrosis (*yellow*), partial necrosis (*green*), viable tumor (*blue*), stroma (*purple*) and adipose tissue (*red*)

unmixing to determine the localization of bright-field stains introducing additional challenges including dedicated hardware [21]. By contrast, utilizing fluorescence based microscopy allows investigators to label multiple aspects of a single sample with enhanced specificity and verify the results simultaneously. Therefore the vascularity mentioned above can be analyzed simultaneously with ER on the tumor (or any) cells. The number of vessels, size of the vessels and lumens, the distance to ER positive or negative tumor cells or any number of additional metrics may be quantified and studied using image analysis [33].

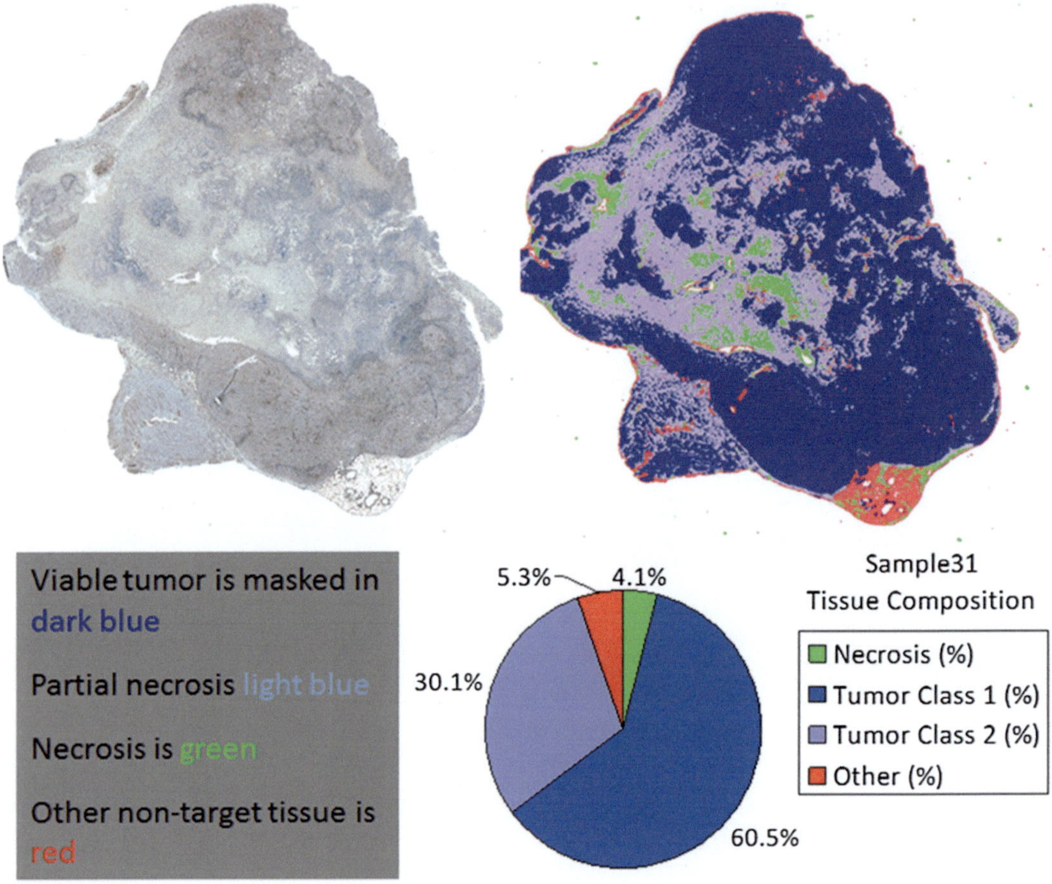

Fig. 1.3 Shows the tumor volume in dark blue and the partial and complete necrosis regions in light blue and green, respectively. Other tissue (e.g. adipose) is in red. This allows samples to be compared quantitatively for necrosis or other tumor microenvironmental features with exquisite detail

Confocal microscopy enables fluorescently tagged samples to be observed with even more precision. With confocal microscopy investigators can see small targets (i.e. proteins, receptors) with exquisite depth specificity. In other words, two proteins may appear to be in a single location but one may be deeper in a cell that the other. The Z plane specificity provided by confocal microscopy allows researchers to overcome this challenge [10]. Image analysis is used to identify, segment, count and determine the localization of these targets which can provide information to the investigator about the location and molecular status of the tumor microenvironment as it relates to the tumor itself (see Fig. 1.4). For example, the heterogeneity of the tumor microenvironment may contribute to the ability of a tumor to grow or progress [12, 37]. This may be observed at a protein level by investigating the molecular differences in different areas of the tumor microenvironment with higher X, Y and Z plane specificity. Finally, super-resolution, as the name implies, is a technique to identify and capture the smallest targets to be captured digitally and measured by image analysis techniques to provide investigators with even finer information about the cells which directly interact with the cancer of study [7].

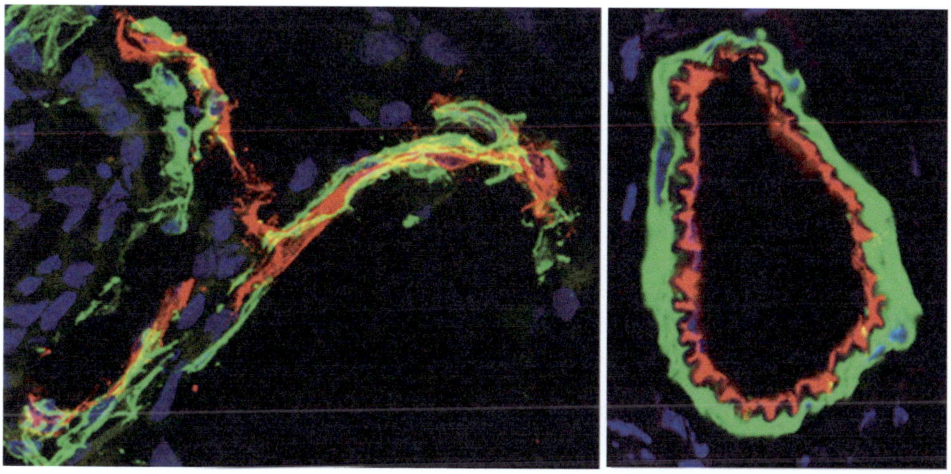

Fig. 1.4 Shows two examples of murine prostate tumor nuclei stained with DAPI in blue and two vascular related cell markers for pericytes in green and basal lamina in red. These three independent channels were acquired sequentially, minimizing crosstalk between the fluorophores ensuring specificity for each target

1.5 Second Harmonic Generation to Evaluate the Tumor Microenvironment

Second harmonic generation (SHG) is a label-free technique which enables imaging of specific tissue types, including collagen [43]. The process itself is a nonlinear optical image acquisition method. In SHG, photons with the same frequency come into contact with biological material and are effectively doubled which also gives SHG the moniker, frequency doubling. When this doubling occurs, the result is photons with twice the frequency and half the wavelength of the original photon, which are emitted and may be captured [32].

Collagen is a principle target of SHG and also prevalent in the tumor microenvironment of many diseases and may be involved in tumor progression [34]. The ability to analyze the differences in collagen layer thickness and/or arrangement is increasingly becoming a scientific area of interest in the study of the tumor microenvironment [16]. An example includes the progression of ductal carcinoma in situ to invasive cancer (see Fig. 1.5).

1.6 Conclusion

Image analysis has been increasingly providing new avenues for repeatable and quantifiable study of histiocytic sections of tumor. A more rapid and accurate interrogation of the tumor microenvironment of digital and whole slide images will enable the extraction of mathematical values from images which can be used to study and model these interactions. Image analysis allows pathologists to now extend beyond the tumor itself and now incorporate study of the tumor microenvironment. In fact, investigators are now looking to other fields of image analysis, including landscape ecology [25], to help provide metrics and measures of the tumor and its microenvironment to investigate cancer from a new perspective.

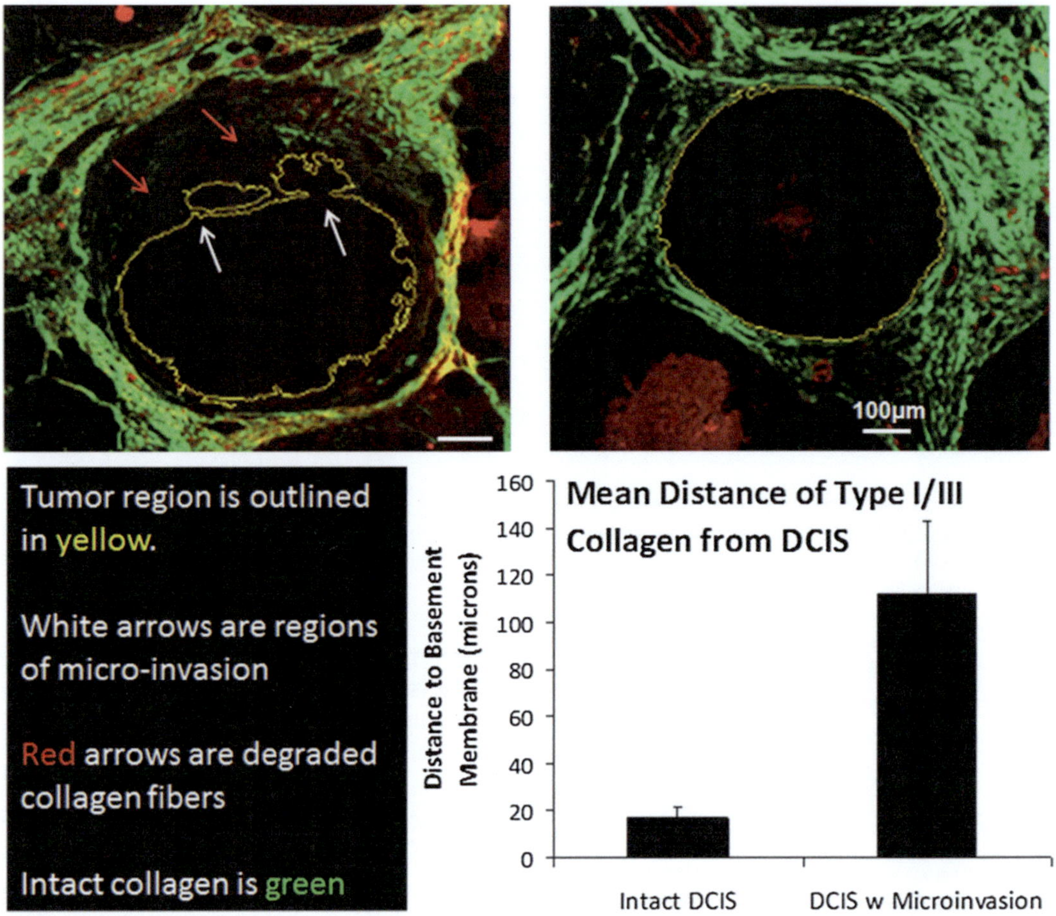

Fig. 1.5 In this figure the tumor regions (*top*) are encircled in the thin yellow line. The microenvironment is imaged with SHG and the shown result on the left illustrates the lack of collagen (*red arrows*) remaining near a microinvasion site (*white arrows*) as compared to a complete and intact basement membrane. In the bottom right 20 intact DCIS lesions and 20 DCIS regions with microinvasion were evaluated. The graph depicts the mean distance of the collagen fibers identified by SHG (I/III) to be statistically much further from the basement membrane in microinvasive samples indicating a notable relationship between collagen presence and microinvasion

References

1. Amat-Roldan I et al (2010) Fast image analysis in polarization SHG microscopy. Opt Express 18(16):17209–17219
2. Anderson A, Chaplain M, Rejniak K (eds) (2007) Single-cell-based models in biology and medicine. Birkhauser-Verlag, Mathematics and Bioscisnce in Interaction (MBI) series
3. Anitei M-G et al (2014) Prognostic and predictive values of the immunoscore in patients with rectal cancer. Clin Cancer Res 20(7):1891–1899
4. Beck AH et al (2011) Systematic analysis of breast cancer morphology uncovers stromal features associated with survival. Sci Transl Med 3(108):108ra113–108ra113
5. Ben-Baruch A (2002) Host microenvironment in breast cancer development: inflammatory cells, cytokines and chemokines in breast cancer progression-reciprocal tumor–microenvironment interactions. Breast Cancer Res 5(1):31
6. Bradford JA et al (2004) Fluorescence intensity multiplexing: simultaneous seven marker, two color immunophenotyping using flow cytometry. Cytometry Part A 61(2):142–152
7. Chien M-P et al (2013) Enzyme-directed assembly of nanoparticles in tumors monitored by in vivo whole animal imaging and ex vivo super-resolution fluorescence imaging. J Am Chem Soc 135(50):18710–18713
8. Deepak RU et al (2015) Computer assisted pap smear analyser for cervical cancer screening using quantitative microscopy. J Cytol Histol S3:010. doi:10.4172/2157-7099.S3-010

9. Dvorak HF et al (2011) Tumor microenvironment and progression. J Surg Oncol 103(6):468–474
10. Egeblad M et al (2008) Visualizing stromal cell dynamics in different tumor microenvironments by spinning disk confocal microscopy. Dis Model Mech 1(2–3):155–167
11. Estrella V et al (2013) Acidity generated by the tumor microenvironment drives local invasion. Cancer Res 73(5):1524–1535
12. Faratian D et al (2011) Heterogeneity mapping of protein expression in tumors using quantitative immunofluorescence. J Vis Exp 56:e3334
13. Glatz K, Pritt B, Glatz D, Hartmann A, O'Brien MJ, Blaszyk H (2007) A multinational, internet-based assessment of observer variability in the diagnosis of serrated colorectal polyps. Am J Clin Pathol 127(6):938–945
14. Gurcan MN et al (2009) Histopathological image analysis: a review. IEEE Rev Biomed Eng 2:147–171
15. Helm J, Centeno BA, Coppola D et al (2009) Histologic characteristics enhance predictive value of American Joint Committee on Cancer staging in resectable pancreas cancer. Cancer 115(18): 4080–4089
16. Iyengar P et al (2005) Adipocyte-derived collagen VI affects early mammary tumor progression in vivo, demonstrating a critical interaction in the tumor/stroma microenvironment. J Clin Investig 115(5):1163
17. Junttila MR, de Sauvage FJ (2013) Influence of tumour micro-environment heterogeneity on therapeutic response. Nature 501(7467):346–354
18. Kayser K et al (2010) AI (artificial intelligence) in histopathology–from image analysis to automated diagnosis. Folia Histochem Cytobiol 47(3): 355–354
19. Karabulut A, Jesper R, Marianne Hamilton T, Finn P, Nielsen HW, Erik D (1995) Observer variability in the histologic assessment of oral premalignant lesions. J Oral Pathol Med 24(5):198–200
20. Kenny PA, Lee GY, Bissell MJ (2007) Targeting the tumor microenvironment. Front Biosci 12: 3468
21. Levenson RM, Cronin PJ, Pankratov KK (2003) Spectral imaging for brightfield microscopy. Biomedical Optics 2003. International Society for Optics and Photonics
22. Li H, Fan X, Houghton JM (2007) Tumor microenvironment: the role of the tumor stroma in cancer. J Cell Biochem 101(4):805–815
23. Lloyd MC et al (2010) Using image analysis as a tool for assessment of prognostic and predictive biomarkers for breast cancer: how reliable is it? J Pathol Inform 1:29
24. Lloyd MC, Alfarouk KO, Verduzco D, Bui MM, Gillies RJ, Ibrahim ME, Brown JS, Gatenby RA (2014) Vascular measurements correlate with estrogen receptor status. BMC Cancer 14(1):279
25. Lloyd MC, Rejniak KA, Brown JS, Gatenby RA, Minor ES, Bui MM (2015) Pathology to enhance precision medicine in oncology: lessons from landscape ecology. Adv Anat Pathol 22(4):267–272
26. Mass RD et al (2005) Evaluation of clinical outcomes according to HER2 detection by fluorescence in situ hybridization in women with metastatic breast cancer treated with trastuzumab. Clin Breast Cancer 6(3):240–246
27. Mavaddat N et al (2010) Incorporating tumour pathology information into breast cancer risk prediction algorithms. Breast Cancer Res 12(3):R28
28. McInerney T, Terzopoulos D (1996) Deformable models in medical image analysis: a survey. Med Image Anal 1(2):91–108
29. Messina JL et al (2012) 12-Chemokine gene signature identifies lymph node-like structures in melanoma: potential for patient selection for immunotherapy? Sci Rep 2:765
30. Nyberg P, Salo T, Kalluri R (2007) Tumor microenvironment and angiogenesis. Front Biosci: A J Virtual Libr 13:6537–6553
31. Ohashi A et al (2005) Quantitative analysis of the microvascular architecture observed on magnification endoscopy in cancerous and benign gastric lesions. Endoscopy 37(12):1215–1219
32. Oka M (1990) Second harmonic generation. U.S. Patent No. 4,910,740. 20 Mar
33. Peng C-W et al (2011) Patterns of cancer invasion revealed by QDs-based quantitative multiplexed imaging of tumor microenvironment. Biomaterials 32(11):2907–2917
34. Provenzano PP, Eliceiri KW, Keely PJ (2009) Multiphoton microscopy and fluorescence lifetime imaging microscopy (FLIM) to monitor metastasis and the tumor microenvironment. Clin Exp Metastasis 26(4):357–370
35. Rejniak KA (2012) Homeostatic imbalance in epithelial ducts and its role in carcinogenesis. Scientifica 132978
36. Robertson AJ, Anderson JM, Swanson Beck J, Burnett RA, Howatson SR, Lee FD, Lessells AM, McLaren KM, Moss SM, Simpson JG (1989) Observer variability in histopathological reporting of cervical biopsy specimens. J Clin Pathol 42(3): 231–238
37. Rojo MG, Bueno G, Slodkowska J (2010) Review of imaging solutions for integrated quantitative immunohistochemistry in the Pathology daily practice. Folia Histochem Cytobiol 47(3): 349–348
38. Sarode VR et al (2011) A comparative analysis of biomarker expression and molecular subtypes of pure ductal carcinoma in situ and invasive breast carcinoma by image analysis: relationship of the subtypes with histologic grade, Ki67, p53 overexpression, and DNA ploidy. Int J Breast Cancer 217060
39. Schindewolf T et al (1994) Evaluation of different image acquisition techniques for a computer vision system in the diagnosis of malignant melanoma. J Am Acad Dermatol 31(1):33–41

40. Serra J (1982) Image analysis and mathematical morphology, vol 1. Academic, New York
41. Song N, Tao LI, Xue-Min Z (2014) Immune cells in tumor microenvironment. Prog Biochem Biophys 41(10):1075–1084
42. Sugimoto H, Mundel TM, Kieran MW, Kalluri R (2006) Identification of fibroblast heterogeneity in the tumor microenvironment. Cancer Biol Ther 5(12):1640–1646
43. Uchugonova A et al (2013) Multiphoton tomography visualizes collagen fibers in the tumor microenvironment that maintain cancer cell anchorage and shape. J Cell Biochem 114(1): 99–102
44. Wetzels RH et al (1989) Detection of basement membrane components and basal cell keratin 14 in noninvasive and invasive carcinomas of the breast. Am J Pathol 134(3):571

Hypoxia in Gliomas: Opening Therapeutical Opportunities Using a Mathematical-Based Approach

2

Alicia Martínez-González, Gabriel F. Calvo, Jose M. Ayuso, Ignacio Ochoa, Luis J. Fernández, and Víctor M. Pérez-García

Abstract

This chapter explores the use of mathematical models as promising and powerful tools to understand the complexity of tumors and their, frequently, hypoxic environment. We focus on gliomas, which are primary brain tumors derived from glial cells, mainly astrocytes and/or oligodendrocytes. A variety of mathematical models, based on ordinary and/or partial differential equations, have been developed both at the micro and macroscopic levels. The aim here is to describe in a quantitative way key physiopathological mechanisms relevant in these types of malignancies and to suggest optimal therapeutical strategies. More specifically, we consider novel therapies targeting thromboembolic phenomena to decrease cell invasion in high grade glioma or to delay the malignant transformation in low grade gliomas. This study has been the basis of a multidisciplinary collaboration involving, among others, neuro-oncologists, radiation oncologists, pathologists, cancer biologists, surgeons and mathematicians.

Keywords

Microscopic mathematical model of glioblastoma • Antithrombotics and cancer • Malignant transformation of low-grade gliomas • Mathematical models of hypoxia in tumors

A. Martínez-González • G.F. Calvo
V.M. Pérez-García (✉)
Mathematical Oncology Laboratory (MôLAB), University of Castilla-La Mancha, Castilla-La Mancha, Spain
e-mail: alicia.martinez@uclm.es; gabriel.fernandez@uclm.es; victor.perezgarcia@uclm.es

J.M. Ayuso • I. Ochoa • L.J. Fernández
Aragón Institute of Engineering Research (I3A), University of Zaragoza, Zaragoza, Spain
e-mail: josayuso@unizar.es; iochgar@unizar.es; luisf@unizar.es

K.A. Rejniak (eds.), *Systems Biology of Tumor Microenvironment*, Advances in Experimental Medicine and Biology 936, DOI 10.1007/978-3-319-42023-3_2

2.1 The Glioma Microenvironment and its Macroscopic Fingerprints

2.1.1 The Hypoxic Tumor Microenvironment

Low oxygenation (hypoxia) constitutes a characteristic physio-pathological feature of about 50–60 % of locally advanced solid tumors. Ample experimental evidence has shown that oxygen concentrations vary significantly, both spatially and temporally, in solid tumors, mostly due to the aberrant neovasculature induced by the rapid cell proliferation. Although sustained exposure to complete oxygen deprivation (anoxia) may result in the formation of necrotic areas within the tumor, viable hypoxic cancer cells often encompass these areas and infiltrate towards regions with higher oxygen levels where they subsequently exhibit substantial growth rates. Near-zero oxygen levels are observed at distances of the order of 150 μm from feeding blood vessels [13, 16] but it is frequently found that, owing to the lack of integrity and increased permeability of the tumor vasculature, anoxia can arise at smaller distances.

Patients with more hypoxic tumors have a significantly poorer prognosis following chemotherapy and/or radiotherapy [29]. At the molecular level, adaptation of tumor cell subpopulations to strong spatio-temporal fluctuations of oxygen availability is mediated by the family of hypoxia inducible factors (HIF). The HIF signaling cascade is controlled primarily through the post-transcriptional modification and stabilization of HIF1-α and HIF2-α subunits. Their activity is increased as a result of genetic alterations that activate oncogenes and inactivate tumor suppressor genes [22, 46]. The subunit HIF1-α turns on the hypoxic response, increases neovascularization, induces inflammation and promotes apoptosis evasion. HIF2-α regulates erythropoietin synthesis and modulates vascular endothelial cell function. HIF1-α and HIF2-α share some redundant functions, but they also display distinctive and even opposing activities in cell growth, metabolism, angiogenesis and nitric oxide homeostasis. They are broadly expressed in many human cancers and they often correlate with poor patient outcome [30, 56].

Identifying hypoxic regions within tumors has attracted considerable research by means of a number of different molecular imaging and other techniques [19, 25]. Nevertheless, determination of robust clinical biomarkers for tumor hypoxia remains a challenge. Several clinical procedures have been investigated, involving HIF-1, carbonic anhydrase IX, glucose transporter 1 (Glut1), and vascular endothelial growth factor (VEGF), but none represent a true gold standard due to the dramatic spatio-temporal heterogeneities encountered in tissue oxygen levels and the great difficulty in completely capturing this heterogeneity. Quantitative approaches based on mathematical models may assist in providing useful biomarkers.

2.1.2 Glioma Patterns of Invasion and Proliferation

Gliomas account for approximately 80 % of all malignant tumors of the central nervous system. These tumors comprise a heterogeneous group of neoplasms that initiate in the brain or the spine. According to the World Health Organization (WHO) classification [34], the most prevalent gliomas are diffuse oligodendrogliomas and astrocytomas (WHO grade II), anaplastic oligodendrogliomas and astrocytomas (WHO grade III) and glioblastoma multiforme (GBM) (WHO grade IV).

An important feature of primary brain tumors is that they neither invade into blood vessels nor infiltrate into the bony calvarium which encloses the brain [7]. That is, brain tumors very rarely metastasize to other ectopic organ tissues, although it is worth mentioning that in high-grade gliomas (HGGs, WHO grades III and IV), tumor cells may spread along blood vessel walls and even through the fiber tracts of the corpus callosum to the contralateral hemisphere. Low grade gliomas (LGGs, WHO grade II) infiltrate slowly through the brain but show mild or no contrast enhancement in post-contrast MRI images, suggesting an intact blood-brain barrier (BBB) and a lack of tumor necrosis. Radial LG growth rates are modest with annual increases in diameter of 2–4 mm/year while HGGs radial growth rates are typically 10 times larger [9]. As astrocytomas progress in time through the pathologic spectrum from the lower end of grade II to the upper end of grade III, their proliferative capacity creeps upward, resulting in a more densely cellular tumor with greater malignant potential [49].

It is believed that cancer cells do not move and proliferate simultaneously. The reason is that the same cytoskeleton structures are used to develop the lamellipodia (cytoplasmic sheets extended at the front of moving cells) and to build the spindle fibers during mitosis. In addition, a gradient of the chemoattractant, naturally present in the extracellular environment, may drive the directional motility which is known as haptotaxis. Therefore, the haptotaxis output node will be part of the life-cycle response just like proliferation, quiescence and apoptosis. This fact is referred to as the proliferation/migration dichotomy (or the *go or grow* hypothesis) and leads to the fact that highly motile cells should exhibit low proliferation rates [18].

The switch between proliferative and invasive phenotype cannot be only mutation driven [24, 37]. It has been suggested that invasive glioma cells are able to revert to a proliferative cellular program and vice versa, depending on the environmental stimuli [18, 31]. It would appear that oxygen may be this key stimulus and thus, for each oxygen level, there exists a dominant (fittest) tumor cell phenotype that corresponds to a certain proliferation/migration ratio [18].

This proliferation/migration dichotomy manifests very clearly in gliomas where it has been studied in detail [18]. However, some features of this dichotomy are also present in other types of cancers [57]. In fact, cells located in the perivascular areas have both oxygen and nutrients and exhibit a high proliferative activity. On the contrary, cells exposed to hypoxia show typically increased migration and slower proliferation to deal with more hostile microenvironmental conditions [9, 12, 60].

2.1.3 Pseudopalisades and Coagulation in Gliomas

Hypoxia plays a central role both in the progression and resistance to therapy in primary brain tumors [44], where it has been proven to play a key role in the biology and aggressiveness of these cancers [14]. In GBM, the most common and lethal type of malignant primary brain tumor in adults, the tumor microenvironment is significantly different from that of normal tissue.

Some of the distinguishing hallmarks of GBM are the presence of a high proliferative index and the disruption of the BBB which, under normal conditions, protects the brain from many common bacterial infections and prevents the entry of potential neurotoxins. There is also an increased vascular permeability, the presence of edema, microvascular hyperplasia and the development of necrotic foci surrounded by hypercellular areas, known as pseudopalisades [26]. Some of them are depicted in Fig. 2.1 subplots (a), (c) with hematoxylin and eosin (H&E) marker.

Pseudopalisades are a key component in the pathophysiology of GBM. In fact, the differential histological diagnosis of GBM against lower grade glioma depends on the presence of tumor necrosis and pseudopalisades [28]. Most pseudopalisading cells are hypoxic and express elevated levels of HIF-1α, which drives the expression of the angiogenic factor Vascular Endothelial Growth Factor (VEGF), promoting neovascularization. The result is a prominent microvascular density, one of the highest among all neoplasias. This microvascular hyperplasia gives rise to immature and leaky vessels, with reduced cell-cell adhesions and decreased pericyte coverage.

Accordingly, the contact between blood and tumor cells activates the intrinsic coagulation pathway [27, 59]. Specifically, 25–30 % of glioma patients suffer from venous thromboembolism (VTE) [51, 52] and it is known that the more tumoral tissue is removed the less-likely are the patients to die from VTE [51]. This fact has led to the consideration of using thromboprophylaxis for glioma patients with higher potential risks of VTE [6, 21, 27, 32]. Furthermore, the

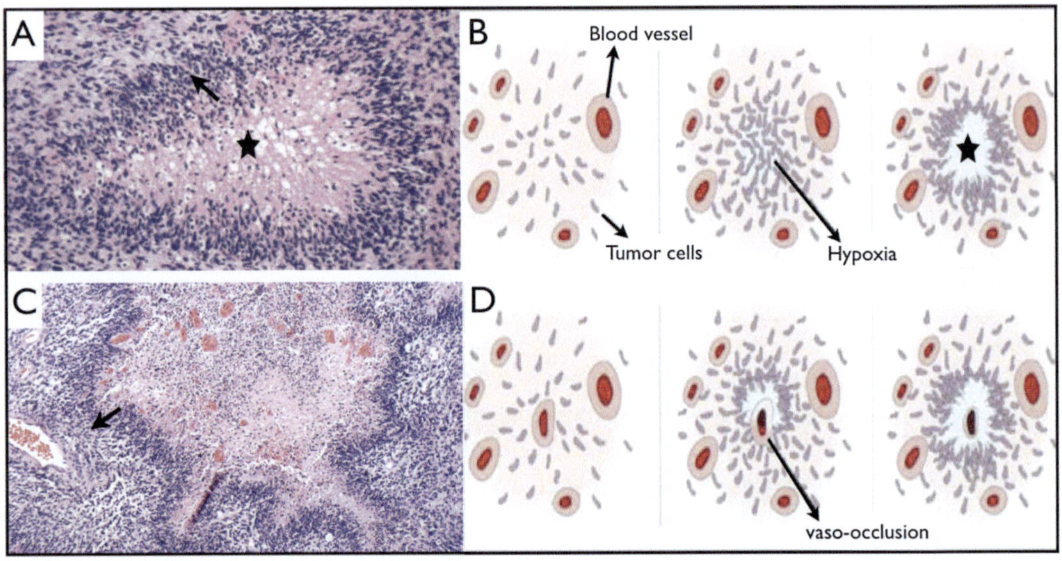

Fig. 2.1 Potential mechanisms of pseudopalisade formation caused by hypoxia-induced migration. The first hypothesis is that tumor cells at greatest distance from vessels become hypoxic beyond a critical point in tumor growth and migrate toward peripheral vessels, leaving a central clear zone (**a** and **b**). Alternatively, a second theory hypothetizes that vascular occlusion could lead to central hypoxia, followed by tumor cell migration toward a viable blood supply (**c** and **d**) (Figure adapted from [10, 48, 49])

extrinsic coagulation pathway is also activated by the action of the tissue factor (TF), a transmembrane protein, though its action is not only pro-coagulant, but also promotes GBM cell invasion and induces neovascularization [5]. In this sense, there have been preliminary clinical trials evaluating the effects of antithrombotics in GBM, all of them aiming at reducing the vaso-occlusive events and, overall, giving satisfactory results without evidence of adverse side effects [50].

Laboratory markers of coagulation activation such as thrombin/antithrombin complex or prothrombin fragments $1 + 2$ support the premise that malignancy is a hypercoagulable state. Inflammatory cytokines (e.g. tumor necrosis factor and interferon), coagulation proteins (e.g. TF and factor VIII), and procoagulant microparticles may be elevated in patients with high malignant tumors [53]. However, the relative contribution of chemotherapeutics, tumor cells, endothelium, and circulating procoagulants in promoting thrombus formation continues to be investigated [15, 20, 32, 61].

2.1.4 How Do Pseudopalisading Structures Arise in GBM?

The fact that GBM is one of the most highly vascularized human tumors seems to contradict that its microcirculation is functionally very inefficient [28]. In GBM, at least five neovascularization mechanisms have been identified: vascular co-option, angiogenesis, vasculogenesis, vascular mimicry and GBM-endothelial cell transdifferentiation [23]. However, its chaotic vasculature leads to a decrease in both the supply of essential nutrients and in the removal of waste products.

The characteristic necrotic regions of GBM are suspected to have their origin from tissue areas subjected to persistent hypoxia (and most likely anoxia). Thus, it is possible that a combination of hypoxic and other intrinsic tumor agents are responsible for the development of pseudopalisading necrosis [48]. In fact, [9] proposed that pseudopalisades in GBM are generated by an actively migrating and hypoxic wave of glioma cells.

One possibility of pseudopalisade formation is that those tumor cells located at farther distances from arteries become hypoxic, after a critical point in the tumor natural history, caused by increased cell density and metabolic demands, and start to migrate toward peripheral and more functional vessels, leaving a central clear zone (see Fig. 2.1 subplots a and b). Alternatively, vascular occlusion or collapse within the neoplasm could lead to central hypoxia, followed by tumor cell migration toward a viable blood supply (see Fig. 2.1 subplots c and d). Previous works have proposed that this last pseudopalisade formation mechanism would be the result of a multistep process [9, 10, 49]. Firstly, in anaplasic astrocytoma or de novo GBM in its initial stages, tumor cells proliferate and infiltrate through the parenchyma and receive oxygen and nutrients via the intact native blood vessels. Secondly, vascular insult occurs as a result of uncontrolled tumor growth causing endothelial injury and vascular leakiness. Both endothelial injury and the expression of procoagulant factors by the neoplasm result in intravascular thrombosis and increased hypoxia in the regions surrounding the vessel [45]. Subsequently, tumor cells begin to migrate away from hypoxia, creating a peripherally moving wave that is seen microscopically as pseudopalisading cells, leading to an expansion of the zone of hypoxia and central necrosis. Meanwhile, hypoxic tumor cells in pseudopalisades secrete proangiogenic factors, such as VEGF and Interleukins, which contribute to the angiogenesis process forming more aberrant vessels that will again eventually suffer vaso-occlusions.

Furthermore, it is believed that cell proliferation is not the cause of pseudopalisade formation since it has been observed that proliferation rates are lower in pseudopalisades than in adjacent astrocytoma tissue for nine glioma lines [9]. Considering that pseudopalisades are mainly formed by hypoxic cells, there exist biological evidences to think that hypoxic cell proliferation indices are lower than for the normoxic cells. In addition, tumor invasion increases in GBM concurrently with pseudopalisade development. For this reason, the phenotypic behavior of pseudopalisading cells seems to be less proliferative and more motile than that of the adjacent normoxic tumor cells.

2.1.5 Proliferation/Migration Dichotomy Validation from Biopsies

Following the pseudopalisade hypothesis formation, cells comprising the pseudopalisade, mainly those close to the necrotic area, will simultaneously show higher motility and lower proliferation (such as Ki67) biomarkers than cells in the adjacent tissue. Since mammalian diaphanous homolog 1 (DIAPH1) is involved in a number of actin-related biological processes (e.g. invasion and metastasis in a number of human cancers) it has been proposed as a motility biomarker for glioma cells [33]. Then, pseudopalisading cells should express higher DIAPH1 and lower Ki67 than cells located in the adjacent tissue. Figure 2.2 shows that those GBMs displaying a more highly marked staining tend to have a larger number of pseudopalisades. This staining is specially intense in those cells located at the internal side of the pseudopalisade. In addition, the results suggest that inflammatory cells are also DIAPH1 positive. This might be due to the fact that they move towards the committed event (e.g. a vaso-occlusion event). Also, endothelial cells tend to be DIAPH1 positive particularly if they are close to the vessels so it is important to distinguish among the different types of cells.

2.2 Mathematical Simulation of Pseudopalisade Formation Following Vaso-Occlusion Events

This Section presents a mathematical model suggested by [35] and discusses its potential to describe the phenomena described in Sect. 2.1.1. The model will be built to simulate the formation of pseudopalisades in GBM according to the vaso-occlusion hypothesis discussed previously

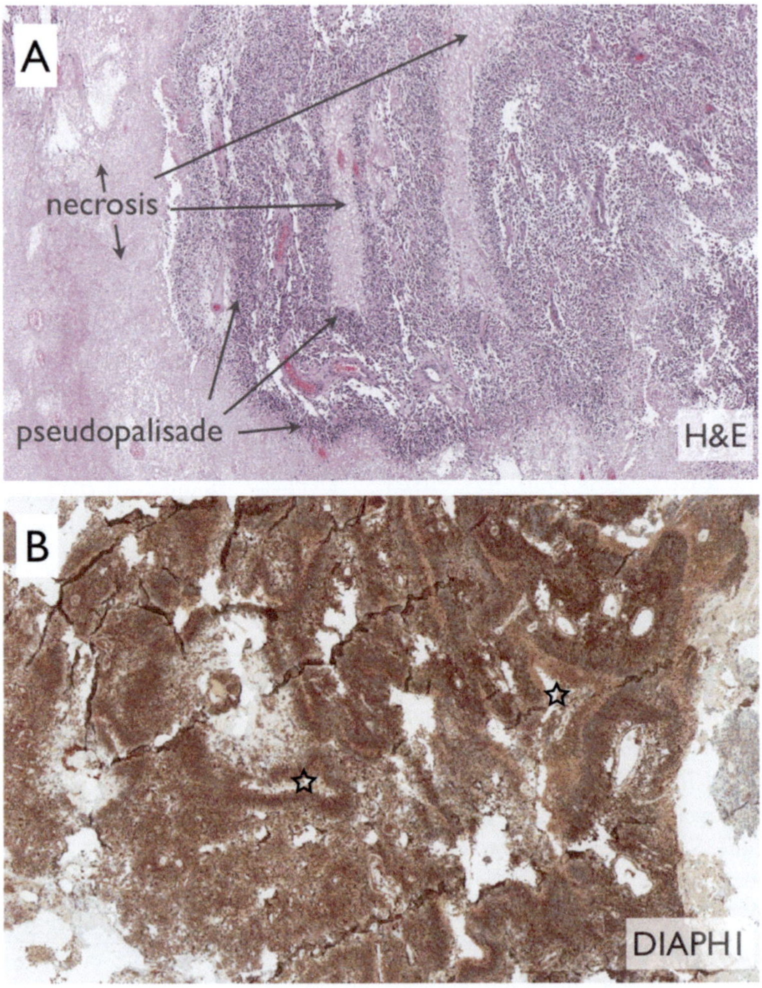

Fig. 2.2 (**a**) Hematoxylin and eosin (H&E) staining shows multiple perinecrotic palisades. (**b**) DIAPH1 staining is associated with higher motility in the internal side (*stars*) of the pseudopalisades (Courtesy of the Pathology Service of Hospital General Universitario de Ciudad Real)

and summarized in Fig. 2.3. The therapeutical implications of that model and also of a subsequent one [36] will be discussed in detail.

2.2.1 The Model

The model is based on Fisher-Kolmogorov type equations and incorporates the interplay between two tumor dominant cell phenotypes, a necrotic core C_d and the oxygen concentration O_2. Specifically, two different coupled tumor cell subpopulations compete for space and oxygen. They correspond to the normoxic phenotype C_n which is more proliferative (higher proliferation rate τ_n) and less motile (lower diffusion constant D_n), and the hypoxic phenotype C_h which is less proliferative (lower proliferation rate τ_h) and more mobile (higher diffusion constant D_h).

In the simulated microenvironment, the populations were embedded within two blood vessels arranged in a linear domain (one dimensional section). One of the vessels, surrounded by tumor cells, suffered a vaso-occlusion event resulting in no oxygen flux from it. The model equations, displayed in Fig. 2.4, were solved using second order finite differences in space with an explicit fourth order Runge-Kutta method in time.

2.2.2 Results

The numerical simulations reveal the formation of a traveling wave of hypoxic cells that qualitatively reproduces the experimentally observed patterns. In the framework of the model, preventing vessels from breaking (or delaying their breakup) by means of an antithrombotic may lead to slower tumor invasion speeds what might be useful for therapeutic purposes (see Fig. 2.5).

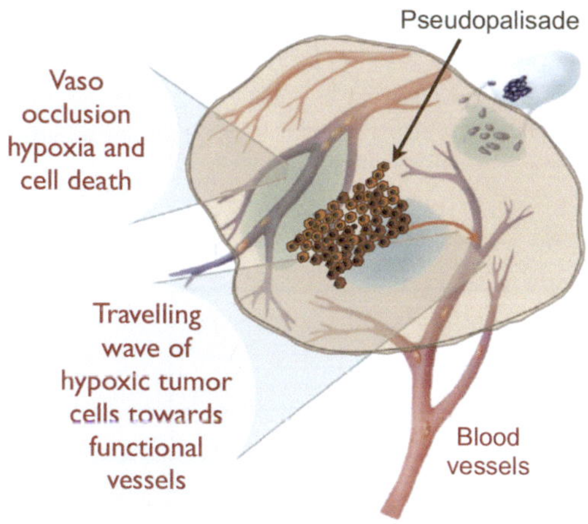

Fig. 2.3 Schematic representation of the mechanism of pseudopalisade formation. Pathologic observations indicate that hypoxia and necrosis within gliomas could arise secondary to vaso-occlusion [49]. Pseudopalisades may represent a wave of tumor cells actively migrating away from central hypoxia after vascular insult (See Sect. 2.1.4)

The model also provides an estimate of palisade timescale formation, lifetime, palisade width and the necrotic core size considering different distances between the vessels (the occluded vessel and the functional one). The model, although biologically simple, supports the hypothesis that large pseudopalisades are more persistent. This prediction has a good agreement with the pathologic observation that pseudopalisades surrounding larger broken vessels are easier to observe.

These results suggest the coexistence of two mechanisms which drive invasion in GBM. One purely diffusive and modulated by a logistic growth (where the only restriction is the maximum carrying capacity of the tissue) obeying the Fisher-Kolmogorov equation, as described by [54] and related works. The second one shows an accelerated invasion caused by micro thrombi formations in the tumor vessels.

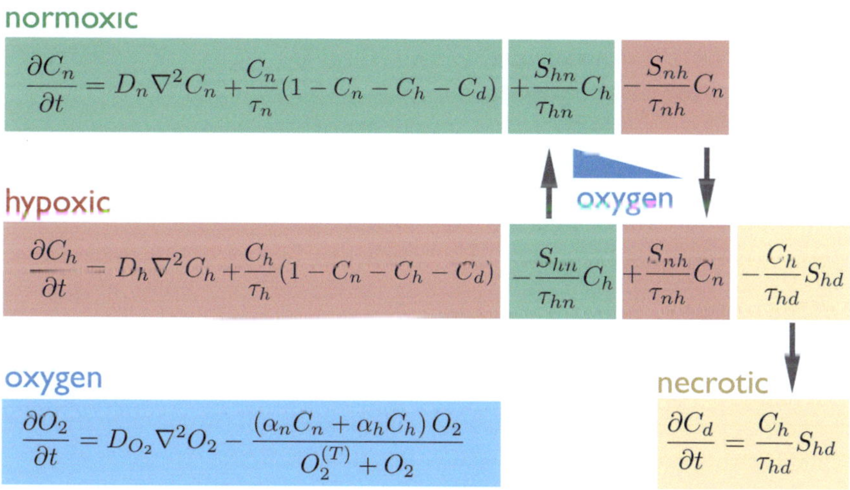

Fig. 2.4 Oxygenation levels influence tumor cell phenotype. Oxygenation drives the phenotypic switching mechanisms coupling the cell populations: Normoxic to hypoxic, hypoxic to normoxic, and hypoxic to necrotic. Thus the oxygen-dependent functions $S_{nh}(O_2), S_{hd}(O_2), S_{nh}(O_2)$ (see [35]) satisfy the conditions, $S_{nh}(0) = S_{hd}(0) = 1, S_{nh}(\infty) = S_{hd}(\infty) = 0, S_{hn}(0) = 0, S_{hn}(\infty) = 1$. High oxygen levels favor the existence of more proliferative phenotypes which are less mobile. On the contrary, cells respond to low oxygen concentrations by expressing less proliferative phenotypes which are more motile. τ_{nh}, τ_{hn} and τ_{hd} are characteristic switching times between phenotypes. Finally, hypoxic cells under persistent anoxia eventually die and give rise to necrosis [9]

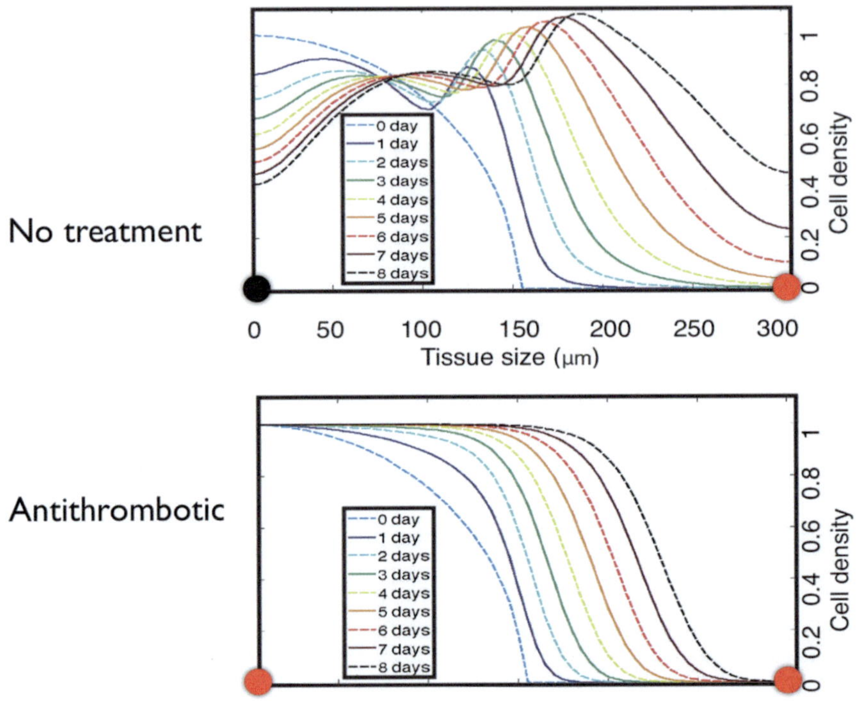

Fig. 2.5 **Simulations of the total tumor cell density evolution in a one dimensional domain surrounded by two blood vessels (represented by dots placed at 0 and 300 μm)**. *Top*: Due to the pressure and release of prothrombotic factors, the vessel on the left collapses and becomes non-functional leading to hypoxia and massive migration of glioma cells to other, better oxygenated regions. *Bottom*: The effect of antithrombotic therapy prevents the thrombotic episode leading to a better oxygen supply and to a slower expansion of the tumor front

2.2.3 Discussion and Therapeutical Implications for GBM Patients

It seems that vaso-occlusive events in GBM may play a key role in accelerating tumor invasion through pseudopalisade formation. Actually, we suggest that during the time window where vaso-occlusions occur any oxygen dependent treatment will have a limited efficacy (e.g. radiotherapy). This fact suggests that the use of chemical agents slowing-down the vessel impairment, might delay tumor progression in GBM patients. In addition, the formation of foci of pseudopalisading necrosis may contribute to the acceleration in the transition from low to high degree of malignancy. We hypothesized that antithrombotic administration could delay malignant glioma transformation by retarding the development of hypoxia.

Thus, this suggests that antithrombotics (AT) may constitute potentially attractive agents for glioma treatment. One example might be low molecular weight heparine (LMWH), that promotes the release of tissue factor pathway inhibitors for preventing VTE and with very low toxicities [42, 43, 47]. It is important to underscore that the macroscopic invasion of GBM can be partially conceived, from a microscopic point of view, as consisting of a large number of intravascular thrombotic events. Unregulated tumor cell proliferation in the vicinity of the vessels might cause these thrombotic events. They give rise to the formation and coalescence of small necrotic foci coupled with subsequent episodes of hypoxic cell migrations in search of nearby functional blood vessels. However, these vessels will eventually suffer new intravascular thrombotic events. Finally, the envelope of many of these small-scale pseudopalisades contributes to the high density regions on larger spatial scales.

Moreover if, as predicted, preventing capillary thrombosis induced by the glioma cells results in tumor growth delay, a favourable synergistic effect with radiation and/or chemotherapy is to be expected. Reducing hypoxia would then lead to an enhanced tumor radio and chemo-sensitivity, that might provide an extra benefit in survival. We have considered this scenario in a recent work [36] and found a substantial effect from thromboprophylaxis related to the tumor invasion delay.

It is worth reminding that VTE, which is the formation of thrombi leading to the occlusion of large vessels, is well recognized as a major cancer complication and known to be a common cause of death in cancer patients. The contribution of chemotherapeutics, tumor cells, endothelium, and procoagulants in promoting thrombi formation continues to be investigated [15, 20, 32, 61]. GBM is one of the most prothrombotic tumors, and the use of AT therapy has been proposed in high VTE risk GBM patients as being potentially beneficial.

The main limitation in using LMWH in GBM patients, specially the post-operative ones, is a potential increase in bleeding risk. However, a limited phase II clinical study of LMWH in high grade glioma patients by [47] has shown that thromboprophylaxis is safe if not started immediately after surgery. A phase III trial with a small number of patients seems to confirm this trend [42].

2.3 Can Anti-thrombotics Delay the Malignant Transformation of Low-Grade Gliomas?

2.3.1 Mathematical Modeling of the Malignant Transformation of Low-Grade Gliomas

Under hypoxic conditions, transcription hypoxia-inducible factors, such as HIF-1α, are expressed and trigger a neovascularization response. As discussed before, this induces endothelial hyperplasia and the formation of a high density of aberrant, non-functional and leaky micro vessels. These vessels contribute to a positive feedback in the coagulation cascade, allowing blood extravasation and contact with tumor cells, resulting in a sustained coagulation.

Hypoxia is only marginal in low grade gliomas (LGG) [60], where the vessel architecture remains essentially intact. Many factors may contribute to their transformation into high-grade gliomas, the so-called malignant transformation (MT). The simplest explanation of the MT of LGGs from a dynamical point of view is that these tumors degenerate after the local development of hyperplasias, the subsequent emergence of hypoxic areas and the triggering of the hypoxic response leading to a more aggressive tumor [39, 41, 55].

Thus, the simplest biological scenario, depicted in Fig. 2.6, to consider as a basis for a mathematical model of the malignant transformation includes necrosis and two populations of tumor cells with LGG and HGG behaviour. LGG cells feature lower proliferation rates and motility than HGG cells. A fraction of HGG-type tumor cells die to their damage to the microenvironment giving rise to necrotic areas as in other GBM tumor-necrosis models [40].

Since cell density and vasculature functionality seem to be very related in gliomas, the model incorporates a malignant and irreversible switch from LGG to HGG cells. This hypothesis is based on the fact that after several episodes of hypoxia, cells tend to accumulate a malignant behaviour. This biological fact is incorporated into the model by assuming that there is a cell density threshold beyond which LGG cells transform into HGG.

These assumptions allow one to partially circumvent many biological mechanisms that remain unknown and, at the same time, enable one to propose a simple therapeutical approach: AT would increase the cell density threshold for the MT. A mathematical model incorporating all the previous hypothesis is presented in Fig. 2.7.

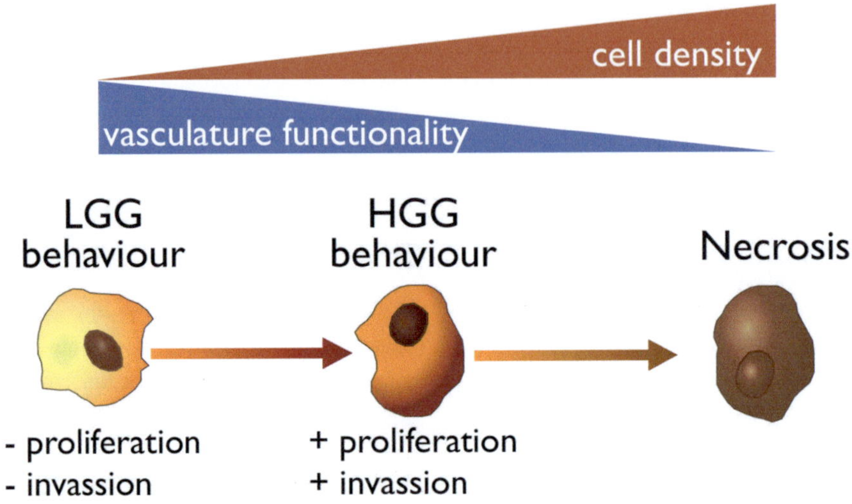

Fig. 2.6 Schematic representation of the malignant transformation scenario considered in the model. The model is based on the hypothesis where LGG cells became malignant when cell density is too high and induces hypoxic events, leading to more aggressive phenotypes

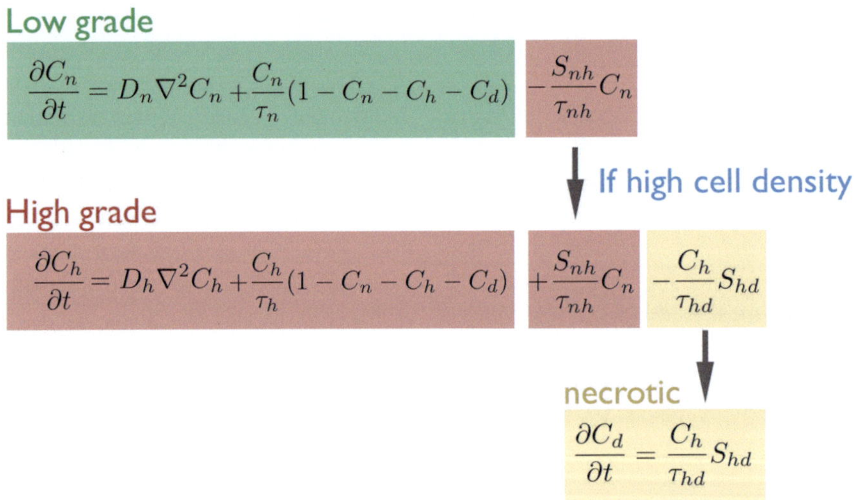

Fig. 2.7 Mathematical modeling of the malignant transformation of low-grade gliomas. Parameters and functions meanings are as in Fig. 2.4

Parameter values for HGG cells in this model are taken to be similar to those used in Sect. 2.2. LGG diffusion and proliferation parameters should be those that yield around 10 times lower infiltration velocity (see e.g. [4, 17]). One approximate formula for the front velocity can be used: $v = 2\sqrt{D\rho}$, where ρ is the proliferation rate. If the switching time from LGG to HGG is fast, then the velocity for tumor invasion will increase as it was studied in [38].

2.3.2 Results

Figure 2.8 exhibits a characteristic simulation comparing an untreated virtual tumor control with a tumor treated with AT from the very beginning. The results suggests that these treatments may lead to a malignant transformation delay of many months, which is a substantial improvement for such a simple therapeutical approach.

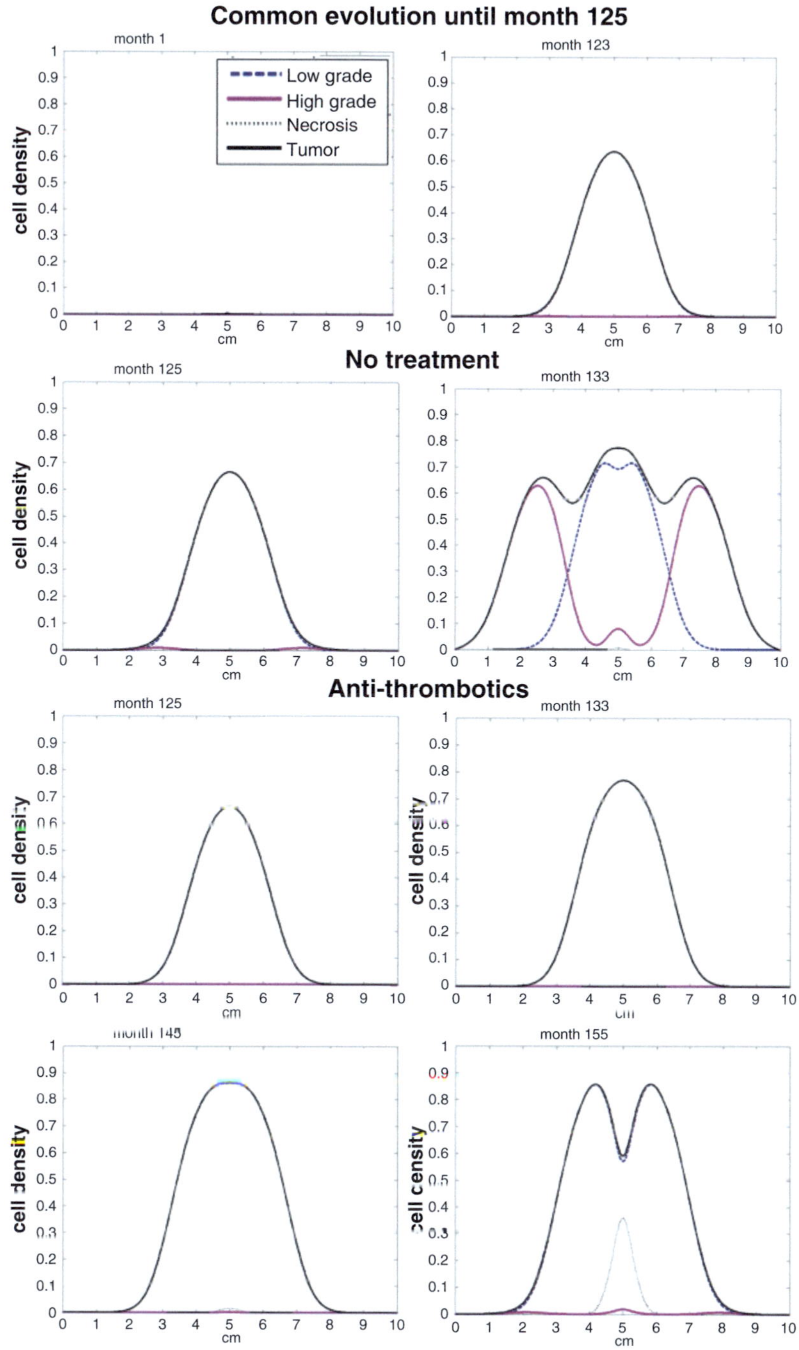

Fig. 2.8 Tumor cell density along one spatial section. LGG and HGG cell densities are represented by the *dashed blue* and *pink lines*, respectively. Shown are the total tumor cell densities (*thick black lines*) and necrosis profiles (*dotted lines*). The results show that LGG suspicious of undergoing a transition to high grade might also benefit from treatment either with vascular normalization (antiangiogenic) therapy or anti-thrombotic agents

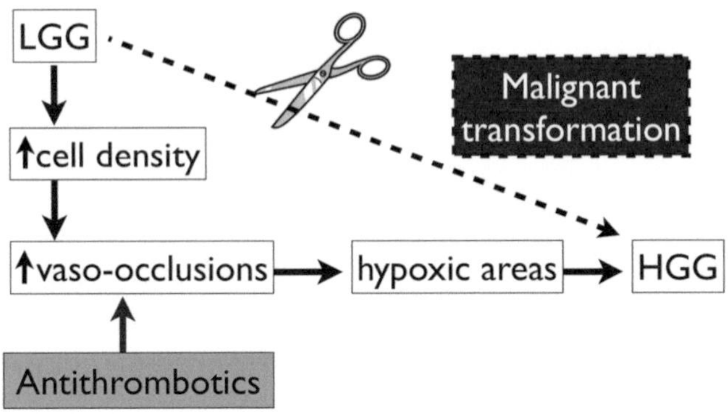

Fig. 2.9 Antithrombotics against the malignant transformation. The goal is to delay the malignant transformation associated with the hypoxic phenotype by preventing the coagulation process

Typically, simulations show a common evolution for treated and non treated virtual tumors for the first months (125 months in Fig. 2.8). After that, cell density is high enough to induce vascular damage, hypoxia, necrosis and an increase in cell malignancy. In contrast, the very same tumors treated with AT maintained its vasculature functionality for larger periods of time delaying malignant transition (around 20 months in the virtual tumor displayed in Fig. 2.8).

2.3.3 Therapeutical Implications

The prediction that AT may result in a delay of the malignant transformation of LGGs into lethal higher-grade tumors has direct therapeutical implications. The only clinical limitation of the proposal is the difficultity in maintaining AT therapy for very long times (i.e. longer than one year), due to potential side effects. Thus, using estimates for the optimal time to start treatment with AT is of great interest from the clinical point of view.

Recent works have suggested that the response of LGGs to either a limited amount of radiation [41] or chemotherapy [8], would allow one to estimate the time to MT. Those estimates may prove useful to specify the times at which AT should begin to be administered when it is really necessary, i.e. around the MT time. Figure 2.9 summarizes the potential benefit from AT in delaying MT.

In summary, the use of AT therapy in gliomas may have a broad range of indications to be explored in more detail. First, AT administered to WHO grade II astrocytic and oligodendroglial patients may help to preclude the malignant transformation into secondary GBM associated with the onset of hypoxia once local vascular damage is produced. Secondly, as discussed in Sect. 2.2.2, it may lead to the reduction of the tumor progression speed. Third, it may have a synergistic effect with conventional cytotoxic therapies as discussed in Sect. 2.2.3.

2.4 Recreating the Glioblastoma Micro-environment In-Vitro: Cells in Microfluidic Devices

2.4.1 In Vivo Versus In-Vitro: The Microenvironment Problem

It is now broadly recognized that classical in-vitro experiments with established cell lines in 2D cultures provide only a very limited information on the behaviour of cancer in real patients. This has lead to an effort to study different kinds of 3D in-vitro experiments were not only the dimensionality of physical space but other features of the microenvironment in which cancer cells evolve are preserved [1, 58].

In the specific case of GBM, the aberrant hypoxic microenvironment is difficult to reproduce in controlled in-vitro experiments. The reason is that hypoxia is not only the result of the growth of cells overcoming the nutrient supply (a situation that is well reproduced in experiments with tumor spheroids). It is the pivotal role played by thrombotic events in GBM progression that is difficult to simulate in simple in-vitro models, and interestingly, even in many animal models [11]. Pseudopalisades are not always present in xenografts of human cells into different animal models, probably because of the different growth behaviour of the human cells in the host tissue and the fact that the vessels are provided by the host.

Microfluidic devices have the potential to provide more realistic microenvironments incorporating more elements of the physiologic GBM microenvironment, while at the same time preserving the controllability advantages of in-vitro experiments [2].

2.4.2 A Microfluidic Device for Mimicking the Hypoxic Glioblastoma Microenvironment

Microfluidic devices have been built to mimick GBM physiopathology and observe pseudopalisading structures in-vitro. A scheme of one of these devices is depicted in Fig. 2.10. GBM cells, are embedded within a collagen matrix, i.e. in a fully 3D microenvironment, and located inside a culture chamber of a microfluidic device. Medium perfusion through the two lateral microchannels plays the role of blood

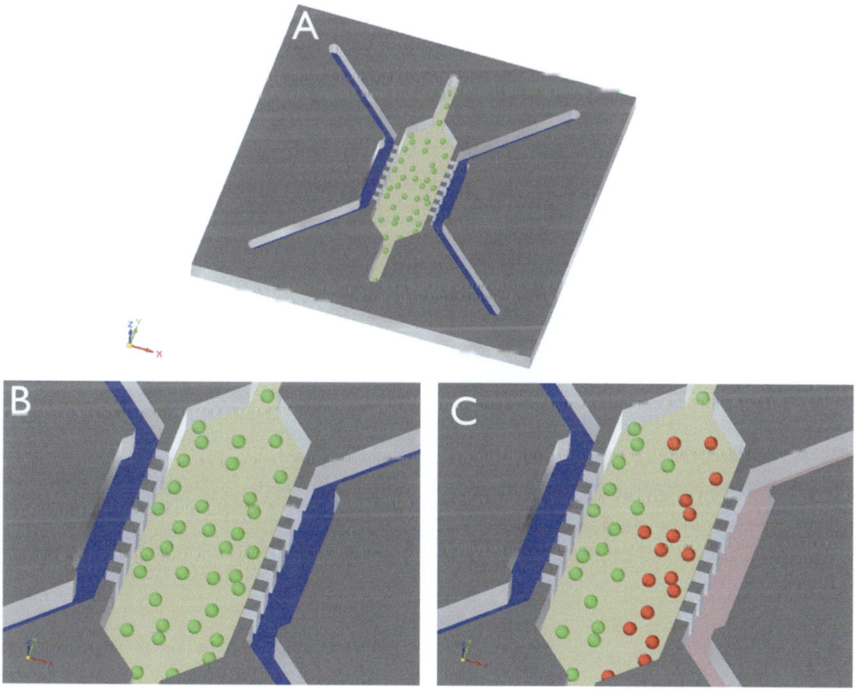

Fig. 2.10 Scheme of the microfluidic device. (**a**) Complete microfluidic device formed by two channels and the culture chamber where cells are embedded within a collagen matrix. (**b**) Device detail where both channels are operative, contributing to the supply of nutrients and oxygen (in *blue*), all cells being in normoxia (in *green*). (**c**) Device detail with one open channel (on the *left* in *blue*) and one closed channel simulating a thrombus (on the *right* in *red*). Cells around the open channel are normoxic (in *green*) and cells around the closed channel become hypoxic (in *red*)

perfusion through blood vessels. In these devices cells are viable for very long times, even longer than one month. Such a medium perfusion can be finely tuned, and thrombosis can be easily reproduced by simply stopping medium flow through any of the channels. When medium flow is stopped on one lateral microchannel, a nutrient/oxygen gradient appears, closely mimicking the spatial gradients arising in the real tumor microenvironment.

These devices allow us to observe the cell dynamics in real time and/or the proliferative status of cells. In the next few years techniques under development will make possible to generate a complete spatio-temporal mapping of the metabolic cell status, oxygen and nutrient densities, etc., thus allowing a quantitative direct comparison with mathematical models.

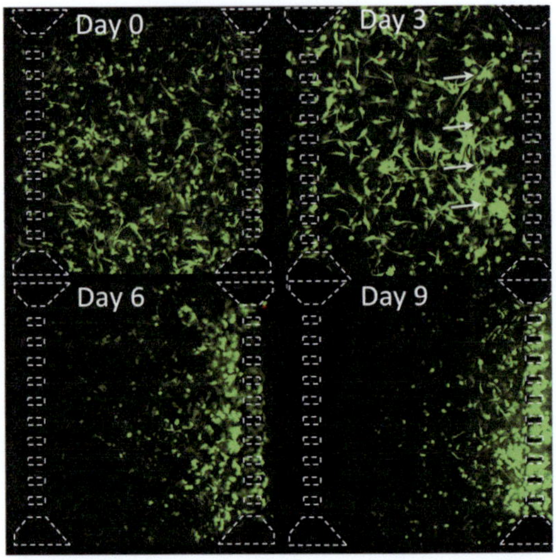

Fig. 2.11 Glioma cell distribution in the microfluidic device. 4×10^6 U-251 cells/ml were seeded in collagen hydrogel at 1.5 mg/ml. Cell viability was assessed using fluorescein diacetate (*green*), *white lines* represent the channels and the arrows show the direction of the cell movement. After day 0 only the channel on the right is operative

2.4.3 Synthetic Pseudopalisades are Generated in Microfluidic Devices

Many realistic situations can be explored by controlling the distances between the channels, their nutrient concentrations and the number of cells seeded in the collagen matrix. Also, it is possible to create chronic hypoxia, not only due to the removal of one of the oxygen sources, but also due to the cell density increase with time. In addition, acute hypoxia appears when one or both of the synthetic capillaries stop contributing oxygen to the medium.

Specifically, synthetic glioma pseudopalisades have been created in this device by seeding U251 human glioma cells and reproducing artificially a vessel-occlusion event. This model has allowed to compare the pseudopalisade characteristic formation time after vessel collapse, and the dynamics and structure of these structures and to compare the results with in silico estimates from mathematical models [3].

Results suggest that when both channels of the microfluidic device work normally and there is a low cellularity, the U-251 glioma cells remain quasi-static within the chamber. This is due to their very slow growth in three-dimensional environments in the absence of other stimuli. However glioma cells migrate when there is only one channel open (the one on the right). After 3 days, glioma cells are already observed to develop structures resembling the pseudopalisades observed in real brain tissue. As time goes on, cells continue moving towards the open channel after 10 days (Fig. 2.11).

2.4.4 Mathematical Model

The mathematical model used to simulate the cell dynamics within the chamber consists of a set of partial differential equations modeling the interplay of three cellular cancer cell phenotypes, the oxygen distribution and necrosis.

The model is based on the one presented in Fig. 2.4 and discussed in Sect. 2.2 where oxygen coming from lateral vessels was the driving force that triggered the celular phenotypic changes between hypoxic and normoxic states. To these two dominant phenotypes, C_h and C_n, based on the go or grow dichotomy [24, 35], we have incorporated a third phenotype C_m accounting for hypoxic cells that arrive to normoxic areas and switch to a more aggressive phenotype when oxygen levels are restores. This more malignant phenotype shows a high proliferative capacity and is essential to explain the phenomenology observed in these devices.Also, in addition to the diffusive random motion, we have included a directional transport term driving the cell's motion towards better oxygenated regions for hypoxic cells. Parameters used in simulations where equal or similar to the ones employed in [36].

The equations governing the interplay between the relevant phenotypes are

$$\frac{\partial C_n}{\partial t} = D_n \nabla^2 C_n + \frac{C_n}{\tau_n}\left(1 - C^T\right) - \frac{S_{nh}}{\tau_{nh}} C_n, \tag{2.1a}$$

$$\frac{\partial C_h}{\partial t} = D_h \nabla^2 C_h + \frac{C_h}{\tau_h}\left(1 - C^T\right) + T_h \nabla O_2 \left(\frac{\partial C_h}{\partial x}\right) \tag{2.1b}$$

$$+ \frac{S_{nh}}{\tau_{nh}} C_n - \frac{S_{hm}}{\tau_{hm}} C_h + \frac{S_{mh}}{\tau_{mh}} C_m - \frac{S_{hd}}{\tau_{hd}} C_h, \tag{2.1c}$$

$$\frac{\partial C_m}{\partial t} = D_m \nabla^2 C_m + \frac{C_m}{\tau_m}\left(1 - C^T\right) - \frac{S_{mh}}{\tau_{mh}} C_m + \frac{S_{hm}}{\tau_{hm}} C_h, \tag{2.1d}$$

$$\frac{\partial C_d}{\partial t} = \frac{S_{hd}}{\tau_{hd}} C_h, \tag{2.1e}$$

where $C^T = \frac{C_n + C_h + C_m + C_d}{C^M}$ and C^M is the carrying capacity of the medium.

2.4.5 Comparison of Synthetic Versus Mathematical Pseudopalisades

Under unrestricted conditions (i.e. with fully functional vessels), the number of cells and their distribution along the chamber is only slightly modified due to the very slow proliferation rates. A typical example is shown in Fig. 2.12a, c for experiments and simulations respectively. In our computer simulations of the mathematical model, after 9 days, all cells had a normoxic phenotype and displayed a similar behaviour. When the nutrient flow along the left channel was disrupted, cellularity remained spatially homogeneous for 3 days, while the nutrient level was being exhausted and cells slowly switched their phenotypes.

Our numerical simulations reveal the formation of a pseudopalisading structure moving towards the active channel after 6 days and an impressive increment in cell density around the right channel at day 9 in excellent agreement with the experimental result (Fig. 2.12b, d).

We have done an extensive parameter value scan and is very interesting to point out that this dynamical behaviour cannot be obtained within the framework of the two-phenotype model of [35], i.e. by assuming that tumor cells revert to their original normoxic phenotypes when reaching areas of higher oxygenation. Then, the incorporation of the third, highly proliferative, phenotype might be essential to reproduce the evolution of the cell density profiles within the chamber.

It is remarkable that later experimental analyses using proliferation markers (Ki67) have shown that the areas of higher oxygenation do not only have a higher accumulation of cells but also a much larger proliferation that confirms the mathematical model predictions and points out to the existence of, at least transiently, more aggressive cell phenotypes after hypoxic episodes [2].

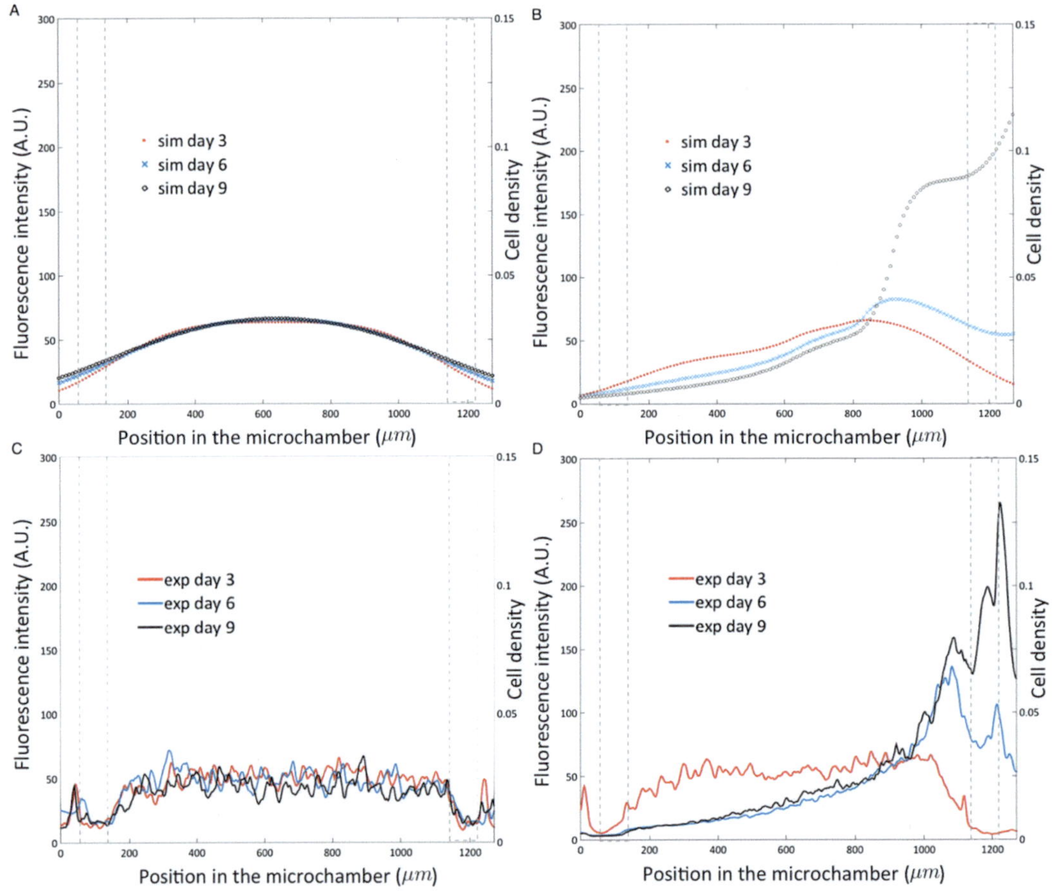

Fig. 2.12 Computer simulations versus experimental data of the cell evolution profiles hypothesis. Simulations of tumor cell density evolution and experimental data of fluorescence intensity within the chamber under unrestricted conditions (**a** and **c**) and under thrombotic conditions (**b** and **d**). *Left Y axis* denotes fluorescence intensity from experiments at days 3 (*red line*), 6 (*blue line*) and 9 (*black line*). *Right Y axis* denotes cell density from simulations at days 3 (*red dots*), 6 (*blue crosses*) and 9 (*black circles*), respectively

2.5 Discussion and Conclusions

In this chapter we have presented another example of the use of mathematical modeling to unveil the intrinsic complexities of the tumor microenvironment. This is a timely topic that is of great relevance due to its potential utility in clinical situations, although it is still not well understood even from the biological point of view. Here we have focused on simple mathematical models trying to describe in the simplest possible way the dynamical interplay of tumor cells and the oxygen distribution for the specific case of gliomas, mainly glioblastoma multiforme.

It is well known that every different cancer type is in some sense a different kind of disease and this applies very specifically to gliomas. These tumors are very infiltrative and do not grow as compact masses, a characteristic that justifies their description using nonlinear reaction-diffusion equations. Moreover, they are not metastatic, which means that the problem remains essentially localized in space within the organ of origin of the tumor. Moreover, glioblastomas are among the most prothrombotic malignancies and the only ones displaying pseudopalisading structures, which play an instrumental role in the progression of these high-grade brain tumors.

We have also tried to provide evidence of the relevance of mathematical models as useful tools to raise hypothesis of potential interest in the clinics. In combination with artificial biological systems, based on microfluidic devices, we have further tested our mathematical models which have lead to sound results. Several of the hypothesis presented here are currently under consideration by clinicians.

Acknowledgements The work of the Mathematical Oncology Laboratory (MôLAB) on the mathematical modelling of the glioma microenvironment and its therapeutical implications is supported by: University of Castilla-La Mancha; Junta de Comunidades de Castilla-La Mancha, Spain (grant number PEII-2014-031-P); Ministerio de Economía y Competitividad/FEDER, Spain (grant numbers: MTM2012-31073 and MTM2015-71200-R); and James S. Mc. Donnell Foundation twenty-first Century Science Initiative in Mathematical and Complex Systems Approaches for Brain Cancer (Special Initiative Collaborative-Planning Grant 220020420 and Collaborative award 220020450). We would like to thank the Pathology Department from Hospital General Universitario de Ciudad Real (Spain) for providing some of the histopathological images and to Dr. Marcial García Rojo from the Pathology Department at Hospital de Jerez de la Frontera, (Spain) for helpful discussions.

References

1. Asghar W, El Assal R, Shafiee H, Pitteri S, Paulmurugan R, Demirci U (2015) Engineering cancer microenvironments for in vitro 3-D tumor models. Mater Today 18:539–553
2. Ayuso JM, Basheer HA, Monge R, Sánchez-Álvarez P, Doblaré M, Shnyder SD, Vinader V, Afarinkia K, Fernández LJ, Ochoa I (2015) Study of the chemotactic response of multicellular spheroids in a microfluidic device. PLoS One 10(10):e0139515
3. Ayuso JM, Monge R, Martinez-González A, Llamazares GA, Berganzo J, Hernández-Laín A, Santolaria J, Doblaré M, Sánchez-Gómez P, Pérez-García VM, Ochoa I, Fernández LJ (2015) Glioblastoma on a microfluidic chip: generating pseudopalisades and enhancing aggressiveness through thrombotic events. Cancer Res 75:B04
4. Badoual M, Gerin C, Deroulers C, Grammaticos B, Llitjos J-F, Oppenheim C, Varlet P (2014) Oedema-based model for diffuse low-grade gliomas: application to clinical cases under radiotherapy. Cell Prolif 47:369–380
5. Bastida E, Ordinas A, Escolar G, Jamieson GA (1984) Tissue factor in microvesicles shed from U87MG human glioblastoma cells induces coagulation, platelet aggregation, and thrombogenesis. Blood 64:177–184
6. Batchelor TT, Byrne TN (2006) Supportive care of brain tumor patients. Hematol Oncol Clin North Am 20:1337–1361
7. Bellail AC, Hunter SB, Brat DJ, Tan C, Van Meir EG (2004) Microregional extracellular matrix heterogeneity in brain modulates glioma cell invasion. Int J Biochem Cell Biol 36:1046–1069
8. Bogdanska MU, Bodnar M, Belmonte-Beitia J, Murek M, Schucht P, Beck J, Pérez-García VM (2016, submitted) A mathematical model of low grade gliomas treated with temozolomide and its therapeutical implications. Math Biosci
9. Brat DJ, Castellano-Sanchez AA, Hunter SB, Pecot M, Cohen C, Hammond EH, Devi SN, Kaur B, Van Meir EG (2004) Pseudopalisades in glioblastoma are hypoxic, express extracellular matrix proteases, and are formed by an actively migrating cell population. Cancer Res 64:920–927
10. Brat DJ, Van Meir EG (2004) Vaso-occlusive and prothrombotic mechanims associated with tumor hypoxia, necrosis, and accelerated growth in glioblastoma. Lab Invest 84:397–405
11. Candolfi M, Curtin JF, Stephen Nichols W, Muhammad AG, King GD, Elizabeth Pluhar G, McNiel EA, Ohlfest JR, Freese AB, Moore PF, Lerner J, Lowenstein PR, Castro MG (2007) Intracranial glioblastoma models in preclinical neuro-oncology: neuropathological characterization and tumor progression. J Neuro-Oncol 85:133–148
12. Elstner A, Holtkamp N, von Deimling A (2007) Involvement of Hif-1α in desferrioxamine-induced invasion of glioblastoma cells. Clin Exp Metastasis 24:57–66
13. Eltzschig HK, Carmeliet P (2011) Hypoxia and inflammation. N Engl J Med 364:656–665
14. Evans SM, Judy KD, Dunphy I, Jenkins WT, Hwang WT, Nelson PT, Lustig RA, Jenkins K, Magarelli DP, Hahn SM, Collins RA, Grady S, Koch CJ (2004) Hypoxia is important in the biology and aggression of human glial brain tumors. Clin Cancer Res 10:8177–8184
15. Furie B, Furie BC (2006) Cancer-associated thrombosis. Blood cells Mol Dis 36:177–181
16. Gatenby RA, Gillies RJ (2004) Why do cancers have high aerobic glycolysis? Nat Rev Cancer 4:891–899
17. Gerin C, Pallud J, Grammaticos B, Mandonnet E, Deroulers C, Varlet P, Capelle L, Taillandier L, Bauchet L, Duffau H, Badoual M (2012) Improving the time-machine: estimating date of birth of grade II gliomas. Cell Prolif 45:76–90
18. Giese A, Bjerkvig R, Berens ME, Westphal M (2003) Cost of migration: invasion of malignant gliomas and implications for treatment. J Clin Oncol 21:1624–1636
19. Gillies R, Gatenby R (2007) Hypoxia and adaptive landscapes in the evolution of carcinogenesis. Cancer Metastasis Rev 26:311–317

20. Green D, Kwaan HC (2009) Coagulation in cancer. Springer, New York/London
21. Hamilton MG, Hull RD, Pineo GF (1994) Prophylaxis of venous thromboembolism in brain tumor patients. J Neurooncol 22:111–126
22. Hanahan D, Weinberg RA (2011) Hallmarks of cancer: the next generation. Cell 144:646–674
23. Hardee ME, Zagzag D (2012) Mechanisms of glioma-associated neovascularization. Am J Pathol 181:1126–1141
24. Hatzikirou H, Basanta D, Simon M, Schaller K, Deutsch A (2012) "Go or Grow": the key to the emergence of invasion in tumour progression. Math Med Biol 29:49–65
25. Horsman MR, Mortensen LS, Petersen JB, Busk M, Overgaard J (2012) Imaging hypoxia to improve radiotherapy outcome. Nat Rev Clin Oncol 9:674–687
26. Huse JT, Holland EC (2010) Targeting brain cancer: advances in the molecular pathology of malignant glioma and medulloblastoma. Nat Rev Cancer 10:319–331
27. Jenkins EO, Schiff D, Mackman N, Key NS (2010) Venous thromboembolism in malignant gliomas. J Thromb Haemost 8:221–227
28. Jensen RL (2009) Brain tumor hypoxia: tumorigenesis, angiogenesis, imaging, pseudoprogression, and as a therapeutic target. J Neurooncol 92:317–335
29. Joiner M, Van der Kogel A (2009) Basic clinical radiobiology, 4th edn. Hodder Arnold, London
30. Keith B, Johnson RS, Simon MC (2012) HIF1α and HIF2α: sibling rivalry in hypoxic tumour growth and progression. Nat Rev Cancer 12:9–22
31. Keunen O, Fack F, Thorsen F, Taxt T, Bartos M, Jirik R, Miletic H, Wang J, Stieber D, Stuhr L, Moen I, Rygh CB, Bjerkvig R, Niclou SP (2011) Anti-VEGF treatment reduces blood supply and increases tumor cell invasion in glioblastoma. Proc Natl Acad Sci USA 108:3749–3754
32. Khorana AA, Francis CW (2008) Cancer associated thrombosis: new findings in translational science, prevention, and treatment. Informa Healthcare, New York
33. Li Z, Xu J, Zhang C, Liu X, Jiang L, Chen F (2013) Mammalian diaphanous-related formin 1 is required for motility and invadopodia formation in human U87 glioblastoma cells. Int J Mol Med 2:383–391
34. Louis DN, Ohgaki H, Wiestler OD, Cavenee WK (2007) World Health Organization classification of tumours of the central nervous system, 4th edn. Renouf Publishing Co. Ltd., Geneva
35. Martínez-González A, Calvo GF, Pérez Romasanta LA, Pérez-García VM (2012) Hypoxic cell waves around necrotic cores in glioblastoma: a biomathematical model and its therapeutic implications. Bull Math Biol 74:2875–2896
36. Martínez-González A, Durán-Prado M, Calvo GF, Alcaín FJ, Pérez-Romasanta LA, Pérez-García VM (2015) Combined therapies of antithrombotics and antioxidants delay in silico brain tumour progression. Math Med Biol 32:239–262
37. Onishi M, Ichikawa T, Kurozumi K, Date I (2011) Angiogenesis and invasion in glioma. Brain Tumor Pathol 28:13–24
38. Pardo R, Martínez-González A, Pérez-García VM (2016) Hypoxic ghost waves accelerates the progression of high-grade gliomas. Commun Nonlinear Sci Numer Simul 39:360–380
39. Pérez-García VM, Pérez-Romasanta L (2016) Extreme protraction for low-grade gliomas: theoretical proof of concept of a novel therapeutical strategy. Math Med Biol. doi:10.1093/imammb/dqv017
40. Pérez-García VM, Calvo GF, Belmonte-Beitia J, Diego D, Pérez-Romasanta L (2011). Bright solitary waves in malignant gliomas. Phys Rev E 84:021921
41. Pérez-García VM, Bogdanska M, Martínez-González A, Belmonte-Beitia J, Schucht P, Pérez-Romasanta L (2015) Delay effects in the response of low grade gliomas to radiotherapy: a mathematical model and its therapeutical implications. Math Med Biol 32:307–329
42. Perry JR, Julian JA, Laperriere NJ, Geerts W, Agnelli G, Rogers LR, Malkin MG, Sawaya R, Baker R, Falanga A, Parpia S, Finch T, Levine MN (2010) PRODIGE: a randomized placebo-controlled trial of dalteparin low-molecular-weight heparin thromboprophylaxis in patients with newly diagnosed malignant glioma. J Thromb Haemost 8:1959–1965
43. Planes A (2003) Review of bemiparin sodium: a new second-generation low molecular weight heparin and its applications in venous thromboembolism. Expert Opin Pharmacother 4:1551–1561
44. Pope WB, Young JR, Ellingson BM (2011) Advances in MRI assessment of gliomas and response to anti-VEGF therapy. Cur Neurol Neurosci Rep 11:336–344
45. Rahman R, Smith A, Rahman C, Grundy R (2010) Antiangiogenic therapy and mechanisms of tumor resistance in malignant glioma. J Oncol 2010(251231):1–16
46. Ranalli NJ, Evans SM, Judy KD (2009) Hypoxia in brain tumors: a review of the literature. Neurosurg Quart 19:1–12
47. Robins H, ONeill A, Gilbert M, Olsen M, Sapiente R, Berkey B, Mehta M (2008) Effect of dalteparin and radiation on survival and thromboembolic events in glioblastoma multiforme: a phase II ECOG trial. Cancer Chemother Pharmacol 62:227–233
48. Rong Y, Post DE, Pieper RO, Durden DL, Van Meir EG, Brat DJ (2005) PTEN and hypoxia regulate tissue factor expression and plasma coagulation by glioblastoma. Cancer Res 65:1406–1413
49. Rong Y, Durden DL, Van Meir EG, Brat DJ (2006) 'Pseudopalisading' necrosis in glioblastoma: a familiar morphologic feature that links vascular pathology, hypoxia, and angiogenesis. J Neuropathol Exp Neurol 65:529–539
50. Schneider T, Mawrin C, Scherlach C, Skalej M, Firsching R (2010) Gliomas in adults. Deutsches Arzteblatt International 107:799–807

51. Simanek R, Vormittag R, Hassler M, Roessler K, Schwarz M, Zielinski C, Pabinger I, Marosi C (2007) Venous thromboembolism and survival in patients with high-grade glioma. Neuro-Oncol 9: 89–95
52. Streiff MB, Segal J, Grossman SA, Kickler TS, Weir EG (2004) ABO blood group is a potent risk factor for venous thromboembolism in patients with malignant gliomas. Cancer 100:1717–1723
53. Svensson KJ, Kucharzewska P, Christianson HC, Sköld S, Löfstedt T, Johansson MC, Mörgelin M, Bengzon J, Ruf W, Belting M (2011) Hypoxia triggers a proangiogenic pathway involving cancer cell microvesicles and PAR-2-mediated heparin-binding EGF signaling in endothelial cells. Proc Nat Acad Sci 108:13147–13152
54. Swanson KR, Rostomily RC, Alvord EC Jr (2008) A mathematical modelling tool for predicting survival of individual patients following resection of glioblastoma: a proof of principle. Br J Cancer 98: 113–119
55. Swanson KR, Rockne RC, Claridge J, Chaplain MA, Alvord EC Jr, Anderson AR (2011) Quantifying the role of angiogenesis in malignant progression of gliomas: in silico modeling integrates imaging and histology. Cancer Res 71:7366–7375
56. Wilson WR, Hay MP (2011) Targeting hypoxia in cancer therapy. Nat Rev Cancer 11: 393–410
57. Wouters A, Pauwels B, Lardon F, Vermorken J (2007) Implications of in vitro research on the effect of radiotherapy and chemotherapy under hypoxic conditions. The Oncologist 12:690–712
58. Xu X, Farach-Carson MC, Jia X (2014) Three-dimensional in vitro tumor models for cancer research and drug evaluation. Biotechnol Adv 32(7):1256–1268
59. Young A, Chapman O, Connor C, Poole C, Rose P, Kakkar AK (2012) Thrombosis and cancer. Nat Rev Clin Oncol 9:437–449
60. Zagzag D, Zhong H, Scalzitti JM, Laughner E, Simons JW, Semenza GL (2000) Expression of hypoxia-inducible factor 1α in brain tumors: association with angiogenesis, invasion, and progression. Cancer 88:2606–2618
61. Zwicker JI, Furie BC, Furie B (2007) Cancer-associated thrombosis. Clin Rev Oncol Hematol 62:126–136

Computer Simulations of the Tumor Vasculature: Applications to Interstitial Fluid Flow, Drug Delivery, and Oxygen Supply

3

Michael Welter and Heiko Rieger

Abstract

Tumor vasculature, the blood vessel network supplying a growing tumor with nutrients such as oxygen or glucose, is in many respects different from the hierarchically organized arterio-venous blood vessel network in normal tissues. Angiogenesis (the formation of new blood vessels), vessel cooption (the integration of existing blood vessels into the tumor vasculature), and vessel regression remodel the healthy vascular network into a tumor-specific vasculature. Integrative models, based on detailed experimental data and physical laws, implement, in silico, the complex interplay of molecular pathways, cell proliferation, migration, and death, tissue microenvironment, mechanical and hydrodynamic forces, and the fine structure of the host tissue vasculature. With the help of computer simulations high-precision information about blood flow patterns, interstitial fluid flow, drug distribution, oxygen and nutrient distribution can be obtained and a plethora of therapeutic protocols can be tested before clinical trials. This chapter provides an overview over the current status of computer simulations of vascular remodeling during tumor growth including interstitial fluid flow, drug delivery, and oxygen supply within the tumor. The model predictions are compared with experimental and clinical data and a number of longstanding physiological paradigms about tumor vasculature and intratumoral solute transport are critically scrutinized.

Keywords

Tumor vascularization • Angiogenesis • Interstitial fluid flow • Drug delivery • Oxygenation • Computer simulation

3.1 Introduction

One of the hallmarks of cancer is angiogenesis, the formation of new blood vessels via sprouting, which fuels tumor growth with additional nutri-

M. Welter • H. Rieger (✉)
Theoretical Physics, Saarland University, Campus E2 6, 66123 Saarbrücken, Germany
e-mail: mwelter@lusi.uni-sb.de; h.rieger@mx.uni-saarland.de

K.A. Rejniak (eds.), *Systems Biology of Tumor Microenvironment*, Advances in Experimental Medicine and Biology 936, DOI 10.1007/978-3-319-42023-3_3

ents [62]. Angiogenesis, vessel cooption (the integration of existing blood vessels into the tumor vasculature), dilatation, and vessel regression remodel the healthy vascular network of the host into a tumor specific vasculature that is different from the arterio-venous blood vessel network of the host tissue [75]. Consequently blood flow, oxygen and nutrient supply, and interstitial fluid flow have tumor specific abnormalities [161] that have dramatic consequences for anti-cancer treatment: (a) tumor vasculature is chaotic, lacking a hierarchical organization, and spatially inhomogeneous comprising regions with low microvascular density (like a necrotic core). As a result, severe hypoxia (deprivation from oxygen) [66] can impede the effectiveness of radiation and chemo therapies [58], and promote invasive growth (migration of tumor cells and penetration of tissue barriers). (b) Tumor vessel walls are leaky, i.e. have a high permeability for blood plasma, and a functioning lymphatic drainage is absent in most malignant tumors, leading to bulk flow of free water in the interstitial space, denoted as interstitial fluid flow (IFF), and a concomitantly elevated interstitial fluid pressure (IFP) [75]. The resulting excessive extravasation of liquid may release most drug prematurely, leading to a retarded delivery into the tumor center, especially in large tumor [74, 76, 81]. Indeed high IFP is regarded as an obstacle in cancer therapy [64, 102]. Therapeutic concepts like vessel normalization via anti-angiogenic therapy have been developed [77] that actually decrease IFP and improve drug penetration in tumors [157].

However, a mechanistic understanding of vascular network formation and various treatment strategies is still lacking and calls for a quantitative analysis of the underlying physics. Drug delivery as well as oxygen supply are determined by blood and interstitial fluid flow, for which reason such an analysis must focus on the relation between the intra- and extra-vascular transport characteristics and the tumor vasculature morphology. Moreover, the analysis must account of the fact that tumor blood vessel networks emerge from, and are connected to the normal, arterio-venous, vasculature of the host.

In this chapter we review the current state of mathematical modeling and simulation of vascularized tumor growth and discuss predictions made by our models for vascular morphology, drug delivery and oxygenation. It is organized as follows: The first section provides an overview of the physiological basics of vascularized tumor growth. It follows a section on obstacles to treatment of cancer. In the subsequent model part we review our work and the related literature, comprising models of vascular network formation, tumor growth, interstitial fluid flow, drug delivery and oxygenation. Then we discuss the various predictions made, limitations of our models, and finally provide an outlook to future work. For further reading on our work, see [11, 90, 165–169]

3.1.1 Physiological Basics

Normal vasculatures are organized in capillaries, small vessels by which most of the solute exchange of nutrients and wastes with blood takes place, and in arterial and venous trees, respectively. Capillaries are organized as homogeneously distributed dense network, the capillary plexus. The walls of capillaries consist mostly of endothelial cells (ECs). This network is supplied by arterial and drained by adjacent arioles and venules, respectively. Arterioles and venules join into larger arteries and veins which eventually join at the heart. Their walls recruit additional cells such as pericytes and smooth muscle cells for reinforcement and control over their diameter. This vascular organization thus minimizes the power required to drive blood and to simultaneously maintain the volume of circulated blood [105]. Normally, maintenance of the vasculature depends on a balance of pro-and antiangiogenic factors such as blood flow and metabolic demand, mediated by a complex biochemical signaling network not yet fully understood. This system adapts the microvascular density (MVD) to the nutrient demand of tissue and regulates development of blood vessels into vascular trees. Components of this system have been studied (see below, in the context of tumors), however the big picture is still elusive.

A solid tumor typically starts off as an avascular multicellular spheroid. It is initially formed, when cells undergo mutations disabling their regulatory circuits for proliferation and apoptosis (programmed cell death) allowing them to divide an infinite number of times. After an initial phase of exponential growth, the radius of a spheroid in nutrient solution continues to grow linearly [20, 39] since proliferation of tumor cells (TCs) is restricted to a few cell layers behind the tumor-tissue interface. Vascularized tumors also show a linear growth regime [38, 67]. TCs beyond an annular outer shell enter a quiescent state due to nutrient and space restrictions or die off (necrosis). Thus a necrotic core develops, and an equilibrium between proliferation and death is established, limiting the size of the spheroid to approximately 1 mm^3. We consider only oxygen as representative of nutrients, which is a common simplification in mathematical models, although tumor metabolism depends on other nutrients and waste products as well. Notably, TCs can switch to a glucose-based metabolism, allowing them to survive hypoxic conditions. Not all tumors start as avascular spheres though. Some types, e.g. glioma brain tumors and breast tumors, incorporate (coopt) the blood vessel network of the host at the beginning of growth [68, 122]. In this process, TCs preferably proliferate around blood vessels, apparently while displacing or destroying cells of normal tissue [37]. The ability to metastasize may develop at a later point in time.

Oxygen in tissue has a high diffusion coefficient of ca. 2 mm^2/s, but it is also bound and consumed which leads to an approximately exponential decrease of the concentration around blood vessels. The range up to which the concentration decreases to zero is typically 100 μm in tumors [25]. In normal tissues it lies between 50 μm (brain) and 150 μm (breast). This diffusion range is thus a major determining factor of the mean intercapillary distance required for adequate oxygen supply. Neither normal cells nor TCs remain viable beyond it. Normal cells as well as tumor cells can respond to hypoxia by releasing chemical compounds known as growthfactors (GFs) which are essential mediator molecules of angiogenic signals. VEGF is a well-known major player [25, 26, 94, 101] but there are many more with various function. They diffuse through tissue where they bind to receptors at blood vessels and collectively they loosen the cell layers of vascular walls, and stimulate ECs to proliferate and to migrate away from their parent vessel. ECs follow GF gradients to the source of GF (chemotaxis) trailed by more ECs that form a new sprout [49, 108, 143]. This process is known as angiogenesis. If the tip encounters another vessel it will fuse with it and mature into a perfused capillary. Otherwise the sprout retracts after some time.

Hence, a hypoxic tumor spheroid might develop a phenotype that enables pro-angiogenic signaling by GFs in an effort to improve its oxygen supply. Like diffusion of oxygen, the angiogenic signal has a finite range. The area where neovascularization is visible in glioma [67] and melanoma [38] is restricted to a 200 μm annular shell around the invasive edge. However, in microscopy images of mammary carcinoma in mice, increased branching and dilation is observed up to ca. 1 mm from the edge [10, Fig.1]. Neovasculature as well as preexisting vessels are coopted when the tumor grows past them. For unknown reasons, tumor vascular network formation is not properly controlled. As a result, dense chaotic vascular excrescence develops (s. Fig. 3.1b), that is very unlike a well ordered normal capillary bed (s. Fig. 3.1a). The additional vessel may provide nutrients required for growth. However, they are often dysfunctional, in some cases even hindering growth [130].

A few 100 μm into the tumor interior, angiogenesis stops and endothelial cells a switch to circumferential growth leading to vaso-dilation. Tumor vessels of Melanoma and Glioma tend to dilate to a maximum radius of ca. 25 μm but no further. Moreover, many vessels undergo a process of regression, until eventual collapse of the lumen and pinch off of blood flow [38, 68]. GFs produced in the tumor interior are partially responsible for the concomitant detachment of supporting cells from the vascular tube, but they also promote ECs survival. Another crucial fac-

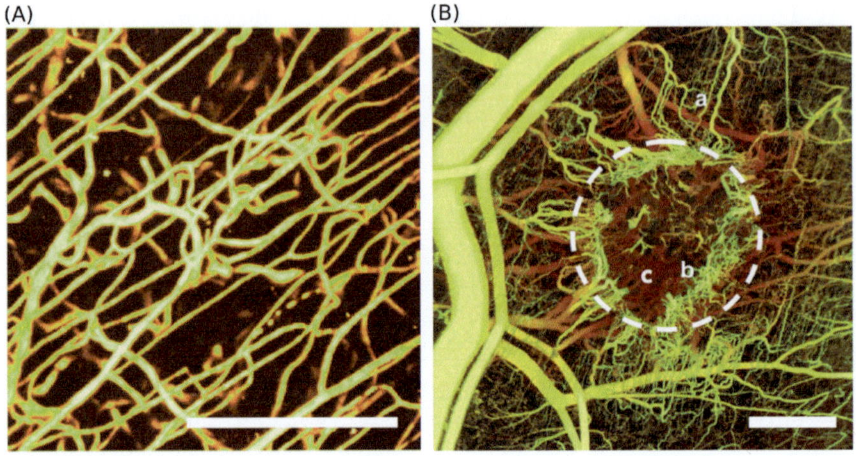

Fig. 3.1 Depth-coded microscopy images of vascular networks: (**a**) A normal capillary network with some supplying and draining arterioles and venules, respectively. Capillaries appear as thin straight segments, which is typical, for instance, for muscle tissue (Scale bar = 100 μm). (**b**) Blood vessel network in a mammary carcinoma bearing mouse (tumor location indicated by *dashed circle*). Vascular remodeling is apparent in proximity of the tumor. Numerous dilated, tortuous vessels proceed from a few parent vessels toward the tumor (*a*). The tumor rim is densely and chaotically vascularized due to excessive branching. The vascular density drops dramatically into the tumor, leaving large regions void of vessels (*c*, *b*; scale bar = 1 mm) (Reprinted from [10] with permission. Copyright 2011 James W. Baish et al.)

tor for survival is blood flow, where Angiopoietins (Ang-1/2) among others act as regulatory molecules [21, 53]. They are expressed by ECs in reaction to the shear stress which is exerted by the blood flow on the vessel wall [7]. Ang-2, a negative regulator of angiogenesis, promoting regression, is frequently overexpressed in tumors [68]. ECs apparently switch from angiogenesis to circumferential growth depending on the sensed direction of the GF concentration gradient [143], which is by the ephB4 guidance molecule [43].

Only few dilated vessels survive this thinning process, leading to a very sparse network of isolated vessels. Viable TCs remain as cuffs around these vessels. Beyond the diffusion range of oxygen, TCs die of hypoxia, whereupon large necrotic regions emerge in the tumor interior. Thus, a normal blood vessel network is progressively transformed into a tumor specific vasculature by the angiogenic activity that is mostly confined to an area around the tumor edge. The result is a compartmentalization into a ca. 200 μm wide band around the periphery where the MVD is elevated to ca. 1.5 times the baseline normal tissue MVD. The MVD decreases sharply into the tumor interior to approximately half of the MVD of normal tissue [38, 67]. Images of experimental tumors are reprinted in Figs. 3.1 and 3.2. Quantitative morphological data from [38] is reprinted in Fig. 3.3.

Normally, only a small amount of blood plasma leaks from blood vessels through nanometer sized gaps between ECs whereupon it becomes part of the interstitial fluid (IF). IF is absorbed into lymphatic channels which eventually feed the liquid back into the blood stream. Leakiness of tumor vessels is caused by huge gaps present in their walls due to missing ECs [25] leaving holes of the size of micrometers. The permeability of the vessel walls therefore increased by two orders of magnitude [83]. Moreover, tumors often lack functional lymphatic vessels, although they can induce lymphangiogenesis similar to regular angiogenesis and can metastasize through lymphatics in the tumor periphery [153]. The lack of lymphatics as well as vascular hyperpermeability lead to the phenomenon of elevated interstitial fluid pressure (IFP), an elevation of the hydrostatic pressure of the IF which approaches

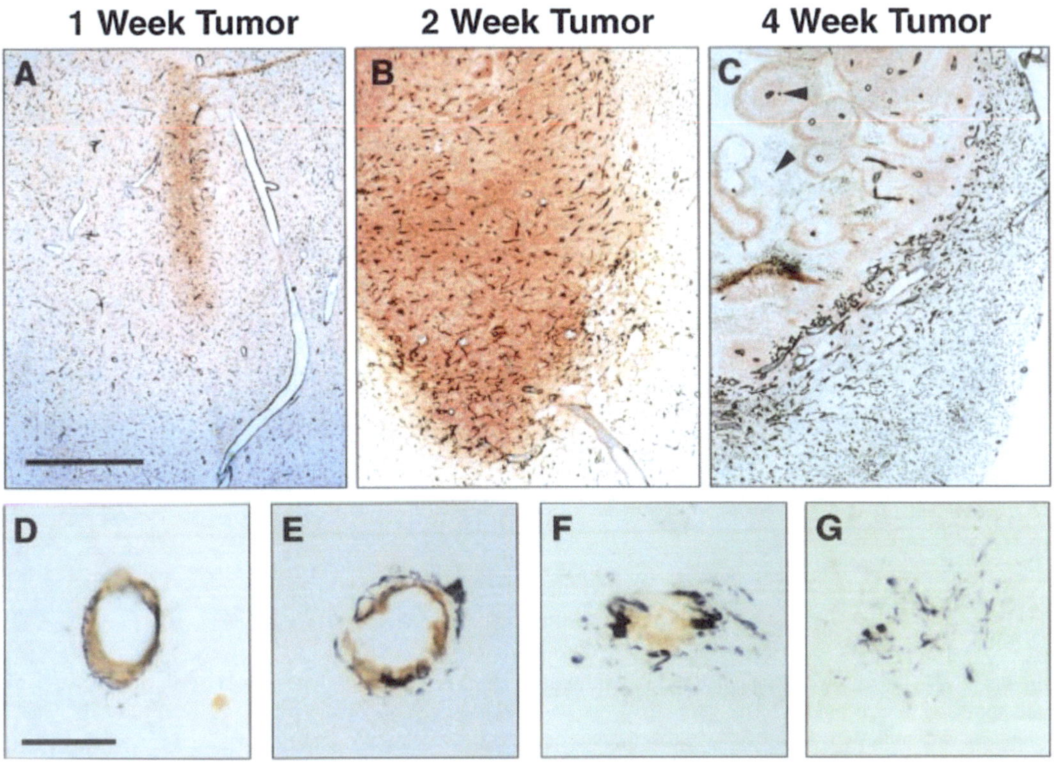

Fig. 3.2 Histological sections of rat glioma brain tumors: (**a–c**) depict the progression of a tumor (viable tumor cells stained *red*; endothelial cells stained *black*; scale bar = 1 mm). Small 1-week tumors exhibit normal appearing blood vessels. After two weeks, decreased density and vaso-dilation are visible. In 4-week tumors, vessels are mostly isolated and have cuffs of viable tumor cells around them. Distal regions are necrotic. The tumor rim is densely vascularized. (**d–g**) depicts regression of a blood vessel with detachment of pericytes and smooth muscle cells (*black*) from the vessel wall (*brown*) (Scale bar = 50 μm; Reprinted from [68] with permission. Copyright 2005 American Association for the Advancement of Science)

the level of blood pressure [152]. The IFP in the tumor interior is relatively homogeneous at levels between 10 and 40 mmHg. Across the tumor boundary it drops down to the level of normal tissue where the IFP is zero in good approximation. The interstitial fluid flows through tissue like water or oil flows through a porous medium, e.g. through rock. In tissues, cells and ECM assume the role of the medium. Consequently, IF flows predominantly in radial direction out of the tumor spheroid. Peak velocities between 0.1 and 0.2 μm/s were measured near the boundary of a 1 cm sized tumor [73]. Elsewhere, velocities are much lower due to shallower IFP gradients. This may drive TCs into the surrounding lymphatics and wash out drug from the tumor.

3.1.2 Obstacles to Cancer Treatment

This section reviews biophysically relevant obstacles to treatment most of which are founded in the peculiarities of tumor blood vessel networks. Current cancer-killing drugs have poor selectivity, i.e. they are toxic to normal cells, too. Therefore, it is not possible to simply increase the dose to compensate for inadequacies of the vasculature [102].

Since tumor vasculatures are heterogeneous, one can find areas in tumors, so called hot-spots, where the MVD is locally increased. The MVD at hot-spots is used as indicator for malignancy and tumor progression with varying success [38]. Therefore a solid understanding of the interac-

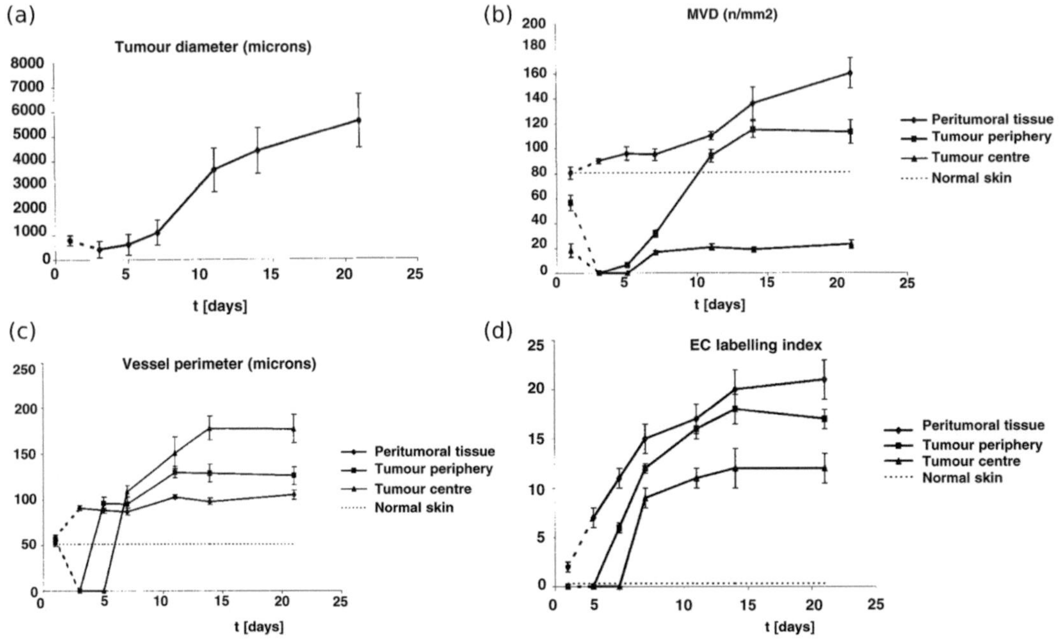

Fig. 3.3 Experimental morphological data of human melanoma in mouse models: The vessel network development was followed during tumor growth from an intradermal inoculation of 10^4 tumor cells until the tumor reached 4–5 mm in diameter. At day 10, tumor growth transitions to a linear regime, consistent the confinement of proliferative activity to an annular shell behind the invasive edge (**a**). (**b–d**) display data for different regions: Tumor center; the tumor periphery – a 100 μm wide band of tumor immediately adjacent to the invasive edge; peritumoral tissue – a 200 μm wide band of host connective tissue immediately adjacent to the tumor periphery. After 15 days, MVD (**b**) and Vessel perimeters (**c**) assume plateau values. Vessels are generally abnormally dilated, and the MVD is high near the invasive edge whereas it stays low in the tumor center. The tumor coopts the dense peripheral vasculature and subsequently dilutes it. Thus the activity of vascular remodeling moves with the invasive edge. EC labeling index (**d**) is essentially the percentage of proliferating endothelial cells (ECs), i.e. the plot indicates angiogenic activity all across the tumor and beyond (Reprinted from [38] with permission. Copyright 2002 John Wiley & Sons, Ltd)

tions between vascular network formation and growth dynamics of the tumor spheroid is required.

The reasons for poor drug delivery are manifold. In addition to premature release and washout due to excessive extravasation, the vasculature is sparse in large areas of the tumor and therefore the efficacy of drugs depends on the ability to penetrate tissue well. However penetration is often poor, instead, strong drug concentration gradients emerge around blood vessels, and persist over long periods of time [121]. Vascular normalization strategies can help [80], but other approaches should be considered, too, such as alteration of tissue permeability.

The discovery of tumor induced angiogenesis [45], and VEGF, sparked the development of a new type of treatment in which the vasculature is targeted with angiogenesis suppressing agents to deprive the tumor of nutrients. This is a so-called anti-angiogenic therapy, today often used concomitantly to other measures, such as chemotherapy. Vascular normalization is a more recent concept, where a balance between excessive pruning and a reduction of angiogenic activity is to be effectuated in order to reduce leakiness and thus improve blood flow [78]. However the underlying mechanisms are still poorly understood. What works for one kind of tumor can have an adverse effect in another type of tumor [102]. Relief of mechanical stress on blood vessels is now also seen as therapeutic opportunity [79] to improve blood flow.

Moreover, the success of ordinary chemo and radiation therapy is tied to the oxygenation status of the tumor. For instance, some chemotherapeutics work poorly in oxygen deprived environments due the chemical reactions involved. Other drugs can only kill cycling (proliferating) cells and are therefore unefficative against tumor cells (TCs) which are quiescent. Hypoxic TCs are also resistant to radiation therapy since oxygen is required so that ionizing radiation can produce DNA damaging compounds [85]. Hypoxia also promotes invasive growth, i.e. the tendency and ability of TCs to migrate increases [113]. Hence, hypoxia is generally associated with poor prognosis [19, 63].

It is possible to obtain important tumor characteristic data such as perfusion, blood volume and hypoxia status from patients using positron emission tomography (PET) and other imaging methods. However the interpretation of raw sensor data requires theoretical models. Moreover the resolution of current methods is limited to a voxel size of ca. 1 mm^3. On the other hand, microscopic information are hardly accessible experimentally. Direct measurements by invasive probes are limited to small sample sizes and may be afflicted with systematic errors [159]. Interstitial fluid flow velocities are measured by invasive microscopy [76], not applicable to humans. Concentration distributions of drugs were measured by microscopy of dissected tumorous tissue [121], exploiting auto-fluorescence. In this regard theoretical models and computer simulation can provide insight into the tumor micro environment in order to foster the understanding of macroscopic phenomena and therapy failures.

3.2 Theoretical Models

This section reviews basic theory and modeling approaches of mathematical models of tumor growth and its microenvironment. See also the Refs. [95, 124, 134, 158] for reviews of recent work.

3.2.1 The Bulk of Tissue

There are two approaches to describe tissues. In continuum mechanics conservation equations are formulated for mass, momentum and sometimes energy, and on the other hand, in particle methods, particles represent either cells or macroscopic sections of tissue and move according to Newtons equations of motion. Fluids and deformable bodies are described in this way, too. But for living tissues, addition and removal of mass and momentum due to growth and death needs to be taken into account. The simplest form of mass conservation to satisfy this is the partial differential equation (PDE)

$$\frac{d\rho}{dt} = -\rho\nabla \cdot \boldsymbol{u} + \alpha, \tag{3.1}$$

where $\rho = \rho(\boldsymbol{x}, t)$ is the density depending on space and time, $d\rho/dt$ is total derivative in time which can be expanded into $d\rho/dt = \partial\rho/\partial t + \boldsymbol{u}\nabla\rho$, ∇ is the Nabla operator, $\boldsymbol{u}$ is the local velocity, and α embodies local sources and drains. The general form of momentum equations is

$$\frac{d(\rho\boldsymbol{u})}{dt} = \nabla \cdot \boldsymbol{\sigma} + \boldsymbol{f}, \tag{3.2}$$

where $\boldsymbol{\sigma}$ is the Cauchy stress tensor and $\boldsymbol{f}$ is the total body force, accounting for gravity for instance. In reality, biological tissues are highly complex materials [162, see the review]. On short time scales, they show elastic behavior which is usually neglected in models of tumor growth. On long time scales, i.e. days, residual stresses are relaxed by rearrangement of ECM fibers and cell adhesion molecules, leading to viscous behavior. Moreover, cells show active responses to stimuli, e.g. migratory behavior. In practice, growing tumors are therefore often modeled like (viscous) liquids, including an isotropic (solid) pressure, friction, and adhesion forces. Inertial forces can be neglected since tissue growth and cell migration happens at very low Reynolds numbers ($Re \ll 1$). Conservation of energy is mostly not considered, assuming a homogeneous constant temperature. The growth of multicellular spheroids [3, 4, 8, 32, 163] and tumors in general

[117, 151, 178] was described using continuum models of a single homogeneous material.

Current state of the art are multi-phase or mixture models where mass, momentum and stress are given as summations over contributions from cells of different types, ECM and water. These phases coexist in space, so that each phase occupies a fraction of the unit volume, given as volume fraction ϕ_i of phase i. The motion of the cell population is often modeled analogous to fluid flow through a porous medium, where the ECM takes the role of the medium. The "flow" thus represents migratory motion in response to solid pressure. Depending on the choice of components and their stress tensors, mixture models describe various growth phenomena, and found numerous applications to study avascular [5,22,32,96,135,171] and vascular tumor growth [18, 29, 71, 97, 148]. Cell-cell adhesion may be modeled by an effective surface tension forces, following [17], allowing the study of growth induced morphological instabilities of the interface between cell populations.

In [169] we introduced a continuum model of the tumor spheroid, closely following [117] and the refs. therein. In principle, a common volume fraction ϕ and a common migration velocity v_ϕ is defined for TCs and normal tissue cells. The interface between TCs and normal cells is defined via an auxiliary function, using the Level Set method [139]. Thus the interface is defined as 0-level of the auxiliary function providing the closest distance from the interface within some proximity of the interface. In real tissues, cell-cell adhesion causes a certain degree of smoothness of the tissue interfaces. This has been neglected, but still our model predicts approximately spherical growth under the assumption of equal motilities of TCs and normal cells. The basic mechanism of tumor expansion of this model is based on an increased tolerance to solid pressure of tumor cells, leading to proliferation whereas proliferation of nearby normal cells is inhibited, eventually leading to apoptosis.

In particle based models, matter is described from the frame of reference that is anchored to a point on a material. In actual computations, a material such as a fluid, is divided into thousands to millions of pieces, represented by particles that move and interact with each other. In biological applications, the particle count is not conserved in general. Instead particles are allowed to replicate or vanish to reflect growth and regression of real tissues. In microscopic systems, particles can be conveniently identified with individual cells. Their time dynamics can be described simply by Newtons equation of motion for each particle as in molecular dynamics simulations, i.e.

$$\frac{\partial m_i v_i}{\partial t} = F_i(x_0, \ldots, x_N, \ldots),$$

where m_i, v_i, x_i denote the mass, velocity and position of the i-th particle, and F_i denotes the force on the particle depending on the current state of the system. These equations must be solved numerically in a discrete time-stepping scheme. In between time steps, an extra step can be added to account for proliferation and death of particles. Continuous space particle models were used to study the growth dynamics of multicellular spheroids [39, 125] and of tissues that are in competition with each other [12]. The dynamics can also be described by stochastic processes and be simulated by Monte-Carlo methods (see below).

In cellular automata models, particles are confined to sites on a lattice. Particles may be able to hop or proliferate to neighboring sites. Due to its simplicity this is a popular approach to study tumor growth [2, 11, 16, 41, 42, 90, 111, 164] and angiogenesis [6, 112, 115, 164]. In the latter case, particles represent pipe segments of a network. It is however more adequate to think of the network as a dynamically changing graph as in mathematical graph theory. The space-time dynamics can be determined by deterministic rules which are applied once per discrete time step, or by stochastic processes, or a mix of both. A stochastic process is formally described by the Master Equation for the rate of change of the probability P_k to find the system in state $k = 1..n$

$$\frac{dP_k}{dt} = \sum_l A_{kl} P_l.$$

The matrix A_{kl} contains the transition rates according to which the system transitions from state l to k with probability $A_{kl}dt$.

Bartha and Rieger proposed a simple particle model of individual TCs [11]. Therein, lattice sites are identified with the potential location of one and only one TC, assuming that TCs cannot move but proliferate to neighboring sites. Given a small initial tumor nucleus, proliferation is consequently confined to the tumor rim, yielding linear growth dynamics of tumor spheroids. Moreover, TCs can be flagged as dead in case that the oxygen concentration becomes too low. Dead TC occupy lattice sites, prohibiting proliferation thereto, but are otherwise inert. Thus the size and spatial distribution of necrotic regions can be analyzed. This model is simple but in conjunction with a model of tumor vascular remodeling it is sufficient to predict realistic morphologies of tumor vasculatures [90, 165, 166]. However, the representation of individual cells in three dimensions at macroscopic system sizes is computationally costly. Therefore coarser grained models are better suited there.

3.2.2 Solutes in the Bulk of Tissue

The simplest general partial differential equation to describe the transport of the concentration $c(\boldsymbol{x}, t)$ of one species is the diffusion-advection-reaction equation

$$\frac{\partial c}{\partial t} + \nabla \cdot (c\boldsymbol{u}) = \nabla \cdot (D\nabla c) + R, \qquad (3.3)$$

where the substance diffuses with diffusion constant D and is carried with the flow of the solute with velocity $\boldsymbol{u}$. The reaction term R can comprise sources and drains, e.g. vessels are sources of oxygen whereas binding and consumption may be represented by a homogeneous drain distribution. In multi-components system, each component i is associated with the concentration c_i each of which is governed by an equation of type (3.3) [117]. Then R (or rather R_i) also comprise transition rates between compartments. This way, drug binding to different intracellular compartments was described in simulations of drug concentrations in tumors [141]. The advection term in calculations of oxygen distributions is usually neglected since oxygen transport is dominated by diffusion due to its low molecular weight. Moreover, it is sufficient to consider quasi stationary distributions where $\partial c/\partial t = 0$ since equilibration times are much shorter than growth processes in tissues [11]. Thus we obtain

$$0 = D\nabla^2 c + R, \qquad (3.4)$$

assuming equal concentration c in all compartments and a constant oxygen diffusion coefficient D.

Balance equations like (3.2, 3.2, 3.3) can only be solved analytically in special cases. Often solutions are calculated numerically with the help of finite difference (FD) [93] or finite element methods (FEM) or some variation thereof. FEM have the advantage that they can be applied straight forwardly to unstructured meshes and therefore work well for arbitrary domain shapes. However, FD methods are easier to implement for regular grids, making them well suited for problems where the expansion of a tumor within a rectangular domain is considered. The application of difference operators leads to systems of linear or non-linear equations in the solution values at grid points. The obtained system matrices are usually sparse, for which many specialized tools are available including direct factorization, fast Fourier transformation, multi-grid, and iterative preconditioned Krylov subspace methods.

3.2.3 Normal Blood Vessel Networks

A model for tumor vascularization must start with the blood vessel network of the healthy tissue surrounding the tumor, since, during growth, the tumor coopts the existing tissue vasculature and generates new vessels via angiogenesis. In early models of angiogenesis, the initial network consisted only of a single parent vessel [6]. These models adequately describe angiogenesis in the rabbit eye model [51]. Essentially, a small tumor on the cornea of the rabbit eye stimulates vascular sprouting in a few large parent vessels from up to 1 mm away. These sprouts branch excessively and form a dense capillary mesh between the tumor and the parent vessels. Similar configurations were considered in later theoretical work [145, 146, 172, 173]. However, in reality, the bulk of tissue is interspersed with vessels which may be coopted by the tumor. Therefore, recent works consider a capillary plexus, often represented by a network of segments which are arranged in a regular pattern, e.g. as square or hexagonal grid, omitting supplying arteries and draining veins [2, 11, 16, 23, 90, 111, 164, 173, 174]. Blood flow is computed assuming a fixed blood pressure at boundaries of the simulation box. In [165] a honeycomb pattern is used in place of a square pattern [11] allowing for more realistic branching angles. Other authors use random arrangements of lines [50] or voronoi cells as basis for vascular networks [128].

Only a few attempts have been made to incorporate physiologically relevant arterio-venous vessel networks. First works focused on algorithmic construction of arterial trees branch by branch [132, 133]. At each step, the existing tree is first geometrically scaled to increase in dimension. This increases the distance between vessels, implying that tissue oxygenation would worsen each step. However a new segment is added according to some optimality criterion in order to supply the voxel in space that is most in need of oxygen. Thus, oxygenation stays approximately constant. This process is repeated until the desired size is reached.

Later, Gödde and Kurz [52] developed a relatively simple lattice based growth model comprising the entire vasculature including arteries, capillaries and veins. Therefore, such an arterio-venous vasculature construction model was implemented for the study of tumor growth [166]. In the following we sketch the construction principle (s. Fig. 3.4): first arterial and venous trees are simultaneously grown by successive attachment of bifurcations at randomly selected tree tips. A bifurcation is simply a Y-shaped arrangement of three segments as depicted in Fig. 3.4h. Lattice sites and bonds can only be occupied once, thus growth terminates eventually when no further free sites are available. Thus the space is divided into areas with only arterial and only venous vessels, respectively (s. Fig. 3.4b). Proper interdigitating trees are obtained by the second stage of the algorithm where vascular trees are remodeled, allowing well perfused branches to expand and weakly perfused branches to regress (s. Fig. 3.4c–e). To this end, each remodeling sweep is preceded with determination of vascular radii, addition of temporary connecting segments (capillaries), and computation of blood flow rates and shear stress f. Capillaries are removed again before the vascular trees are altered, however, they are added again for the final output. An overview of our implementation is given below, but a definition in every detail is beyond the scope of this chapter.

As input, the locations of tree roots and their type, i.e. arterial or venous, are given and mark the starting sites for growth (s. Fig. 3.4a). In previous work their selection was arbitrary, i.e. we considered a wide range of configurations using single pairs of nodes, two pairs [166], or occupation of entire side faces of the cuboid simulation domain [169]. In pseudo code, the first stage of random growth reads as follows

```
nodes = rootNodes // a list
while nodes not empty:
  nd = RemoveRandomItem(nodes)
  // return removed item
  newNodes
  = TryAppendBifurcation(nd)
  // return list of nodes;
  may be empty
  nodes += newNodes  // append
```

TryAppendBifurcation probes orientations along the axes of the lattice, taking already occupied sites into account, and picks an admissible configuration randomly if there are any, adding it to the network. The loop terminates when no more space is available, i.e. the list *sites* is empty. Such as state is depicted in Fig. 3.4b. The second stage is more involved due to the dependence on blood flow. Hence, we define

```
function
  CapillariesRadiiAndBloodflow():
  ComputeRadii()
  AddCapillaries()
  ComputeFlow()
```

The function *ComputeRadii* traverses each vascular tree in a simple depth first traversal and

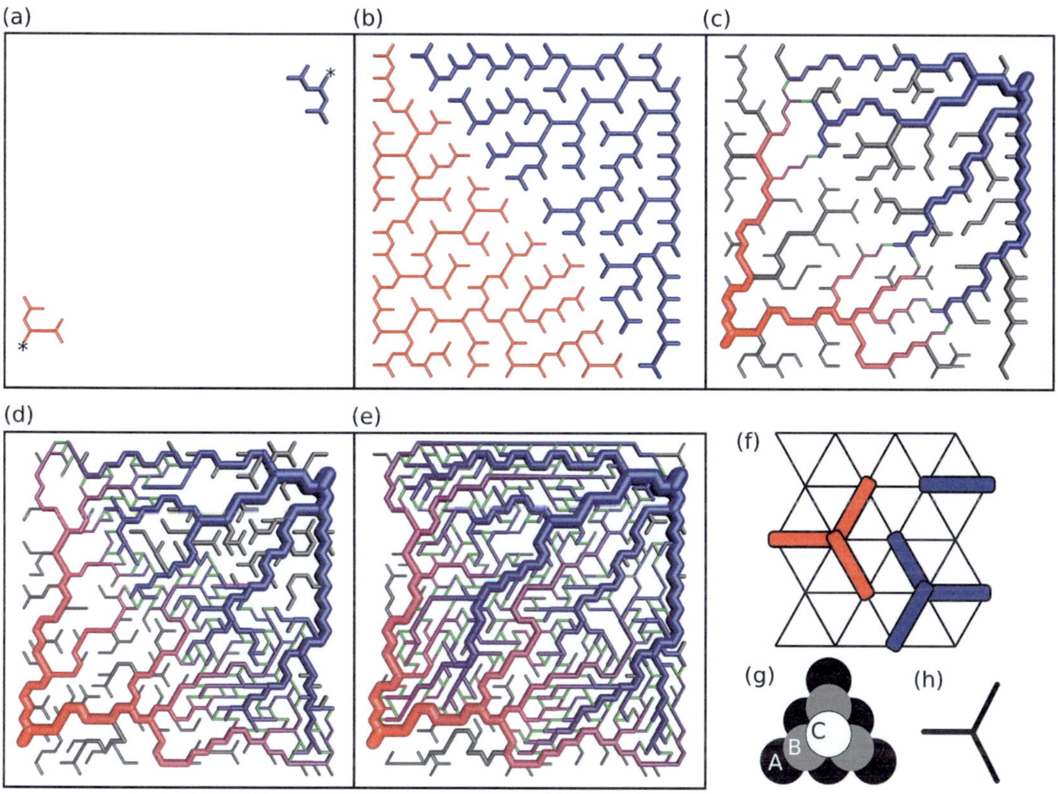

Fig. 3.4 Arterio-venous blood vessel network synthesis: (**a**) The configuration after two steps of the first growth stage. The initial state comprised only two nodes (*). On each side, arterial (*red*) and venous (*blue*), three tripods, as the one depicted in (**h**), were added, creating four tip nodes, respectively. Vessel segments occupy lattice bonds as shown in (**f**) as *red* and *blue bars* and a lattice in the background. (**b**) Both trees were expanded by successive addition of tripods to tip nodes. Eventually, exclusion of occupied sites prevents further additions. This situation is also depicted in small in (**f**). (**c**) At each iteration of the second stage, capillary interconnections (*green*) are inserted where arterial and venous side are separated by only one lattice bond. Then radii are determined and blood flow is computed, arriving at a configuration as depicted in (**c**). Uncirculated branches (*dark grey*) emerge where no connections are made. (**d**) The state after 40 iterations, where weakly, or uncirculated branches cleared space for growth of other branches. (**e**) The result after 1000 iterations. (**g**) Stacking order of FCC lattices for the extension to three dimensions. The layers *A*, *B*, and *C* consist of triangular lattices as depicted in (**f**) which are shifted against each other. Vessel radii in panels (**a**–**e**) are magnified by a factor of four. In panels (**c**–**e**) vessels are color coded by blood pressure (except capillaries)

determines the radius of each segment starting from tree tips up to root nodes. Murray's law is utilized to determine the radius of a parent branch r_p when the radii of child branches $r_{c,1}$,$r_{c,2}$ are known, which states that $r_p^\alpha = r_{c,1}^\alpha + r_{c,2}^\alpha$, with an exponent α between 2.7 and 3 depending on the tissue. The radii of the arterial and venous tree tips are all equal, respectively. Each of the two is an input parameter. The function *AddCapillaries* loops over all nodes of the network, and attempts find neighbors of opposing type (arterial or venous), to which, if admissible, a capillary segment is added. This is carried out again under the exclusion of overlap with other segments. Moreover, (i) in general, at any point, at most three segments are allowed to join at a node. Potential additions of capillaries violating this rule are rejected. (ii) We found it useful for promotion of growth to allow capillaries between vessels of a radius up to a limit of 5 to 20 μm, rather than creating only tip-to-tip connections as done in Ref. [52]. *ComputeFlow* computes blood pressure, flow rates, and shear stress f associated with nodes and vessels as discussed in Sect. 3.2.5 (see below). The main loop of the second stage of the algorithm is as follows

```
for iter = 0 to maxIter:
  CapillariesRadiiAndBloodflow()
  RemoveCapillaries()
  event = dictionary()
  // map nodes to events
  for each Node nd in network:
    event[nd]
    = DetermineRemodelingEvent()
    // does the work
  for each Node nd in network:
  // in random order
    if events[nd] == REGRESSION:
      Remove(nd)
    else if events[nd] == GROWTH:
      added
      = TryAppendBifurcation(nd)
      if not added: // space is
        occupied
        TryAppendSingleSegment(nd)
// prepare final output
CapillariesRadiiAndBloodflow()
```

Segments marked as capillaries have to be removed again, which is carried out by *RemoveCapillaries*. The function *DetermineRemodelingEvent* determines whether a node is marked for *REGRESSION*, *GROWTH*, or for another event denoted *NONE*, indicating no change. *Remove(nd)* also removes adjacent vessel segments. However, since only tree tips are allowed to regress there is only one such segment. Moreover, we found the attempt to insert a single segment to help with grow into crowded spaces. Hence *TryAppendSingleSegment* acts analogous to *TryAppendBifurcation* but adds just a single segment.

The following definition of *DetermineRemodelingEvent* is to some degree arbitrary. However the essential mechanism is growth of perfused branches while others regress. First, let p_G, p_R, and p_N be probabilities for growth, regression and no change. We define them differently for nodes that are perfused ($q > 0$ in at least one adjacent segment) on the one hand, and nodes that are unperfused on the other hand ($q = 0$ for all adjacent segments). For unperfused nodes we simply define

$$p_G = p_{G,x} \tag{3.5}$$

$$p_R = 1 - p_{G,x} \tag{3.6}$$

$$p_N = 0, \tag{3.7}$$

where $p_{G,x}$ determines the rate of regression and is chosen less then $1/2$ to obtain $p_G < p_R$. Figure 3.4e was obtained with $p_{G,x} = 0.4$. Thus unperfused nodes may clear space for growing branches. To define probabilities for circulated branches, let f_{max} be the maximal shear stress taken over all segments, and f be the shear stress average of segments at the considered node. Hence we define the growth "signal" $f_{sig} = f/(f + \epsilon_1 f_{max})$, where $\epsilon_1 \ll 1$ is small number. Taking $\epsilon_1 = 10^{-2}$ one obtains a rapidly increasing function in f which approaches nearly one (0.99 for $\epsilon_1 = 10^{-2}$) for $f = f_{max}$ (see below). The probabilities are defined using f_{sig} as follows

$$p_G = f_{sig}^{\beta} \tag{3.8}$$

$$p_R = (1 - f_{sig})^{\beta} \tag{3.9}$$

$$p_N = 1 - p_G - p_R, \tag{3.10}$$

where β is an exponent larger or equal to one. As a result the growth probability p_G never assumes the value one, which is useful in two-dimensional cases where very well perfused vessels would otherwise form bottlenecks. Moreover taking $\beta > 1$ stabilizes moderately perfused vessels, for which then $p_N > 0$ is obtained. One of the corresponding events is preliminarily picked using tower sampling. However certain conditions need to be fulfilled to be admissible. To grow, a node has to have less than three adjacent segment. To regress, the node has to be a tip node, having only one attached segment. If any of these conditions is not fulfilled, *NONE* is assigned to the node.

Finally we want to add some remarks. First, by setting appropriate values for the lattice constant and capillary radii, the *MVD* and vascular volume *rBV* of generated networks can be adjusted. Secondly, in two dimensions large areas may be left void, depending on selection of root nodes. Such cases were rejected in Ref. [166].

Moreover we found it helpful to vary $p_{G,x} \leq 1/2$ in proportion to the local concentration of growth factors. A corresponding distribution may be incorporated into the model for instance adopting the simplified model in Ref. [11]. This model variant has the advantage that it does not require manual tuning of $p_{G,x}$. Biologically it is justified since vascular sprouts grown by angiogenic signaling via growth factors are also initially not perfused. In the model, unperfused vessels that remain from the first stage of the algorithm, can guide growing branches into a proper tree-like morphology when contact is made with perfused branches.

The extension to FCC lattices seems over complicated, but FCC lattices can be represented by layers of two-dimensional triangular lattices which are offset from each other according to the well known-stacking order ABCABC... (s. Fig. 3.4g). We actually organize sites as sites of a three-dimensional cubic lattice that is spatially distorted to coincide with the conceptual FCC lattice. Neighbors of a given site must correspond to the FCC lattice. Therefore, exploiting translational symmetry, we store precomputed neighbor lists, for a total of six of exemplary sites. Generation of arterio-venous initial networks was carried out again using Y-junctions as structural elements of growth, following the original proposal [52]. Additional rotational degrees of freedom simply add to the number of probed configurations of which one admissible is picked for addition to the network.

3.2.4 Tumor Vascular Remodeling

Bartha and Rieger [11] originally considered a model of tumor vascularization including the processes angiogenesis, cooption, vessel dilatation, regression and collapse (s. Fig. 3.5a). Its basic ingredients are as follows: Vessel segments representing a vascular network, mathematically described as a graph, occupy bonds on a lattice. Junctions (nodes) coincide with sites of the lattice. Various properties are associated with vessels and nodes, such as blood pressure p, radius r, blood flow rates q, and shear stress exerted by the blood flow on vessel walls f. Furthermore, there is a concentration distribution of VEGF representative of all GFs, as well as a tissue oxygen concentration distribution. The system state at $t = 0$ comprises an initial network as described above, and small tumor spheroid in the center of the system. The growth of the spheroid depends on the local oxygen concentration, allowing cell proliferation if the concentration is sufficiently high. If the oxygen concentration drops to hypoxic levels then GFs are locally produced and diffuse into tissue. Diffusion through tissue can be modeled by reaction-diffusion equations (3.4), however simplified models were used [11].

The spatio-temporal evolution of the network is determined by stochastic and continuous processes, reflecting sprouting angiogenesis, vessel dilation, collapse, and regression, respectively (s. Fig. 3.5b–f). In practice, time is advanced in discrete steps of length $\Delta t = 1\,\text{h}$, and these processes are defined approximately as simple local updating rules:

Angiogenesis: A new segment is added with non-zero probability, connecting the current site $\boldsymbol{x}$ and a distant site $\boldsymbol{x}'$ under the following conditions: Both sites are occupied by circulated vessels, the GF concentration at $\boldsymbol{x}$ is sufficiently high, the distance of $\boldsymbol{x}$ to other branching points is at least $d^{(br,min)}$, the distance $|\boldsymbol{x}-\boldsymbol{x}'|$ is small enough, and no site on the path is occupied by TCs. These conditions reflect lateral inhibition of sprouting (for a modeling approach see [15]), finite growth length of sprouts [108], and the switch to circumferential growth within tumors [43].

Dilation: There is a non-zero probability that the radius r of a vessel segment is increased by the amount corresponding to the addition of 10 μm (diameter of an EC) to its circumference, under the following conditions: the local GF concentration is sufficiently high, the segment is located within the tumor, and r is smaller than the upper limit $r^{(max)}$. The latter condition accounts for observations in real tumors [38, 67], however the mechanism that limits dilation is unknown. It should be emphasized that this process is particular important for blood flow characteristics within the tumor since the blood flow varies with the fourth power of the radius and only modest vessel radius increase by a factor of 2 or 3 has leads to an extreme increase in blood flow.

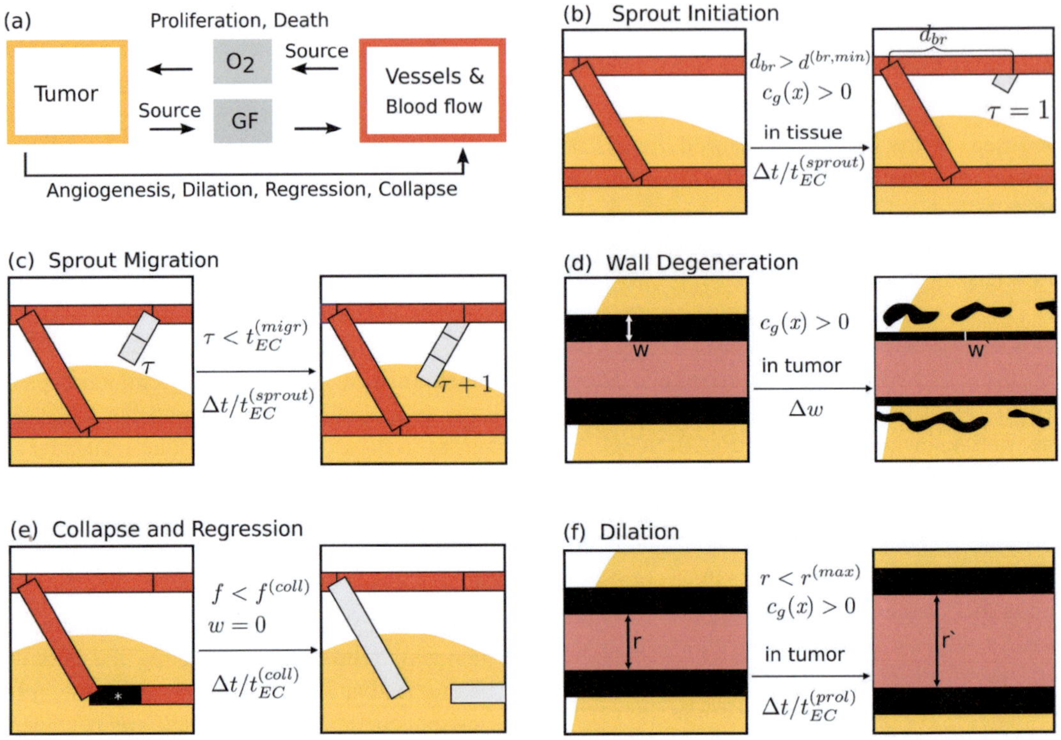

Fig. 3.5 Model of tumor vascular network remodeling: Following [11, 166], tumor and vascular network interact via concentration distributions of growthfactors (GF) and oxygen (O2), where tumor cells are sources of GF and the vascular network is the source of O2 (**a**). Blood flow is computed after alterations of the vascular network to reflect the changes in blood pressure, flow rates and shear stresses. Tumor cells can proliferate in response to a sufficient O2 supply, and will die to O2 deprivation. The dynamical processes of network remodeling are illustrated in (**b–f**), showing the state of vessel segments (*red bars*) before (*left*) and after (*right*) the respective transition. Preconditions are indicated above the center arrows, and transition probabilities are denoted below, respectively. Panel (**b**) depicts the start of a new sprout (*shaded*). A preexisting segment is split at the branching point. The path length on the network to the next branching point d_{br} must be larger than the lower limit $d^{(br,min)}$. Moreover, a sufficient GF concentration c_g must be present and sprouting is not allowed within the tumor mass (*yellow*). The new segment is initialized with an associated lifetime of $\tau = 1$. Panel (**c**) depicts the further extension of the sprout from (**b**). Additional segments inherit τ from the parent segment. Moreover τ is incremented, globally, for all sprouts once per time step. Panel (**d**) depicts the degradation of vessel walls (*black*). The variable w represents the strength of the vessel wall, depicted as varying thickness. It decreases continuously according to the rate Δw, resulting in a value of w' at the next time step. In (**e**) an unstable vessel (*) is removed, representing occlusion of blood flow and complete disintegration. Such event is assumed to happen only to vessels with maximally degenerate walls, $w = 0$ and low wall shear-stresses f, where $f < f^{(coll)}$. The emerging dead ends (*shaded bars*) trivially have $f < f^{(coll)}$, and therefore collapse rapidly, resulting in a long ranged effect. (**f**) depicts the dilation of tumor vessels. Their radii increases at a rate that is given by the area added to the lumen surface assuming division of endothelial cells of the wall every $t_{EC}^{(prol)}$ hours

Collapse: A vessel segment surrounded by TCs has a non-zero probability to be removed if the wall shear stress f is less than the threshold f_{crit}. This process reflects the dependence of vessel survival and maturation on blood flow [68].

Regression: An uncirculated vessel segment has a non-zero probability to be removed if the oxygen concentration is less than a threshold, reflecting complete disintegration of unperfused sections of the vasculature.

One time step comprises the application of these rules at all sites occupied by the network and subsequent recomputation of blood flow, oxygen distribution, and time propagation of other model components such as the tumor spheroid. The probabilities are given as fractions $\Delta t/\tau_{proc}$ of the time step Δt and process specific time constants τ_{proc}, requiring $\Delta t < \tau_{proc}$.

In [11] the network was represented by sequences of 10 μm wide pieces, representing endothelial cells that occupy lattice sites. However, it is much more computationally efficient to associate vessel segments with a series of lattice bonds and allow for segments longer than a single bond. However we still use a basic lattice constant $h_{(tum)}$ of 10 μm, to allow for a sufficiently high resolution of the tumor neovascular plexus. To conserve memory we store pointers to segments in a hash table [156] using a pair of sites indices as key. The lattice constant $h_{(gen)}$ of the initial network synthesis model corresponds is normally larger than $h_{(tum)}$. This is well defined since for proper choice of $h_{(gen)}$, e.g. 100 μm, the location of initial vessels coincides with bonds and sites of the finer $h_{(tum)} = 10$ μm lattice.

In subsequent work [165] we considered a minor extension to the angiogenesis process where sprouts grow over a period of time. Instead of creating a "bridge" instantly, a sprout segment is added and extended with additional segments in subsequent time steps until a timer t_{sp} associated with the sprout runs out. This allows for emulation of tip splitting by sprouting off of a growing sprout and fusion of sprouting branches [165].

To make the model applicable to arterio-venous initial networks, we incorporated a stability variable w associated with segments [166], reflecting the wall strength of vessels, allowing for thick vessels to be more resistant to collapse. The wall strength, w, is continuously decreased at rate Δw until zero, and only then a segment is allowed to collapse (be deleted).

3.2.5 Computation of Blood Flow and Hematocrit

Circulated vessel, i.e. vessels which are perfused at rates $q > 0$, can be determined with the help of the biconnected component graph algorithm [69]. To robustly handle general cases, including arterio-venous networks, we first make an ad hoc augmentation to the network: all boundary (root) nodes are connected temporarily to an extra node which is added. Then the set of perfused vessel is the biconnected component that comprises all edges for which a loop, without repetition of nodes or edges (simple cycle), exists which they are part of and which also runs through the extra node (see also [156]). For partially remodeled square or other regular networks, the augmentation can be omitted, using any of the boundary nodes instead.

Depending on the application, blood flow can be considered on various scales, from computation of the velocity field on micrometer scale to bulk perfusion measured in *ml blood ml tissue*$^{-1}$*min*$^{-1}$ as obtained for instance by positron emission tomography (PET). For pipe networks of tumors models, blood flow is approximated as ideal laminar flow, where the flow rates q define the blood volume throughput per time through each pipe. Blood pressure, p, is associated with nodes. Thus, q is determined by Poiseuille's law

$$q = \frac{\pi r^4}{8\eta} \frac{\Delta p}{l}, \tag{3.11}$$

where r is the vessel radius, η the viscosity, l the length, and Δp denotes the blood pressure difference between the ends of the segment. Conservation of mass requires that the flow into a node equals the flow out of the node, analogous to Kirchoff's laws of electricity, i.e.

$$\sum_i q_i = 0, \tag{3.12}$$

where i indexes vessels adjacent to a given node under consideration. Together with boundary conditions, a system of equations is obtained which is sparse and can be solved with direct factorization or preconditioned conjugate gradient.

Blood contains red blood cell (RBC) causing non-Newtonian behavior, i.e. its viscosity $\eta(r, H)$ depends on the vessel radius r and on blood hematocrit H, where hematocrit is the blood volume fraction of RBCs. The viscosity is commonly expressed by the decomposition into the product of the blood plasma viscosity η_{plasma}, which is constant, and a correction factor, the relative viscosity $\eta_{rel}(r, H)$. Pries et al. [120] derived a well-known phenomenological formula for $\eta_{rel}(r, H)$ which is easy to incorporate. The distribution of hematocrit is sensitive to blood flow rates, i.e., at bifurcations, RBCs tend to flow into the faster perfused vascular branch which is the well-known phase separation effect. Pries and Secomb [118], developed a phenomenological formula that describes this effect in dependence on flow rates of the vessels at the bifurcation. This allows the computation of the hematocrit in downstream branches. Under assumption of given flow rates, the hematocrit distribution can thus be propagated downstream through the network. By iteration, a self-consistent solution for the hematocrit distribution and blood flow rates can be computed [118], which is utilized in several works [2,16,111,115,165,167]. However, as a first approximation, it may be sufficient to consider a constant prescribed hematocrit as in [11, 168, 169]. This is justified because in spite of a wide value range of the relative viscosity ($\eta_{rel}(r = 25\,\mu\text{m}, H = 0.15) \approx 1.5$, $\eta_{rel}(r = 5\,\mu\text{m}, H = 0.6) = 8$), the dependence of the flow resistance on on r^4 plays a much greater role.

3.2.6 Time Dependent Intravascular Tracer Concentration

Rather than computing stationary concentration distributions, we are interested in following an injected bolus of some substances during the transit through the vascular network. For this purpose, Mc Dougall and Anderson [146] already adopted a method from petrol engineering, originally developed to predict solute transport through porous rock. It allows for computation of time dependent concentrations of a tracer c associated with segments of a vascular networks. Essentially tracer flows into nodes where it accumulates, amounting to mass m. From there it is distributed downstream in proportion to the flow rates q of downstream vessels. This procedure, akin to the upwind-differencing scheme for advection equations, is applied repeatedly in time steps of length Δt. The amounts of substance from upstream vessels, added into downstream nodes, is accordingly $\Delta m = cq\Delta t$. Thus, given a time dependent inlet concentration $c_{in}(t)$, the method yields concentrations $c(t)$ of each vessel. Transvascular loss was not considered although the method would be straight forward to extend to take this into account.

We applied the method to networks created by our tumor growth simulation for regular [165, 168] and arterio-venous initial networks [166]. However, in our network model, the assumption that network edges are of constant length, and short compared to their radius is violated, leading to an amplified propagation velocity. Therefore we track the position of the interface that separates clean blood from tracer "contaminated" blood and moves with the velocity of blood flow. Similar models were developed for the simulation of capillary rise in network models of porous materials [1] and are widely used there, e.g. in [123].

3.2.7 Interstitial Fluid Pressure

Interstitial fluid flow (IFF) is modeled as liquid flowing through a porous medium [24, 81, 83, 137, 138, 173, 174, 176], where tissue cells and the fibers of the extracellular matrix assume the role of the medium. Fluid and medium are described in general within the framework of mixture theory with the help of distributions of their local volume fraction and their velocity distributions. However, the medium is often assumed rigid. The volume fraction of the liquid is identified with the porosity ϵ which describes the amount of space available per unit volume within the medium. This space is filled by definition with the liquid. Assuming rigidity and (quasi) stationary flow, the system is characterized by the spatial velocity field of the liquid, $v(x)$, where x is the space coordinate. The velocity v is determined by the gradient of the IFP p_i according to the well-known Darcy's Law

$$v = -K\nabla p_i, \tag{3.13}$$

where the permeability constant K is the product of an intrinsic permeability constant of the medium, the porosity and the inverse fluid viscosity. Usually, K is obtained directly from experimental data for a specific tissue type. Assuming incompressibility and constant permeability, the mass conservation equation obtained is a Poisson equation in p_i:

$$\nabla \cdot v = -K\nabla^2 p_i = Q, \tag{3.14}$$

where $Q = J_v + J_l$ was added to represent sources and drains with contributions from vessels, J_v, and lymphatics, J_l. We adopted this simple approach to determine IFP and IFF in vascular networks of simulated tumors [169]. Some authors consider IFF within a fully coupled mixture model, where v is the relative velocity between the IF and a moving cell population [171]. Other authors incorporate IFF into models of tumor growth and allow compression of blood vessels due to elevated IFP [174]. Penta and Ambrosi used data of a simulated microscopic volume [114] to predict IFF in macroscopic systems. Zhao et al. [176] used imaging data of real tumors as basis for simulations using a continuum model.

3.2.8 Transvascular Fluid Exchange

The net transvascular liquid flux J_v is driven predominantly by the difference of blood to interstitial fluid pressure. This is expressed by Starling equation

$$J_v = L_p S\left[p_v - p_i + \sigma_T(\pi_v - \pi_i)\right], \tag{3.15}$$

where L_p is the hydraulic permeability of vessel walls, S is the vascular surface area within a given control volume, p_v is the blood pressure, p_i is the interstitial pressure, σ_T is the average osmotic reflection coefficient and π_v and π_i are the osmotic pressures of plasma and IF, respectively [81]. The osmotic term $\sigma_T(\pi_v - \pi_i)$ represents forces generated by dissolved substances and can be considered as a constant offset from p_v at an experimentally determined value. This model of liquid exchange is straight forward to apply if the vascular network is considered as homogeneous phase [81].

Otherwise (3.14) may be taken as definition of a local source strength of a spatially varying IFP distribution. This is facilitated by letting the vessel network occupy the same lattice used for discretization of (3.14) as done in Refs. [24, 173, 174]. Then each node of the vessel network j corresponds to a discretization site of (3.14), so that the flux between them is directly proportional to (3.15) with suitable choice of S corresponding to the surface area of vessels adjacent to node j. Using the standard finite difference stencil for the Laplace operator in (3.14) one obtains a combined system of equations, equivalent to Kirchhoff's laws. The same strategy can be used to simulate drug delivery [138, 141] and oxygenation [33, 44, 86, 103, 142]. We add that drainage due to lymphatics J_l is in all of the literature known to us modeled as continuous sink density analogous to (3.15).

More generally, vessels can be considered as line-like sources akin to the Dirac δ distribution [14, 70], a concept which has been formulated mathematically rigorously for the solution of elliptic equations with Dirac terms by finite element methods [34] and applied to IFF [28]. We can thus replace (3.15) by the distribution

$$J(y) = \int_\Gamma L_p 2\pi r(\tilde{p}_v - p_i)\delta(\boldsymbol{x} - \boldsymbol{y})dx \tag{3.16}$$

where $\boldsymbol{x}, \boldsymbol{y}$ are spatial coordinates on the network and in the bulk of tissue respectively. Γ is the set of one-dimensional curves (or line segments) that describes the vascular network, $\tilde{p}_v$ is the effective blood pressure including the osmosis terms, and r is the vessel radius. The permeability L_p, blood pressure $\tilde{p}_v$ and radius r can vary depending on the position on the network $\boldsymbol{x}$.

The latter approach was taken by us to simulate IFF in simulated tumors grown within synthetic arterio-venous vasculatures [169]. We took inspiration from immersed boundary methods [116] and replaced the Dirac δ distribution with a smoothed kernel δ_ϵ of width $\epsilon > 0$ to allow for resolution of the source distribution J on a grid of finite cell size. Thus the source distribution of vessels is "smeared" over nearby grid cells, very similar to the method used by [14].

3.2.9 Interstitial Drug Transport

Spatio-temporal distributions of macro-molecules were studied theoretically with the help of homogeneous compartment models in spherical symmetry, incorporating diffusion and interstitial fluid flow [13, 81]. In a similar way [176] albumin concentrations were simulated in a continuous but non-symmetrical tumorous tissue. In a theoretical study of drug transport in tumors, [141] the discrete nature of blood vessels was accounted for on the basis of a tumor grown in an square-patterned initial network ($t = 0$).

We followed [141] in the development of a simple model of drug transport guided by data for Doxorubicin, a common chemotherapy drug [169]. In this model, the local drug concentration is divided among an extracellular compartment with concentration $s_1(\boldsymbol{x})$ [169, Eqn. 20] and an intracellular compartment with concentration $s_2(\boldsymbol{x})$ [169, Eqn. 21] where drug is bound immobile. The extracellular concentration s_1 is subject to diffusion and advection with the liquid velocity v_l according to (3.3). Vessels are sources and drains of drug (s. Sects. 3.2.8, and 3.2.10) comprising diffusive and advective transvascular flux densities [169, Eqn. 23]. Lymphatics can sink drug by advection, assuming that the drug concentration within lymphatics is approximately equal to the concentration in tissue. Consequently, drug diffusion into lymphatics is neglected. Both compartments 1 and 2 exchange drug via rates k_{12} and k_{21} depending on assumed trans-membrane diffusion coefficient and cell surface area. For simplicity, degradation of drug molecules is neglected. In future this should be straight forward to add, provided experimental data. The initial condition is a system clean of drug. Drug is inserted via the vasculature where the intravascular concentration $s^v(t)$ is homogeneous in space and follows a exponentially decaying pulse in time, imitating an injection.

We applied the model to study drug transport in tissues supplied by tumor vascular networks embedded within synthetic initial arterio-venous networks [169]. A cohort of tumors was considered. We first simulated tumor growth and then considered drug transport for stationary final ($t = 800\,\text{h}$) configurations. Interstitial fluid velocity distributions $v_l(\boldsymbol{x})$ were determined prior to computation of drug concentrations.

3.2.10 Oxygen Transport

Oxygen diffuses across the blood tissue interface with a net flux that depends on the difference of oxygen partial pressure (PO2) at the vessel wall and within blood [65]. As oxygen diffuses into tissue, its concentration in blood is reduced, leading to a gradient across the micro-vasculature of ca. 100 mmHg at the arterial side and 40 mmHg at the venous side. The coupling of transvascular oxygen flux with the tissue PO2 therefore poses a difficult problem for the computation of intravascular and tissue oxygen distributions.

This problem has been solved for simple configurations where single, straight artificial capillaries are considered. Based on original ideas of Krogh [87], current sophisticated theoretical models achieve very good agreement with experimental data [65, 104, 106, 107].

For many applications it may be sufficient to simply consider a constant blood *PO*2. Then the tissue PO2 distributions P_t can by computed by solution of the reaction diffusion equation (3.4). This is very common approach in the literature on models of tumor growth. In other works the tissue oxygen distribution is analyzed in detail based on stationary configurations of disjoint collections of lines or points (in two-dimensions) representing sources of oxygen [33,35,44,86,88,89,103,142]. The limitation of such models is however that the PO2 in each source must be given as input.

In tumor however, low flow rates may lead to depletion of intravascular oxygen over short distances, making it necessary to model intravascular PO2 variations. However due to the complexity of tumor blood vessel networks, intravascular oxygen distributions are hard to predict without actually simulating them. Some authors attacked this problem [46, 55–57, 70, 126, 127, 136, 160, 167] and computed self-consistent solutions of the equations for intravascular advection of oxygen and diffusion of oxygen in tissue for systems comprising realistic blood vessel networks. For numerical methods, see [57, 136, 167].

To cope with the computation of intravascular PO2 distributions in complex networks compromises must be made (see [54] for a review). Most importantly, vessels are treated as one-dimensional line segments and intravascular PO2 variations in the radial direction are neglected. Instead, the average over the cross-sectional area is considered, $P(x)$, depending only on the position on the center line x. This is justified because radial variations of intravascular oxygen concentrations are relatively small as revealed by theoretical calculations [104].

In the modeling of intravascular oxygenation it is crucial to take into account that oxygen is, for the most part, bound to hemoglobin in red blood cells (RBCs). The steady state of the binding and unbinding processes is described in good approximation by the Hill-curve [54]

$$S(P) = \frac{P^n}{P^n + P_{50}^n}, \qquad (3.17)$$

where P is the partial pressure of oxygen, $S(P)$ is the fraction of oxygen bound relative to the maximal capacity, c_0 is the concentration of oxygen in RBCs at full saturation, n is the Hill exponent and P_{50} denotes the partial pressure of oxygen where $S(P_{50}) = 1/2$. Hence, the total concentration of oxygen c is given by

$$c = \alpha P + H c_0 S(P), \qquad (3.18)$$

where H is the hematocrit and $\alpha = \alpha_p + H\alpha_{rbc}$ is the effective solubility in blood and α_p, α_{rbc} the solubility in plasma and RBCs, respectively.

In large scale network models it is infeasible to compute all microscopic details of spatio-temporal intravascular PO2 distributions and outward diffusion. Instead, the net transvascular flux per blood-tissue interface surface area j_{tv} is determined by the effective, network dependent, mass transfer coefficient (MTC) γ, similar to L_p of Eq. (3.15)

$$j_{tv} = \gamma(P - P_t), \qquad (3.19)$$

where P_t is the $PO2$ at the inner wall of the vessel lumen, and P is the average partial pressure in blood. Note that γ represents an effective radial diffusion coefficient of oxygen in blood. L_p of the Starling equation, on the other hand, represents the permeability of the wall. In small vessels, blood tends to form an RBC-rich core and a RBC-free boundary layer. For larger vessels ($r > 100\,\mu$m), the discrete nature of RBCs plays a lesser role. Therefore the MTC is function of the vessel radius r, hematocrit H, and blood oxygen saturation S [65]. The functional dependency $\gamma(r, H, S)$ can be obtained from single capillary simulations and experiments. Moreover, since vessels are much longer than their diameter it is reasonable to assume that the tissue PO2 is homogeneous over the vessel circumference [136]. Thus, integration yields a transvascular oxygen flux per length amounting to $2\pi r j_{tv}$, The change of the oxygen flux along the vessel axis is therefore simply given by the

$$q\frac{dc}{dx} = -2\pi r j_{tv}, \qquad (3.20)$$

where q is the blood flow rate, and x denotes the longitudinal space coordinate on the vessel axis. In order to determine the oxygen distribution across an entire network, assumptions must be made on the distribution at vessel junctions, e.g. instant equilibration of the partial pressure of oxygen flowing into a junction. With the help of mass balance equations, the concentration of outflowing oxygen can be computed. Thus the solution for the oxygen concentration can be propagated downstream assuming a known tissue PO2 distribution and a given PO2 at the inlets (see [136, 167]).

Locally, at the blood-tissue interface, j_{tv} is also subject to Fick's law $j_{tv} = -\alpha \nabla P$, in addition to (3.19). This relation can be utilized to obtain boundary conditions for a diffusion equation that determines the tissue PO2 [57].

However in this chapter we want to consider the network as volumetric sources of oxygen $J_{tv}(\boldsymbol{x})$. This is well-defined since the oxygen flux into tissue is already known from (3.19). Therefore, J_{tv} may be formulated with the help of the Dirac δ distribution in analogy to (3.16) [136, 167].

The tissue oxygen concentration $c_t = \alpha_t P_t$ is determined by the diffusion equation for the partial pressure P_t

$$0 = \alpha_t D \nabla^2 P_t - M(P_t) + J_{tv}, \qquad (3.21)$$

where D is the diffusion coefficient of oxygen in tissue, $M(P)$ is the partial pressure dependent consumption rate. A good approximation of $M(P)$ is the well-known Michaelis-Menten relation

$$M(P) = M_0 \frac{P}{P + P'_{50}}, \qquad (3.22)$$

which tends to zero for small P, assumes the value $M_0/2$ for $P = P'_{50}$ and goes asymptotically to the maximal consumption rate M_0. For some problems like tumor oxygenation it is usually assumed that the oxygen concentration is rather low, i.e. $P_t < P'_{50}$. Then it is sufficient to use a linear approximation $M(P) \approx -\lambda P$ for some rate coefficient λ. In physiological conditions, where $P > P'_{50}$, $M(P)$ is often approximated by zero order kinetics $M(P) \approx M_0$.

Discretization of the model equations yields a complex system of non-linear equations. Following [14, 136] we developed a new numerical scheme based on finite differences which is sufficiently efficient, allowing us to study three-dimensional networks in a simulation box of ca. $0.5\,\mathrm{cm}^3$ at reasonable accuracy [167]. Our method was applied to study the relation of vascular morphology to clinical data of tissue blood oxygen saturation in human breast cancers. Hsu and Secomb [70, 136] formulated a solution to the system of equations with the help of a Green's function method. Their method was applied to study oxygenation by various small network sections obtain from animal models as well as synthetic human brain vasculatures [126].

Methods developed for the study of oxygen distributions are also applicable to distributions of other substances like drugs which may be simpler since oxygen adds the complication of hemoglobin binding which leads to nonlinear systems of equations.

3.3 Discussion of Model Predictions

Current state of the art models of vascularized solid tumor growth and capillary network remodeling predict the morphological compartmentalization of tumor blood vessel networks in good agreement with experimental data of melanoma and glioma [38, 67, 68]. From the obtained configurations, of which one is shown in Figs. 3.6 and 3.7 conclusions can be drawn on the mechanisms of vascularization. Further conclusions, using model extensions, can be drawn for interstitial fluid flow and solute transport, as discussed in the following.

3.3.1 Vascular Morphology and Compartmentalization

Typical vascular compartmentalization is characterized by dense chaotic vascular sprouting within an annular shell of a width amounting to ca. $200\,\mu\mathrm{m}$ around the invasive edge, and a sharp decrease of vascular density into tumor spheroid. The normal vasculature is progressively transformed while the invasive edge moves forward, leaving predominantly isolated vessels behind. The ingredients, to obtain such characteristics from theoretical models comprise an expanding tumor spheroid, an initial capillary network, blood flow, a growthfactor concentration distribution, an oxygen concentration distribution, and processes reflecting co-option, angiogenesis, vaso-dilation, regression and collapse [11, 90]. The basic mechanism of this remodeling was identified as shear stress correlated collapse. Dilatation causes a decrease in flow rates and shear stress since the blood volume that the tumor vasculature conducts per time is limited by the

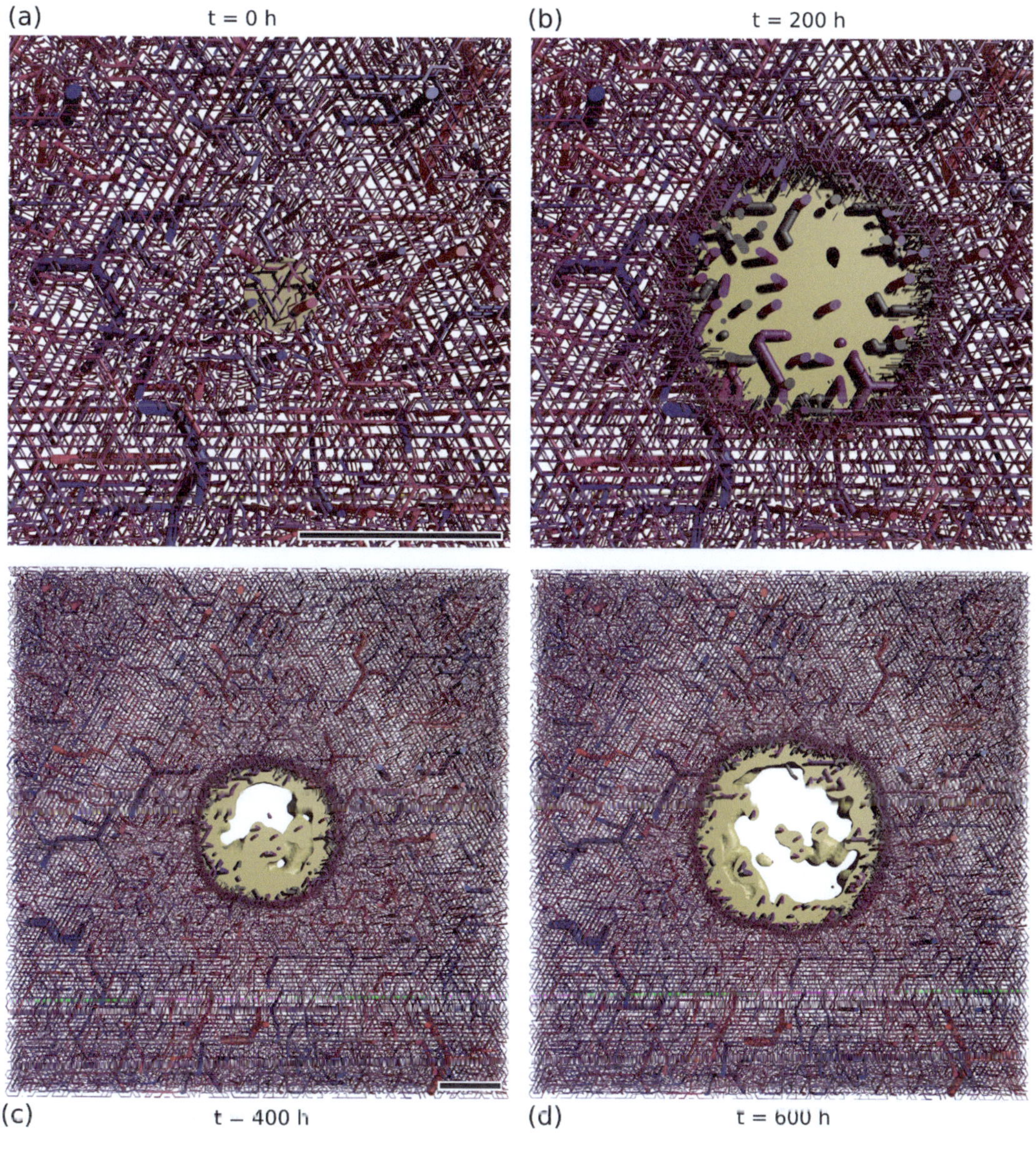

Fig. 3.6 Simulated tumor growth and vascular remodeling: The image sequence shows the temporal evolution of the vascular network and of the viable tumor mass (*yellow*). It is a three-dimensional system, computed for [169], of which a 400 μm thick slice through the system origin is shown. The tumor mass is cut in a slice only half as thick to show the vascular network in its interior. Blood vessel are represented by cylinders, color coded by blood pressure (*red*: approximately 10 kPa, or 75 mmHg, *blue*: 0 mmHg). (**a**) At $t = 0$ h the simulation is initialized with a small tumor nucleus in the center and a pre-generated vasculature of the host. The oxygen consumption of tumor cells is elevated compared to normal tissue, leading to a drop of the tissue oxygen concentration, secretion of diffusing GF and stimulation of angiogenesis. (**b**) As a result, at $t = 200$, the vascular density (MVD) has increased near the tumor rim. Unperfused segments (*dark gray*), i.e. dead ends, are visible. Some of them are newly extending angiogenic sprouts. Others pertain to vessel segment chains where one segment has been removed according to the vascular regression and collapse process, pinching off blood flow. Angiogenesis, dilation and regression act mostly near the expanding tumor-tissue interface, transforming the host vasculature into a typical compartmentalized tumor network. (**c**) A necrotic core emerges as a result of hypoxia and drastically decreased vascular density. Since only viable areas are shown, the necrotic core appears as hollow interior. (**d**) Isolated vessels emerge that have cuffs of viable tumor cells (TCs) around them (Scale bar: 1 mm)

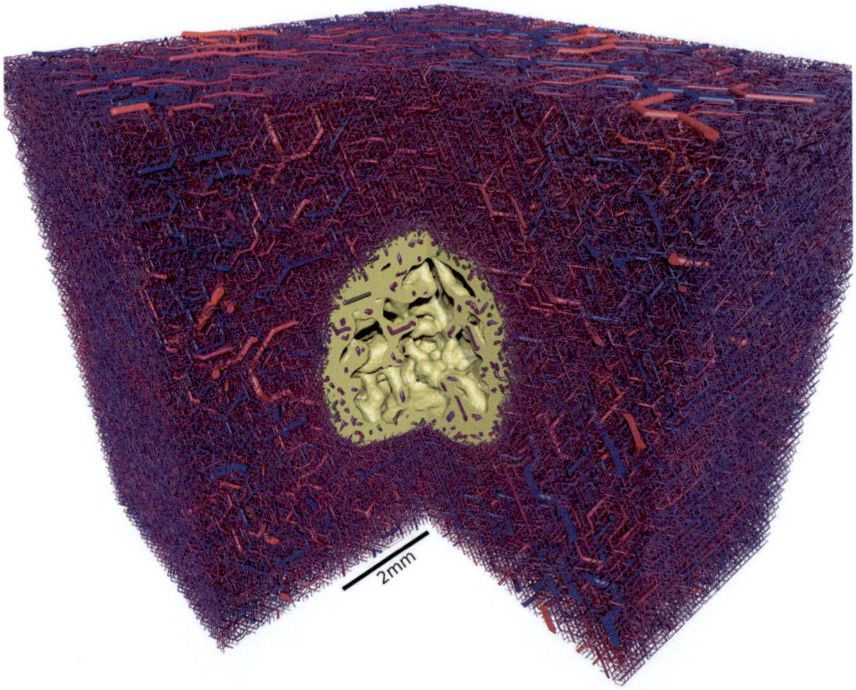

Fig. 3.7 Final simulated tumor and tumor blood vessel network: Depicted is a visualization of the final state of the simulation shown in Fig. 3.6 at $t = 700$ h, where the simulation is stopped. The full simulation cube of 8 mm lateral length is shown, where a quadrant is cut out, so that the tumor spheroid and its interior can be seen. The tumor vasculature exhibits the typical compartmentalization found in melanoma and glioma [37, 67]. It is connected to the bulk of the surrounding vascular network which appears solid, but actually fills only ca. 10 % of the available volume. It is spatially homogeneously distributed and consists of arterial and venous trees and interconnecting capillaries. Configurations such as this are the basis of further studies of interstitial fluid pressure and drug transport [169] and tumor oxygenation [167]

flow resistance of the surrounding vasculature. This leads to removal of segments according to the collapse rule, redirecting blood flow to other vessels. As a result, blood flow and shear stress is stabilized above the critical collapse threshold in surviving vessels. Remaining dead ends are rapidly removed by the regression process.

In synthetic capillary-only initial networks (CNs), vessels of identical diameter are laid out in regular square or hexagonal patterns. However, it is hardly possible to select realistic blood flow boundary conditions for such networks of macroscopic size beyond a few hundred micrometers. For instance, imposing a homogeneous blood pressure gradient yields tumor vascular networks where tumor vessels survive preferably in the direction parallel to the imposed gradient [11, 165]. The explanation is simply that vessel segments of linear chains that run, on average, perpendicular to the gradient, lie on approximately equal blood pressure potentials and therefore no significant blood flow can occur, resulting in collapse of these vessels.

In reality, the capillary plexus is however supplied and drained by adjacent arterioles and venules which exhibit irregular spatial configurations. Therefore, there is no global flow direction, which is why arterio-venous initial networks (AVNs) abolish this artifact in model predictions [166]. In AVNs blood flow depends on only a few boundary conditions at in-and outlets for which experimental reference values for pressure or blood flow can be used. Models based on synthetic arterio-venous networks predict vascular morphologies which obey realistic compartmentalization of MVD and radii.

However, in addition to dilated capillaries, the tumor center also exhibits higher-caliber vessels co-opted from the initial network. Such vessels exhibit a radius r larger than the maximal dilation threshold r_{max} and are therefore not subject to dilation. As a result, predictions of flow rate q are a factor of 10 larger than predicted for CNs.

Model predictions of average quantities such as radial distributions of *MVD*, blood flow, oxygenation and tumor density are robust against model alterations, as studied in Refs [165, 166]. This is true in particular for the rather drastic alteration of the introduction of arterio-venous blood vessel networks (AVNs) [166]. Other model variations, such as calculation of blood flow in conjunction with varying hematocrit, or use of spatially varying collapse probabilities, do not change predictions qualitatively [165]. The parameters vessel collapse probability $p^{(col)}$, wall degradation rate Δw, critical collapse shear stress $f^{(col)}$ and contact inhibition length of angiogenesis $d^{(br,min)}$ correlate with the *MVD* obtained for the tumor center. The MVD at the invasive edge is determined by the MVD of the original network and the contact inhibition length $d^{(br,min)}$. A certain invariance against model details is expected and even required, because it would be implausible if the results were dependent on a specific abstraction of the biological reality (within reasonable accuracy).

Our model predicts that MVD of the tumor interior, MVD at the tumor periphery, and tumor expansion speed are uncorrelated if the peripheral blood vessel network can support the metabolic demand of tumor cells required for growth [11]. Growth within AVNs additionally leads to clustering of vessels in clusters of differing size and density depending on the initial network configuration [166]. The density of such hot-spots is used as a diagnostic tool [38]. However these results suggest that it rather unreliable. A recent meta-study [110] of clinical data comes to the same conclusion. Correlations between MVD and the outcome of the disease is likely due to metastases which was not considered.

We add that we considered the line density L_D, the summed lengths of vessel segments within a given region per volume of this region, as a measure for *MVD*. It is however not the same as the histological MVD because vessels in parallel to the cutting-plane which contribute to L_D cause L_D to overestimate the MVD by a factor of approximately two.

3.3.2 Fractal Properties of Tumor Vasculatures

Following [11], fractal dimension numbers were computed for vascular networks. Fractal dimension d_f is an extension of the conventional dimension to self-similar (fractal) objects. For instance a line has $d_f = 1$, but a fractal curve within the two-dimensional plane can have d_f between 1 and 2 depending on how densely it permeates space. $d_f = 2$ corresponds to a solid object like a disc. For real objects of finite size several approximative metrics exists, e.g. the number obtained by box-counting [98]. Useful model systems are percolation clusters: In conventional percolation, sites of a lattice are randomly occupied with probability p. At some critical probability p_c, a percolating cluster forms that spans across the lateral size of the considered domain. The dimension of this cluster is exactly known $d_f^{perc} = 1.891$ [149]. Similarly, a system spanning cluster can be created from an invasive growth process into a heterogeneous matrix, the dimension of which is known and amounts to $d_f^{inv-perc} = 1.81$ in two dimensions [47].

Gazit [48] measured the dimension of photographs of tumor vascular networks and obtained $d_f^{exp} = 1.89 \pm 0.04$, in good agreement with $d_f^{inv-perc}$ and therefore hypothesized that fractal properties of tumor vascular networks emerge from angiogenic sprouting into a heterogeneous extracellular matrix. Bartha and Rieger [11] obtained $d_f = 1.85$ by box-counting from the entire vasculature that was changed by tumor vascular remodeling. Since no ECM heterogeneities were modeled, it was hypothesized that the mechanism leading to the fractal properties is a random dilution process similar to conventional percolation. Later simulations predicted d_f between 1.6 and 1.9, correlated with the tumor MVD, where the parameters critical collapse shear-stress $f^{(col)}$ and collapse probability $p^{(col)}$ were varied, with similar results respectively [165]. Fractal dimensions of tumor vascular networks obtained from simulations based on three-dimensional AVNs [168] yielded $d_f = 2.51 \pm 0.03$, in good agreement with percolation theory and [90] where tumors in three-dimensional CNs are considered. However, accurate measurement of fractal dimensions of real and simulated tumor vascular networks is hardly possible due to their limited size [11]. Moreover different methods were used: theoretical values were determined by two-or three-dimensional box-counting, whereas Gazit [48] considered two-dimensional projections of real vasculatures. We conclude that fractal dimension is mostly a function of MVD and that it is not a reliable means to determine mechanism of vascularization in tumors.

Morphological analysis was approached from another angle in [166], where frequency distributions of (i) local MVD, (ii) area of clusters of necrotic tissue, and (iii) area of hot-spots of high MVD were computed for tumors grown in two-dimensional AVNs. Predicted distributions show good agreement with a power law, and exhibit all the same exponent of -1.4. Such an algebraic decrease, in contrast to an exponential decrease, is known for systems at a critical threshold where systems undergo a phase transition. In the case of percolation, the critical threshold at $p = p_c$ marks the transition from isolated clusters to a single connected region. Bartha and Rieger [11] suggested that the tumor vasculature is driven automatically into a state akin to the critical percolation cluster by the mechanism of shear-stress correlated vascular collapse. As a result vessels permeate through the entire tumor, robust against moderate variations in $f^{(col)}$ and $p^{(col)}$. These predictions are experimentally testable and, if confirmed, would support that real vascular networks of the interior of tumors are the result of a dilution process rather than the result of sprouting growth into a heterogeneous environment.

3.3.3 Interrelation of Initial and Emergent Tumor Vasculature

High-caliber arterioles and venules ($>50\,\mu$m radius) protruding into the tumor form a backbone of stable vessels in-between which thinner vessels form short and straightforward paths [166]. The flow resistance decreases in proportion to $1/r^4$ with radius r and is therefore, in comparison to capillaries, extremely low in such high-caliber vessels. Therefore, in analogy to electrical networks, the blood pressure (voltage) drop across them is also low. In zero-th order approximation the blood pressure is constant, i.e. high-caliber vessels act like a pressure boundary condition for adjacent capillaries. Short, directed paths, have a survival advantage as discussed above in the context of CNs. The distribution of tumor vessels thus becomes dependent on initial ($t = 0$) vascular networks and is generally heterogeneous and anisotropic.

Assuming an arteriole runs near a venule then a large spatial blood pressure gradient is present. When a tumor grows near this area, a connection (short cut) is formed by angiogenesis, imposing the spatial gradient onto blood flow through this newly formed vessel which is the more stable the steeper the gradient. We attempted to quantify this dependence by correlation of tumor MVD ($t > 0$) versus the magnitude of blood pressure differences in-between vessels of initial ($t = 0$) AVNs. For this purpose, an auxiliary "pressure" field $p(\boldsymbol{x})$ was computed as function of space $\boldsymbol{x}$ at $t = 0$ that interpolates approximately the blood pressure p_v in spaces between vessels [166] and is determined by $\nabla^2 p + a \cdot (p_v - p) = 0$, where a is zero in empty space and $a \gg 1$ at sites coincident with vessels. We plotted the magnitude of the gradient $||\nabla p||$ as local averages taken over small boxes versus the local MVD at $t = 1200\,\text{h}$ Predicted correlation coefficients ranged from 0.2 to 0.5 per simulation. A correlation coefficient of 0.9 was obtained for averages over entire tumors of a cohort of simulations [166]. This finding may eventually be useful for model validation by experiments, should it become possible to scan real three-dimensional vasculatures of host tissues prior inoculation with TCs.

Somewhat different vascular configurations are indeed observed in real tumors of the same tissue, e.g. breast tumors [40]. Predictions outlined above suggest that heterogeneity of the initial vascular network has a strong impact on the emerging tumor vasculature rather than heterogeneity of the ECM through which vascular sprouts grow as originally proposed by Gazit [48].

3.3.4 Blood Flow and Blood Borne Drug Transport

McDougall et al. [145, 146] first considered conduction of a tracer substance through tumor vascular networks using a simulation model of a time-dependent intravascular concentration distribution that was previously used in geo-engineering. Following them, intravascular tracer conduction was studied in stationary tumor vascular networks based on CNs [165] and AVNs [166,168]. Simply, a pulse, or a constant infusion, is applied at inlet vessels, which is from there propagated down-stream through the vascular network. The unspectacular model predictions show tracer flowing through networks within a duration of seconds (AVNs) and ca. 1 min (CNs).

McDougall et al., on the other hand, consider a model system based on angiogenesis experiments on the cornea of a rabbit eye [51] (rabbit-eye model). There, the tumor is not connected to an extensive vascular network, but instead, a single parent vessel spawns a few sprouts (angiogenesis) which travel a long distance of ca. 1 mm and branch into a dense network permeating a tumor spheroid and adjacent tissue. It was concluded that the tumor vasculature conducts drug poorly and that most drug bypasses the tumor. The cause of this contradiction appears to be dilution of the tracer concentration within the dense network near the tumor and much lower flow rates leading to transit times of the order of 10–30 min. Moreover a vascular adaptation model was considered, leading to formation of shunts that bypass the tumor network [145].

A good perfusion is consistent with several clinical studies of human tumors based on PET measurements [72, 92, 99, 170] where elevated perfusion *rBF* by factors of 4.7 to 5.2 were observed [167, Tbl. 4]. Blood flow velocities in tumors predicted by our model are of the order of 1 mm/s, similar to blood flow in normal human micro vessels [100]. Our model predicts arterio-venous short-cuts within the tumor, i.e. vaso-dilation gives rise to mostly very well perfused vascular threads connecting arterioles with venoules [165, Fig. 8]. Such shunts were suggested in the experimental literature where "flow hotspots" are frequently found in tumors of patients [129]. However, it is well-known that blood flow in animal models is can be severely reduced to only 0.1 to 1 mm/s. The exact causes for discrepancy are presumably vessel compression and excessive blood plasma extravasation [82,175], both of which were not considered here.

3.3.5 Interstitial Fluid Flow

Interstitial fluid flow (IFF) in tumorous tissue has been considered theoretically for some time, for instance within the framework of continuum models [76, 83, 176]. More recent models incorporate a discrete tumor vasculatures, e.g. based on the rabbit-eye model [172, 173], and remodeling of capillary networks (CNs) [174].

Welter and Rieger [169] considered IFF and extravascular drug transport in tumors grown within synthetic arterio-venous networks. Predicted interstitial fluid pressure (IFP) distributions exhibit an average radial profile that increases sharply from the tumor edge into the tumor center (s. Fig. 3.8). There, the IFP approaches a plateau value asymptotically, close to the level of blood pressure, amounting to ca. 6.5 kPa (49 mmHg) [169, Fig. 4 and 5]. This is expected due to the high vascular permeability, implying a small pressure drop across the vessel wall. The plateau value lies above experimentally observed mean values taken over various human tumors, but it is still lower than the absolute maximal observed IFP [83, Tbl. 1]. Since the IFP is generally assumed to rise very close to the level of blood pressure, this is rather indicative of overestimated blood pressure. IFF distributions,

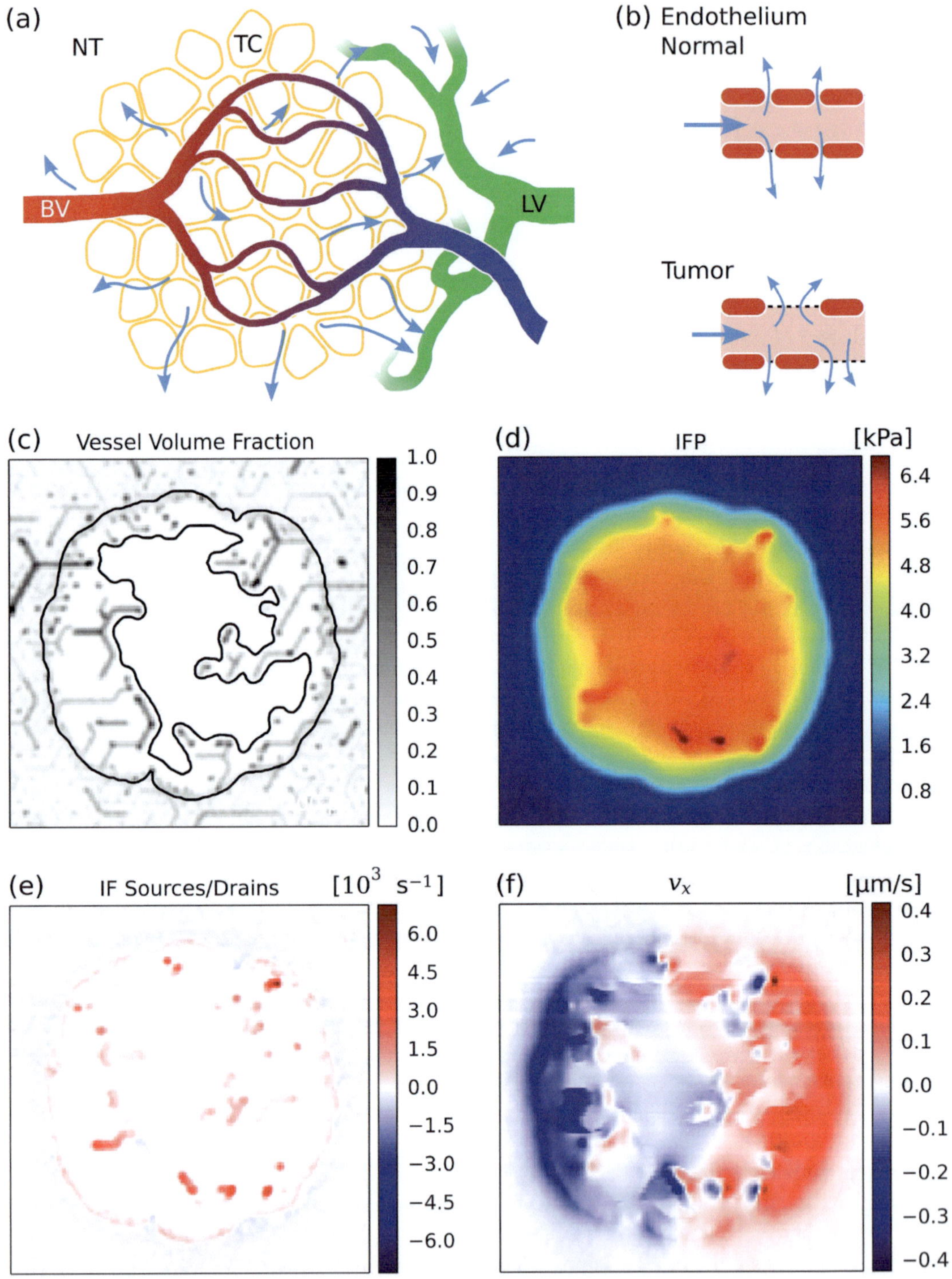

Fig. 3.8 Interstitial fluid flow: (**a**) Sketch of the transport of interstitial fluid from blood vessels (BV) through tumor tissue into lymph vessels (LV) outside the tumor. Interstitial fluid (IF) escapes through gaps in-between endothelial cells (**b**; *top*), which line the lumen of blood vessels, into extracellular space. These spaces also contain adhesion molecules and a network of fibers composed of various proteins such as collagen. Pores and fibers pose a resistance to the flow of the IF akin to the flow of water or oil through a porous rock. IF is absorbed into lymphatic channels from where it is brought back into the blood stream. In normal tissue, a large resistance to transvascular

i.e. the scaled negative gradient of the IFP, follow trivially and exhibit the expected sharp rise at the tumor rim amounting to a maximal value of 0.2 μm/s in good agreement with the literature. As a novel prediction due to the discrete AVN model used, the IFP and IFF distributions exhibit heterogeneity, i.e. they vary spatially in-between vessels of different blood pressure values. Thus vessels are predicted to drain the interstitial fluid in some instances [169, Fig. 4C].

It was often suggested that an elevated IFP poses a barrier to drug delivery [64, 76, 102, 174]. However, the reason for this cannot simply be a decreased transvascular hydrostatic pressure gradient that drives extravasation according to the Starling equation (3.15). To the contrary, in standard modeling approaches (s. Sect. 3.2.7), interstitial fluid flow is analogous to an electrical current flowing through a chain of resistors, of which one resistor, namely leaky tumor vessel walls, is particularly small (s. Fig. 3.9). Thus, an increase in leakiness, i.e. an elevation of IFP, would actually increase the liquid flux throughout the tumor, as predicted by our model. However, the analysis was restricted to good perfusion, where only a negligible liquid fraction escapes in spite of leakiness. Otherwise the way through tumorous tissue into lymphatics could presumably present an alternative well conducting pathway, draining downstream vessels of blood plasma, resulting in reported low flow velocities [82].

The recent theoretical work [174], using CNs and a sophisticated model of tumor growth that incorporates vessel compression due to IFP, comes to similar conclusions about the role of various permeabilities However, it was concluded that IFP is a barrier with little supporting numerical evidence, i.e. no simulation of actual drug transport was performed.

3.3.6 Interstitial Drug Transport

Experimentally, penetration experiments are performed for homogeneous cells layers and genetic causes for drug resistance are examined. However direct observation of drug distribution in tumors is difficult due to a lack of suitable markers. As a result there is only little experimental where spatial distributions of drug were measured [102, 121, 177] and quantitative data is scarce.

In oder to shed light on barriers to drug delivery, transport through tissue by advection and diffusion after extravasation must be taken into account. For this purpose, we analyzed a simple model, according to which, we computed time-dependent concentration distributions of drug in simulated tissues containing a vascularized tumor, grown in three-dimensional AVNs [169]. The considered tumors were static, and obtained by simulations guided by melanoma and glioma. In addition to concentration distributions, we also computed maps of time-independent metrics of doses delivered to the intracellular compartment: the local maximal concentration s_2 taken over time (ICMAX) and the time integral of s_2, respectively (ICAUC). The computation was stopped after a simulated time of 96 h.

Anti-cancer drugs come in a variety of kinds, from light molecules e.g. Cisplatin or Doxorubicin (≈543 g/mol) to heavy nano particles and viruses as carrier systems. We considered a base case guided by Doxorubicin since it is experi-

Fig. 3.8 (continued) flow leads to a large drop of the hydrostatic pressure across the vessel wall, so that the interstitial fluid pressure (IFP) approximately assumes the reference value of zero purported by the lymphatic system. In tumors, the IFP is elevated to approximately the level of blood pressure due to extremely large gaps in vascular walls (**b**; *bottom*) and lack of functional lymphatics [64]. The IFP measured in human tumors ranges from 0 to 94 mmHg, depending on the type of tumor [83]. Cuts through three dimensional simulation data sets are shown in the following panels: (**c**) Fractional volume of blood vessels per voxel volume. The interface to the distribution of viable tumor cells (TCs) is shown as contour. (**d**) Interstitial fluid pressure. (**e**) IF source and drain density in units of liquid volume per tissue volume and time. (**f**) X-component of the IF velocity v. Varying blood pressures and the presence of necrotic regions, of which we assumed a 10× increased permeability for IF, lead to a dissymmetrical IFP distribution. The IFF distribution is discontinuous as a result of the change in permeability

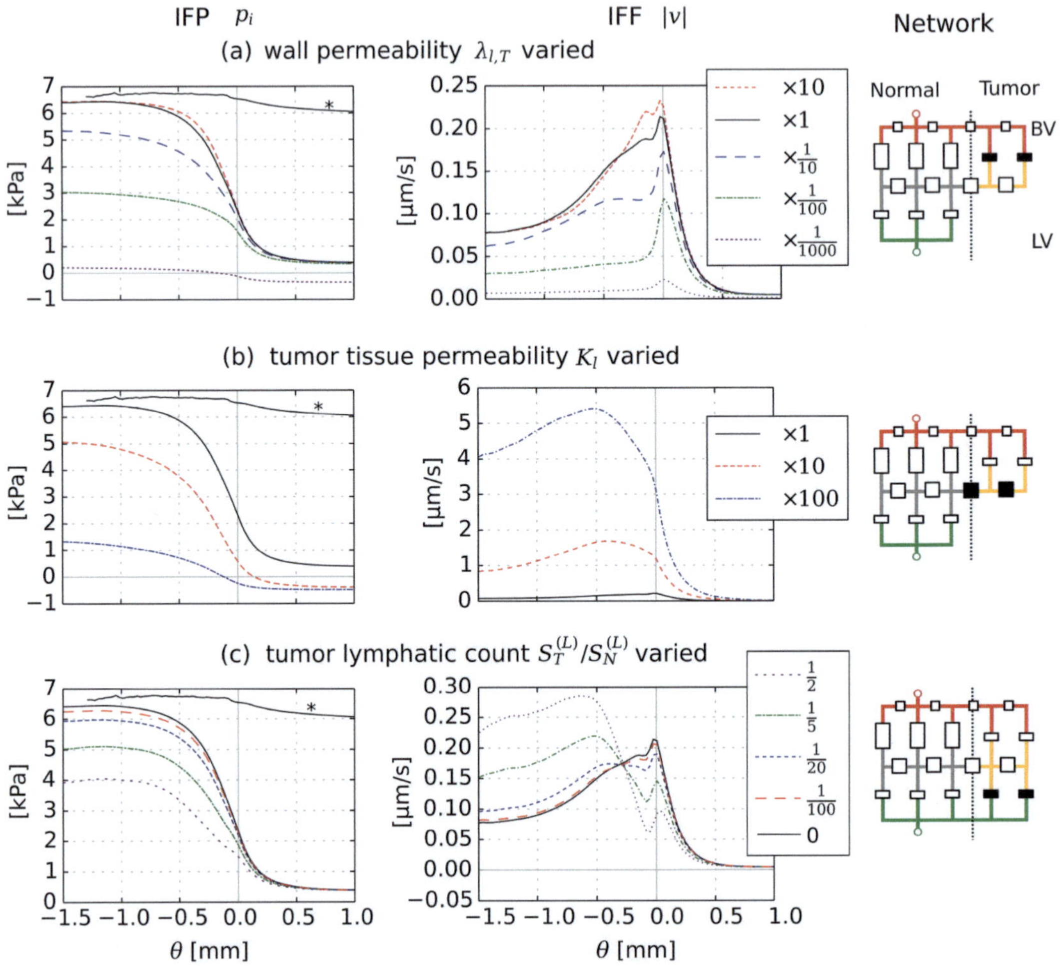

Fig. 3.9 Radial distributions of IFF and IFP as result of parameter variations. *Left column* shows the IFP and the *center column* shows the IFF. The curves are obtained from averages over annular shells and over a cohort of 15 simulated tumors. The *curve* marked with an *asterisk* shows the average blood pressure. Our model of IFF is analogous to an electrical network, where the IFP is the electrical potential. The *right column* shows this in simplified schematics. In each of them vessels are on the *top* (BV; *red*), the *middle* represents interstitial space (*shaded* and *yellow*), and lymphatics (LV; *green*). *Outlined boxes* represent various resistances, or permeabilities, in the system. *Solid black boxes* indicate the varied parameter. The relative deviation from the original base case parameter values is given in the figure legends, except in (D). The considered cases are as indicated in the sub-figure heading: (**a**) Variation of the upper vessel wall permeability bound $\lambda_{l,T}$ (case iv in [169]). (**b**) Variation of the interstitial permeability coefficient K_l (case v in [169]). (**c**) Variation of the amount of tumor lymphatics $S_T^{(L)}/S_N^{(L)}$, where the legend shows $S_T^{(L)}/S_N^{(L)}$ directly (case vii in [169])

mentally relatively well studied and widely used [121, 177].

Our model predicts that, in general, the dose delivered is subject to a compartmentalization similar to the vascular density (MVD), where metrics ICAUC and ICMAX likewise reflect the distribution of the MVD [169, Fig. 9]. Hence the average dose within the center of the tumor spheroid is significantly lower than in normal tissue, unless stated otherwise, and doses are highest at the tumor edge. This result provides an additional explanation of the incompletely understood success of combination therapies of anti-angiogenic agents and chemotherapy, whereas a single drug fails to improve survival [78]. TCs behind the tumor edge might be killed by high doses of chemotherapeutics, effective against cycling cells, whereas the TC population of the tumor center is reduced by necrosis caused by hypoxia. A monotherapy might leave the one or the other part of the TC population unaffected. The mechanism by which combination therapies are known to act is suppression of the activity caused by vascular growth factors, leading to a decreased vascular permeability. This allows overall better delivery of chemotherapeutics due to improved perfusion [78]. We also considered the case of a prolonged infusion which yielded similar results but with higher magnitudes of concentrations and doses.

Variation of the various permeabilities showed that average doses delivered with the permeability. Doses showed the highest sensitivity with respect to interstitial hydraulic and diffusive permeabilities which were varied simultaneously. See Fig. 3.9b, where a 10× increase in permeability leads to a similarly drastic increase in extravasation and IFF, not as obtained by other cases. As a result, more drug is delivered into tissue as well. However, the mechanism only works under the assumption that blood flow is sufficently high that it is not disturbed much by extravasation. As suggested before [174], this could be exploited for therapy. However an increased IFF could aggravate tumor invasion and metastatic dissemination [140]. Moreover, angiogenic normalization therapy, i.e. a reduction of permeability and pruning of vessels [78], might be ineffective or even detrimental for tumors where blood flow is negligibly impaired.

Doxorubicin and lighter molecules have the advantage that diffusion helps to distribute a substantial dose homogeneously around blood vessels regardless of IFF. This was demonstrated in recent simulations of another group [141] where very smooth and homogeneous concentration distributions of the more diffusive drug Cisplatin (300 g/mol) arising from extravasation from a CN were predicted.

Since the diffusion coefficient decreases with the molar mass of the solute, transport of drugs like nano-particles is strongly advection dominated. Simulation of the flow of such particles predicted interstitial drug concentrations that follow the stream of interstitial fluid in significant concentrations through the largest parts of the tumor spheroid, starting from the initial insertion through the vasculature (s. Fig. 3.10). As a result small isolated islands were predicted to exist right behind the invasive edge of the tumor where no significant dose had been delivered within the time frame of the simulation of 96 h. Presumably, this discrepancy to earlier work [76] is caused by the discrete nature of the blood vessel network considered allowing for flow in-between vessel of different blood pressure levels. Thus radial flow component vanishes by chance at some places as dictated by the random configuration of the vascular network. This suggests that a monotherapy with agents of high molar mass would be prone to recurring cancer.

3.3.7 Oxygen Distribution

Extremely good perfusion of tumor vessels cannot be assumed for tumors in general. This necessitates consideration of spatially varying substance concentrations because a substantial fraction may be lost during the transit through the tumor. Maps of tissue and intravascular oxygen partial pressure (PO2) distributions were calculated previously for small system volumes of the order of 0.1 mm^3 [54, 136]. Our recently developed computational method allows for computation of PO2 distributions in macroscopic simulation boxes of ca. 0.5 cm^3 on standard hardware (i7-2600K, 3.4 GHz, 4 GB Ram) within hours to a fair degree of accuracy [167]. It is still computationally expensive, however, few simplifying assumptions need to be made for the vascular network. Computed distributions of PO2 and blood oxygen saturation are shown in Fig. 3.11a, b.

Critical to performance and accuracy is the regularization of the singular source term (3.16), containing Dirac δ distributions, in conjunction with an efficient numerical scheme for the solution of the diffusion equation. In general the method should yield a sparse system matrix to enable numerical solutions in $O(n \log n)$ time in the number of unknowns n. In future, adaptive

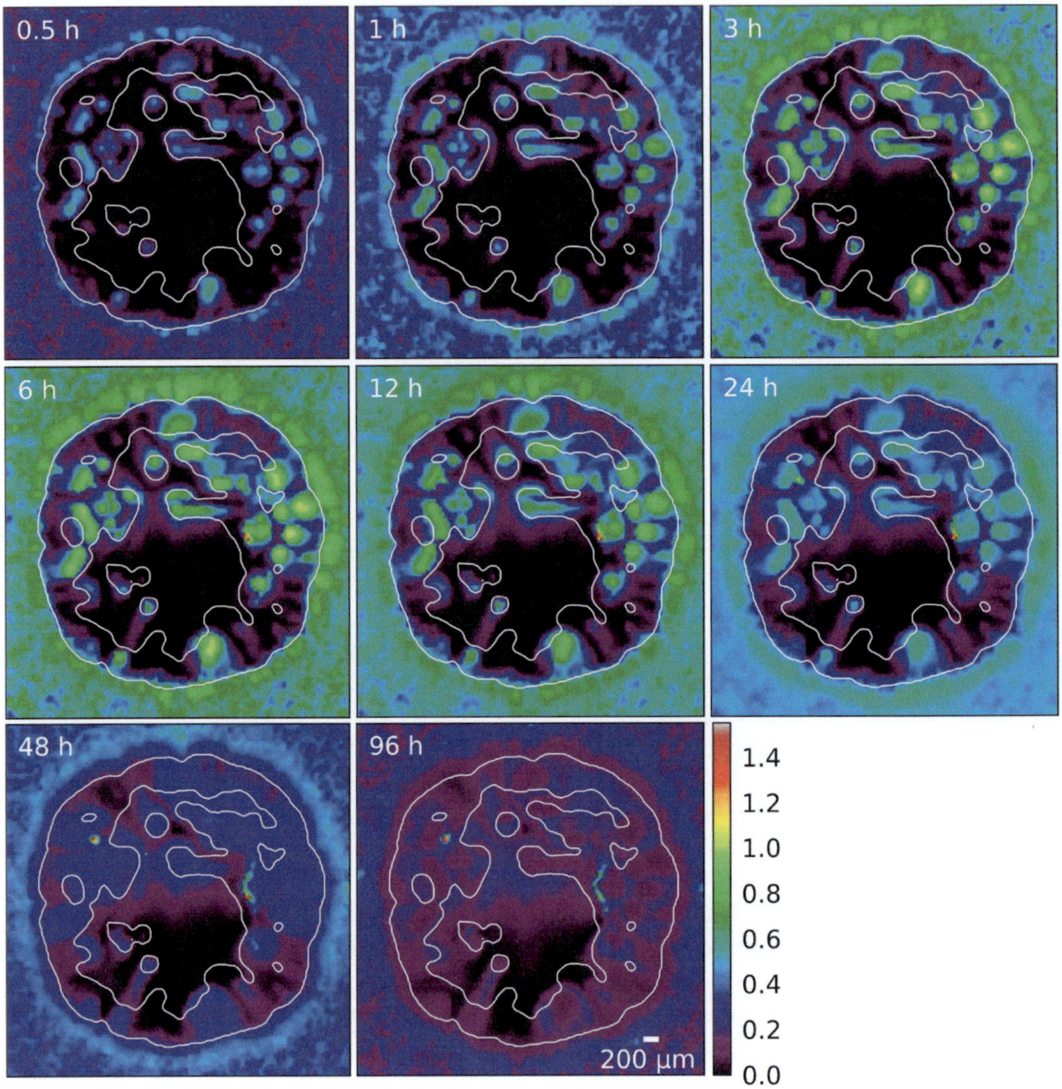

Fig. 3.10 Snapshots of the spatio-temporal distribution of a macro-molecular tracer concentration. The tissue and network configuration was obtained from simulated tumor growth and vascular remodeling of a synthetic arterio-venous vasculature [169]. Each panel shows a horizontal cut through the origin of the simulation box, showing the entire extent of 8 mm width. The distribution was calculated as solution of an advection equation for extra-vascular tissue. Vessels were sources of tracer which extravasates with the IF, assuming a spatially constant intravascular concentration $s_v(t)$. In time an exponential decrease of $s_v(t)$ was assumed, modeling a short injection and the subsequent clearance period. Moreover, the tracer was assumed inert, i.e. there were no sinks except by back flow into vessels. Since macro-molecules are hardly diffusing, the injected tracer is transported with the flow of the IF. The flow varies locally in direction and magnitude due to the coupling of the IFP to varying levels of blood pressure. As a result the tracer distributions is very heterogeneous. Frequently, as in this example, areas are predicted at the tumor rim that receive no significant dose

tesselation of the tissue domain may be used to increase accuracy [34].

The computation of intravascular PO2 distributions [167] was applied to the case of breast tumors for which several groups measured hemoglobin concentrations c_{Hb} and average blood oxygen saturations Y in large cohorts of patients [60, 144, 154, 155]. They determined

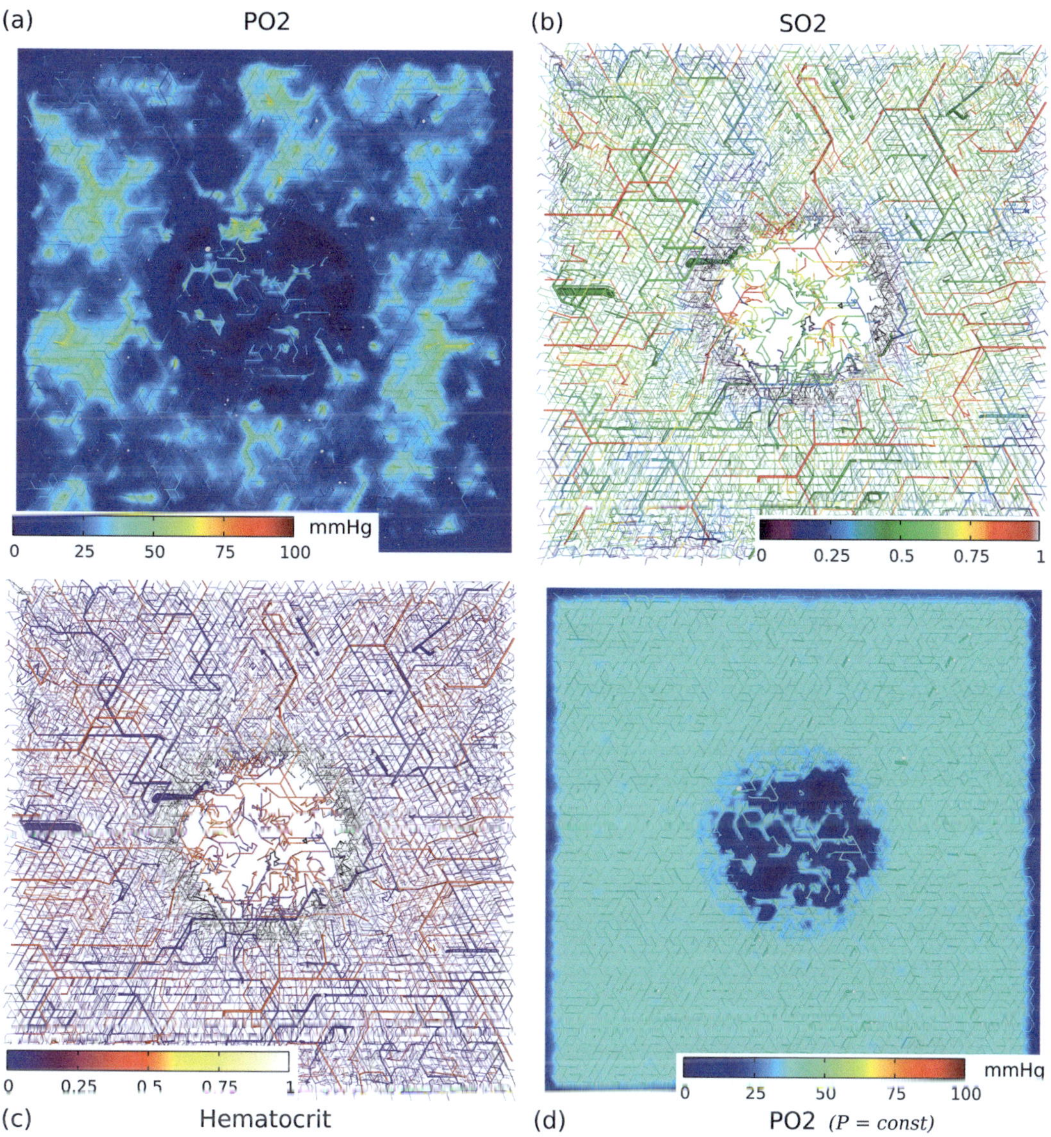

Fig. 3.11 Blood oxygen saturation and oxygen partial pressure. (**a**) Shows simulated intra-and extravascular distributions of partial pressure of oxygen (PO2) P, and P_t, respectively. (**b**) shows the corresponding blood oxygen saturation (SO2). (**c**) shows the hematocrit distribution within the same network. (**d**) shows the oxygen partial pressure obtained from a simplified model where the intravascular oxygen partial pressure was held constant. Data shown was computed for networks obtained by simulation of tumor growth and vascular remodeling [167]. A spherical region of approximately 2 mm radius was changed by the tumor. Each panel shows a horizontal cut through the simulation box. The entire extent of 8 mm width is shown. The vessel network is visualized as collection of cylinders, color coded by respective intravascular distributions. Only a slab, truncated 100 μm above and below the central plane, is shown (cross sectional areas: *light grey*). (**a**) and (**d**) show in addition extravascular tissue PO2 distributions

average concentrations of total hemoglobin c_{Hb}, oxyhemoglobin c_{HbO}, deoxyhemoglobin c_{HbD}, and tissue blood oxygen saturations $Y = c_{HbO}/c_{Hb}$ within normal and tumorous tissue sections. Obtained tumor hemoglobin concentrations $c_{Hb}(tumor)$ were always larger than hemoglobin concentrations in normal tissue $c_{Hb}(normal)$. This is already explained by an increase in regional blood volume rBV due to vaso-dilation. However, the blood oxygen saturation in tumors $Y(tumor)$ was sometimes larger or smaller than the blood oxygen saturation in normal tissue $Y(normal)$, divided approximately to equal numbers among patients. Moreover, high hemoglobin concentrations were correlated with high saturations, but tumors with low hemoglobin concentration exhibited a wider range of blood oxygen saturations.

We considered tumor vascular networks ($t = 800\,\mathrm{h}$) obtained from simulation of tumor growth and vascular remodeling. A large cohort of 90 different networks was simulated emulating a cohort of patients. Regional blood volume rBV, perfusion rBF, hemoglobin concentrations c_{Hb}, oxy-hemoglobin concentrations, deoxy-hemoglobin concentrations, oxygen saturation Y, as well as tissue and vascular oxygen partial pressure distributions P, and P_t were calculated for initial networks and final tumor networks ($t = 800\,\mathrm{h}$). Transient behavior such as transient hypoxia due to temporary occlusion of blood vessels was not considered. In our base case (BASE), different initial ($t = 0$) vascular configurations lead to a spread in tumor oxygen saturations, but it did not predict the clinically observed dependency of $Y(tumor)$ versus $Y(normal)$ since predicted $Y(tumor)$ were always larger than $Y(normal)$. Therefore we considered a phenomenological ad-hoc extension of the model by vaso-compression. On average, taken over the cohort of tumors, the proposed alteration results in a reduction of the radii of arterioles and venules that are thicker than the maximal dilation radius $r^{(max)}$, whereas the radii of smaller vessels are not much affected on average. This modified model predicts saturations Y in good agreement with mammography data (see Fig. 3.12). The reasons for this better agreement are a reduction of blood flow, thus draining a greater fraction of the supplied O2 in order to meet metabolic demand. Moreover, Y is the volume weighted average of the local blood saturation $S(P)$. Therefore compression reduces the weight of dilated arterio-venous shunts which generally exhibit a high saturation S. Cases for which drastically increased metabolic oxygen consumption rates M_0, decreased maximal dilatation radii $r^{(max)}$, and stochastic variations thereof were considered, failed to predict the clinically observed distributions. Therefore our results suggest that a decreased tissue blood oxygen saturation relative to baseline normal tissue of the same patient is indicative of vessel compression which could be exploited in therapy. The clinical data might imply that tumor vascular networks that exhibit higher saturations than normal are vastly different in their vascular architecture than networks that exhibit low saturations. However, our simulations suggest that these networks nevertheless share the traits of typical tumor vascular networks as outlined in the introduction.

For models of tumor growth, it is a very convenient approximation to consider a model of oxygenation where the intravascular PO2 distribution is constant. Then only the tissue PO2 distribution needs to be calculated. There are however qualitative difference in its predictions that one should be aware of (s. Fig. 3.11d). The constant-PO2 approximation fails to predict local oxygen depletion in vessels threading the tumor center. More importantly it over-estimates the PO2 in the neo-vascular plexus around the tumor periphery. The depletion of oxygen predicted by the full model is a consequence of a redirection of most hematocrit into the tumor center due to the phase separation effect (s. Fig. 3.11c). Therefore the densely vascularized capillary plexus around the invasive edge is deprived of red blood cells and thus the oxygen capacity of blood therein drastically reduced.

3.4 Limitations and Outlook

Although current models produce predictions that are in many respect in good agreement with experiments, there are some severe limitations. For one, many predictions were obtained by first simulating tumor growth by a simplified model. Then additional quantities relevant for tumor growth were computed, e.g. interstitial fluid flow or intravascular oxygen distributions [167,169]. Other works have other limitations, e.g. oversimplified

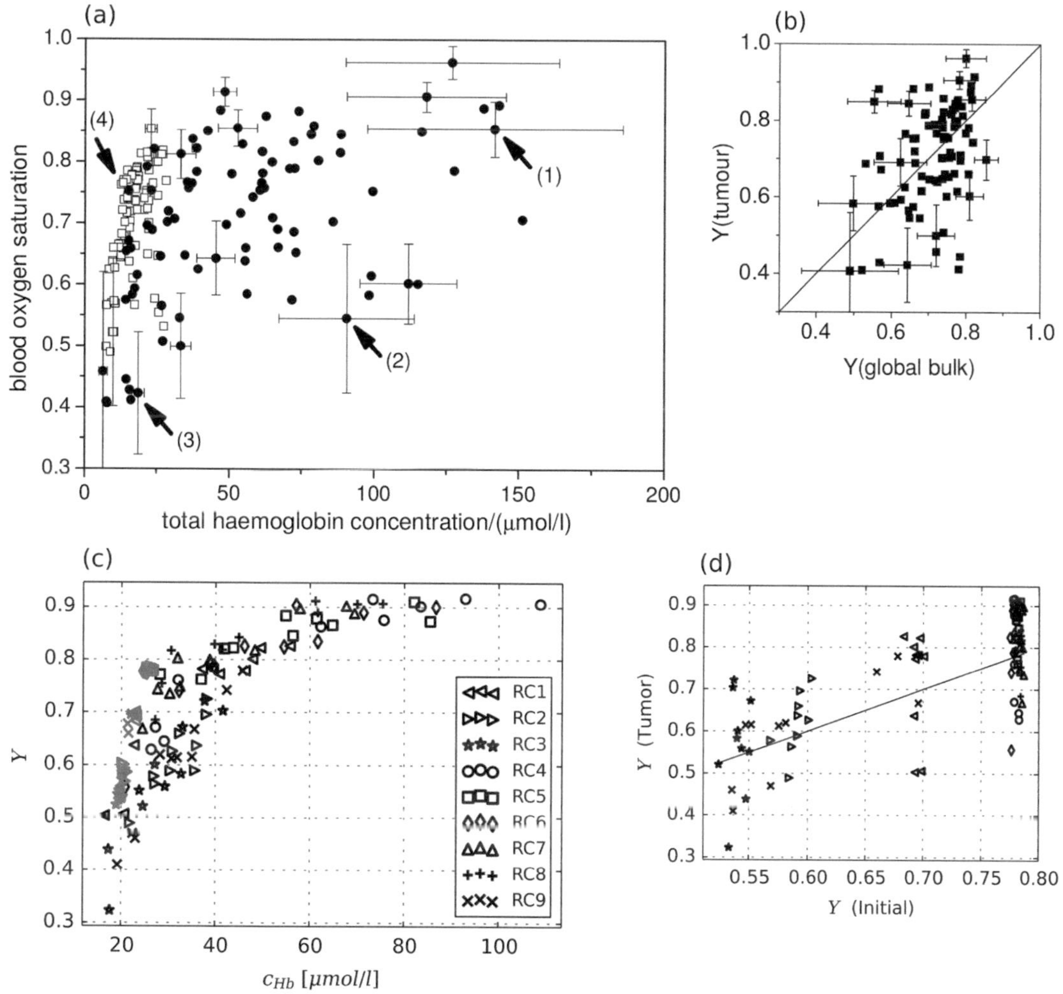

Fig. 3.12 Tissue oxygen saturation: clinical versus simulation data. (**a**) Total hemoglobin concentration c_{Hb} versus tissue blood oxygen saturation Y of tumors (*solid circles*) and healthy breast tissue (*open squares*) for 87 patients, obtained by optical mammography. (**b**) Tissue blood oxygen saturation Y of tumors versus those of corresponding healthy breast tissue for the same group of patients. (Reprinted by permission of IOP Publishing from [60] Figs. 3b, 5a. All rights reserved) (**c**, **d**) Analogous data obtained from simulated tumor vascular remodeling, guided by data for breast cancer, and computation of intra and extravascular oxygen concentration distributions [167] of which examples are shown in Fig. 3.11a, b. Markers in (**c**) correspond to initial tissue (*grey*) and the tumorous areas at $t = 600$ h (*black*). A cohort of 90 tumors was simulated, each using a different initial ($t = 0$) vascular network. Each initial network was grown from one of nine root node configurations denoted RC1–RC9. Depending on the number of root nodes, which is equivalent to the number of arterial and venous trees in the network, varying vascular volumes *rBV* and blood flow rates *rBF* are obtained, introducing significant data scatter. The data shown was predicted assuming vaso-compression of high-caliber vessels that penetrate into the tumor (Case CMPR of Ref. [167])

vascular networks of host tissue [95, 131, 141, 174]. It may be worth to develop an integrated model combining all aspects into time dependent simulation of tumor growth. This may be particularly important for the study of pharmacokinetics where IFF, drug transport, oxygenation and tumor growth are tightly coupled.

Furthermore, To obtain a more faithful cohort of initial blood vessel networks, a systematic analysis of initial networks could be carried out.

Experimental data of blood volume, *rBV*, perfusion *rBF*, and so on, could be used to select a cohort of networks that satisfies experimentally observed statistical distributions. Current networks are unrealistic in some aspects, e.g. they contain no anastomosis, i.e. cross-links between vascular trees [100].

Therefore it would be ideal to use real scanned and digitized blood vessel networks (s. Fig. 3.13 for examples). In principle it is possible to automatically reconstruct networks from voxel data. However, current state of the art microscopy methods can only see through a tissue slab up to a maximal depth of ca. 250 μm [27]. Recently, data obtained from micro computed tomography (μ-CT) was used by Stamatelos et al. [147] to reconstruct large parts of the vascular system of an animal-model breast tumor. However, it is questionable if all capillaries were captured since the resolution of the scanner was only 8 μm, and many dead ends were in the reconstructed network. Similar results were obtained for other cranial [61] and coronary [91] blood vessel networks. A data base of many large scale networks of normal tissues and corresponding tumor networks at different growth stages could be built. Not only would this eliminate the need to construct artificial initial networks, but it would also allow for a detailed comparison between model predictions and real tumor networks.

A major limitation of our model is the restricted applicability to only well perfused tumors. However in animal models blood flow velocities are generally lower amounting to 0.1–1 mm/s [9, 109]. The prediction of good perfusion is inherent to our basic model of vascular remodeling since tumor vessels can only dilate, not shrink, leading to well conducting arterio-venous shunts. The prediction of high flow rates allows for neglect of blood plasma loss due to extravasation, which we justify by a simple worst case estimation of lost plasma amounting to ca. 0.1 %. Therefore, extravasation of plasma cannot be the only cause for low blood flow, but rather it likely aggravates the situation if vessels are constricted by solid pressure for instance.

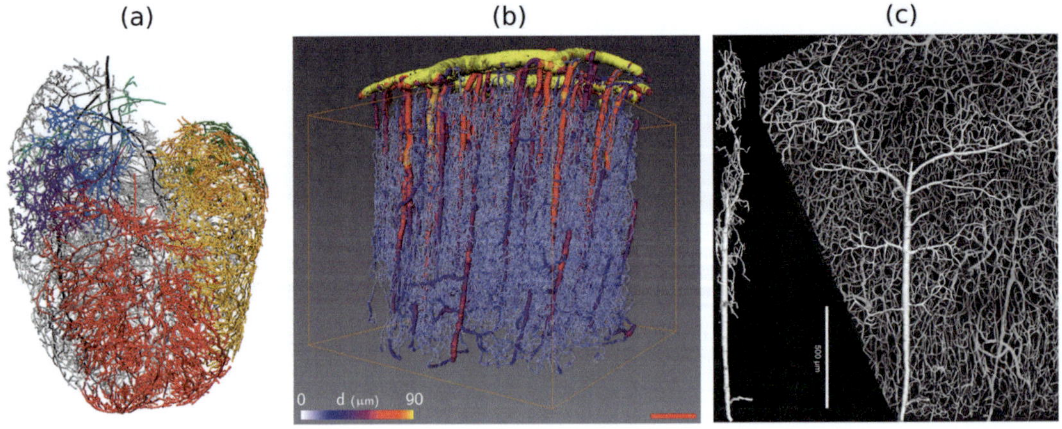

Fig. 3.13 Reconstruction of blood vessel networks from imaging data: (**a**) A coronary vascular network of a rat based on micro-CT images. Various subnetworks are distinguished by random colors. (Reprinted with Permission from [91] Fig. 12e. Copyright 2007 Elsevier Science) (**b**) A section of a cortical blood vessel network after reconstruction based on micro-CT images. Vessels are color coded according to their diameter *d*. (Reprinted with permission from [61] Fig. 1c. Copyright 2010 Nature Publishing Group) (**c**) Tissue slice of the human cerebral cortex. The *left hand side* shows a side view on a large vein with adjacent branches. The reconstruction is based on depth-coded confocal microscopy images. (Reprinted with permission from [27] Fig. 3. Copyright 2006 Taylor & Francis LLC) At the present day, vascular networks of host tissue, in which tumor growth is simulated, are algorithmically synthesized which involves uncertainties and likely model artifacts. The incorporation of such scanned networks, possibly on larger scales than the ones shown, would allow for more accurate model validation and results to be obtained

Consequently, it would be worthwhile to explore extensions to vascular dilation and regression processes rather than limiting blood flow by ad-hoc shrinkage of arterial radii, as done in [167]. One such possible extension already exists in the vascular adaptation model [119]. Essentially, a shrinking-tendency is balanced by a wall shear-stress dependent growth signal. Moreover, compression of blood vessels is insufficiently understood. Forces involved were studied quantitatively, separately (see Refs. below). However, their interplay is not understood or studied much. Obviously, the deformation of vessel walls is governed by a balance of forces which are tensile and compressive stress within the vessel wall, blood pressure, interstitial fluid pressure, and solid pressure. Solid pressure compresses vessels [30, 31, 150] and there is evidence that an elevated IFP aids in compression of vessels [36,59]. There is, to our knowledge, no predictive model of the response of the vessel wall that takes these factors into account. A physical consideration based on first principles e.g. with the help of a elasto-plastic mechanical model of vascular walls in combination with a mechanical model of tissue could help elucidate the forces involved and ultimately yield better predictions of blood flow.

With an ad-hoc extension to emulate compression, our model predicts regional blood flow *rBF* that is about a factor of 5–10 above measured data from breast tumors [167, Tbl. 4]. This apparent deficiency might be founded in the size of the considered tumors (4 mm in diameter, simulated, versus centimeter sized real tumors), since in tumor xenografts [84] blood flow *rBF* of experimental tumors (2–0.3 ml/g/min) correlates negatively with size (0.1–10 cm^3 tumor volume), consistent with predictions of our model. The reasons for this size dependency are currently unknown. However, it suggests that either normal vasculatures can only provide a constant blood flow rate per surface area of the tumor spheroid into the tumor, or that the abnormal organization of the tumor vasculature is only affecting blood flow velocities in tumors much larger than theoretically studied.

References

1. Aker E, Jørgen Måløy K, Hansen A, Batrouni G (1998) A two-dimensional network simulator for two-phase flow in porous media. Trans Porous Media 32(2):163–186. doi:10.1023/A:1006510106194
2. Alarcon T, Byrne H, Maini P (2003) A cellular automaton model for tumour growth in inhomogeneous environment. J Theor Biol 225:257–274. doi:10.1016/s0022-5193(03)00244-3
3. Ambrosi D, Ateshian GA, Arruda EM, Cowin SC, Dumais J, Goriely A, Holzapfel GA, Humphrey JD, Kemkemer R, Kuhl E, Olberding JE, Taber LA, Garikipati K (2011) Perspectives on biological growth and remodeling. J Mech Phys Solids 59(4):863–883. doi:10.1016/j.jmps.2010.12.011
4. Ambrosi D, Mollica F (2004) The role of stress in the growth of a multicell spheroid. J Math Biol 48(5):477–499. doi:10.1007/s00285-003-0238-2
5. Ambrosi D, Preziosi L (2009) Cell adhesion mechanisms and stress relaxation in the mechanics of tumours. Biomech Model Mechanobiol 8(5):397–413. doi:10.1007/s10237-008-0145-y
6. Anderson A, Chaplain MAJ (1998) Continuous and discrete mathematical models of tumorinduced angiogenesis. Bull Math Biol 60(5):857–899. doi:10.1006/bulm.1998.0042
7. Ando J, Yamamoto K (2013) Flow detection and calcium signalling in vascular endothelial cells. Cardiovasc Res 99(2):260–268. doi:10.1093/cvr/cvt084
8. Armstrong NJ, Painter KJ, Sherratt JA (2006) A continuum approach to modelling cell-cell adhesion. J Theor Biol 243(1):98–113. doi:10.1016/j.jtbi.2006.05.030
9. Baish JW, Netti PA, Jain RK (1997) Transmural coupling of fluid flow in microcirculatory network and interstitium in tumors. Microvasc Res 53(2):128–141. doi:10.1006/mvre.1996.2005
10. Baish JW, Stylianopoulos T, Lanning RM, Kamoun WS, Fukumura D, Munn LL, Jain RK (2011) Scaling rules for diffusive drug delivery in tumor and normal tissues. Proc Natl Acad Sci USA 108(5):1799–1803. doi:10.1073/pnas.1018154108
11. Bartha K, Rieger H (2006) Vascular network remodeling via vessel cooption, regression and growth in tumors. J Theor Biol 241(4):903–918. http://dx.doi.org/10.1016/j.jtbi.2006.01.022
12. Basan M, Prost J, Joanny JF, Elgeti J (2011) Dissipative particle dynamics simulations for biological tissues: rheology and competition. Phys Biol 8(2):026,014. doi:10.1088/1478-3975/8/2/026014
13. Baxter LT, Jain RK (1989) Transport of fluid and macromolecules in tumors. I. Role of interstitial pressure and convection. Microvasc Res 37(1):77–104. doi:10.1016/0026-2862(89)90074-5
14. Beard DA (2001) Computational framework for generating transport models from databases of microvascular anatomy. Ann Biomed Eng 29(10):837–843. doi:10.1114/1.1408920

15. Bentley K, Gerhardt H, Bates PA (2008) Agent-based simulation of notch-mediated tip cell selection in angiogenic sprout initialisation. J Theor Biol 250(1):25–36. http://dx.doi.org/10.1016/j.jtbi.2007.09.015
16. Betteridge R, Owen MR, Byrne HM, Alarcon T, Maini PK (2006) The impact of cell crowding and active cell movement on vascular tumour growth. Netw Heterog Media 1(4):515–535. doi:10.3934/nhm.2006.1.515
17. Brackbill JU, Kothe DB, Zemach C (1992) A continuum method for modeling surface tension. J Comput Phys 100(2):335–354. doi:10.1016/0021-9991(92)90240-Y
18. Breward CJ, Byrne HM, Lewis CE (2003) A multiphase model describing vascular tumour growth. Bull Math Biol 65(4):609–640. doi:10.1016/s0092-8240(03)00027-2
19. Brown JM, Wilson WR (2004) Exploiting tumour hypoxia in cancer treatment. Nat Rev Cancer 4(6):437–447. doi:10.1038/nrc1367
20. Bru A, Albertos S, Subiza JL, Garcia-Asenjo JL, Bru I (2003) The universal dynamics of tumor growth. Biophys J 85(5):948–2961. doi:10.1016/s0006-3495(03)74715-8
21. Buchanan CF, Verbridge SS, Vlachos PP, Rylander MN (2014) Flow shear stress regulates endothelial barrier function and expression of angiogenic factors in a 3D microfluidic tumor vascular model. Cell Adh Migr 8(5):517–524. doi:10.4161/19336918.2014.970001
22. Byrne H, Preziosi L (2003) Modelling solid tumour growth using the theory of mixtures. Math Med Biol 20(4):341–366. doi:10.1093/imammb/20.4.341
23. Byrne HM, Owen MR, Alarcon T, Murphy J, Maini PK (2006) Modelling the response of vascualr tumours to chemotherapy: a multiscale approach. Math Models Methods Appl Sci 16(supp01):1219–1241. doi:10.1142/S0218202506001522
24. Cai Y, Xu S, Wu J, Long Q (2011) Coupled modelling of tumour angiogenesis, tumour growth and blood perfusion. J Theor Biol 279(1):90–101. http://dx.doi.org/10.1016/j.jtbi.2011.02.017
25. Carmeliet P, Jain R (2000) Angiogenesis in cancer and other diseases. Nature 407:249–257. doi:10.1038/35025220
26. Carmeliet P, Jain RK (2011) Molecular mechanisms and clinical applications of angiogenesis. Nature 473(7347):298–307. doi:10.1038/nature10144
27. Cassot F, Lauwers F, Fouard C, Prohaska S, Lauwers-Cances V (2006) A novel three-dimensional computer-assisted method for a quantitative study of microvascular networks of the human cerebral cortex. Microcirculation 13(1):1–18. doi:10.1080/10739680500383407
28. Cattaneo L, Zunino P (2014) A computational model of drug delivery through microcirculation to compare different tumor treatments. Int J Numer Method Biomed Eng 30(11):1347–1371. doi:10.1002/cnm.2661
29. Chaplain MA, Graziano L, Preziosi L (2006) Mathematical modelling of the loss of tissue compression responsiveness and its role in solid tumour development. Math Med Biol 23(3):197–229. doi:10.1093/imammb/dql009
30. Chauhan VP, Boucher Y, Ferrone CR, Roberge S, Martin JD, Stylianopoulos T, Bardeesy N, DePinho RA, Padera TP, Munn LL, Jain RK (2014) Compression of pancreatic tumor blood vessels by hyaluronan is caused by solid stress and not interstitial fluid pressure. Cancer Cell 26(1):14–15. doi:10.1016/j.ccr.2014.06.003
31. Chauhan VP, Martin JD, Liu H, Lacorre DA, Jain SR, Kozin SV, Stylianopoulos T, Mousa AS, Han X, Adstamongkonkul P, Popovi? Z, Huang P, Bawendi MG, Boucher Y, Jain RK (2013) Angiotensin inhibition enhances drug delivery and potentiates chemotherapy by decompressing tumour blood vessels. Nat Commun 4:2516. doi:10.1038/ncomms3516
32. Ciarletta P, Ambrosi D, Maugin GA, Preziosi L (2013) Mechano-transduction in tumour growth modelling. Eur Phys J E Soft Matter 36(3):23. doi:10.1140/epje/i2013-13023-2
33. Daşu A, Toma-Daşu I, Karlsson M (2003) Theoretical simulation of tumour oxygenation and results from acute and chronic hypoxia. Phys Med Biol 48(17):2829–2842. doi:10.1088/0031-9155/48/17/307
34. D'Angelo C (2012) Finite element approximation of elliptic problems with dirac measure terms in weighted spaces: applications to one- and three-dimensional coupled problems. SIAM J Numer Anal 50(1):194–215. doi:10.1137/100813853
35. Degond P, Mas-Gallic S (1989) The weighted particle method for convection-diffusion equations. Part 1: the case of an isotropic viscosity. Math Comput 53(188):485–507. doi:10.2307/2008716
36. DelGiorno KE, Carlson MA, Osgood R, Provenzano PP, Brockenbough JS, Thompson CB, Shepard HM, Frost GI, Potter JD, Hingorani SR (2014) Response to Chauhan et al.: interstitial pressure and vascular collapse in pancreas cancer-fluids and solids, measurement and meaning. Cancer Cell 26(1):16–17. doi:10.1016/j.ccr.2014.06.004
37. Döme B, Hendrix MJ, Paku S, Tovari J, Timar J (2007) Alternative vascularization mechanisms in cancer: pathology and therapeutic implications. Am J Pathol 170(1):1–15. doi:10.2353/ajpath.2007.060302
38. Döme B, Paku S, Somlai B, Tímár J (2002) Vascularization of cutaneous melanoma involves vessel co-option and has clinical significance. J Pathol 197(3):355–362. doi:10.1002/path.1124
39. Drasdo D, Höhme S (2005) A single-cell-based model of tumor growth in vitro: monolayers and sphereoids. Phys Biol 2(3):133–147. doi:10.1088/1478-3975/2/3/001
40. Du J, Li FH, Fang H, Xia JG, Zhu CX (2008) Microvascular architecture of breast lesions: evaluation

with contrast-enhanced ultrasonographic micro flow imaging. J Ultrasound Med 27(6):833–842
41. Enderling H, Anderson AR, Chaplain MA, Beheshti A, Hlatky L, Hahnfeldt P (2009) Paradoxical dependencies of tumor dormancy and progression on basic cell kinetics. Cancer Res 69(22):8814–8821. doi:10.1158/0008-5472.can-09-2115
42. Enderling H, Hlatky L, Hahnfeldt P (2009) Migration rules: tumours are conglomerates of selfmetastases. Br J Cancer 100(12):1917–1925. doi:10.1038/sj.bjc.6605071
43. Erber R, Eichelsbacher U, Powajbo V, Korn T, Djonov V, Lin J, Hammes HP, Grobholz R, Ullrich A, Vajkoczy P (2006) EphB4 controls blood vascular morphogenesis during postnatal angiogenesis. EMBO J 25(3):628–641. doi:10.1038/sj.emboj.7600949
44. Espinoza I, Peschke P, Karger CP (2013) A model to simulate the oxygen distribution in hypoxic tumors for different vascular architectures. Med Phys 40(8):081,703. doi:10.1118/1.4812431
45. Folkman J (1971) Tumor angiogenesis: therapeutic implications. N Engl J Med 285(21):1182–1186. doi:10.1056/nejm197111182852108
46. Fraser GM, Goldman D, Ellis CG (2013) Comparison of generated parallel capillary arrays to three-dimensional reconstructed capillary networks in modeling oxygen transport in discrete microvascular volumes. Microcirculation 20(8):748–763. doi:10.1111/micc.12075
47. Furuberg L, Feder J, Aharony A, Jossang T (1988) Dynamics of invasion percolation. Phys Rev Lett 61(18):2117–2120. doi:10.1007/978-94-009-2653-0 17
48. Gazit Y, Berk DA,Michael Leunig LTB, Jain RK (1995) Scale-invariant behavior and vascular network formation in norma and tumor tissue. Phys Rev Lett 75(12):2428–2431. doi:10.1103/physrevlett.75.2428
49. Gerhardt H, Golding M, Fruttinger M, Ruhrberg C, Lundkvist A, Abramsson A, Jeltsch M, Mitchell C, Alitalo K, Shima D, Betsholz C (2003) Vegf guides angiogenic sprouting utilizing endothelial tip cell filopodia. J Cell Biol 161(6):1163–1177. doi:10.1083/jcb.200302047
50. Gevertz JL (2011) Computational modeling of tumor response to vascular-targeting therapies-part I: validation. Comput Math Methods Med 2011:830,515. doi:10.1155/2011/830515
51. Gimbrone M, Cotran R, Leapman S, Folkman J (1974) Tumor growth and neovascularization: an experimental model using the rabbit cornea. J Natl Cancer Inst 52(2):413–427. doi:10.1093/jnci/52.2.413
52. Gödde R, Kurz H (2001) Structural and biophysical simulation of angiogenesis and vascular remodeling. Dev Dyn 220(4):387–401. doi:10.1002/dvdy.1118
53. Goettsch W, Gryczka C, Korff T, Ernst E, Goettsch C, Seebach J, Schnittler HJ, Augustin HG, Morawietz H (2008) Flow-dependent regulation of angiopoietin-2. J Cell Physiol 214(2):491–503. doi:10.1002/jcp.21229
54. Goldman D (2008) Theoretical models of microvascular oxygen transport to tissue. Microcirculation 15(8):795–811. doi:10.1080/10739680801938289
55. Goldman D, Bateman RM, Ellis CG (2004) Effect of sepsis on skeletal muscle oxygen consumption and tissue oxygenation: interpreting capillary oxygen transport data using a mathematical model. Am J Physiol Heart Circ Physiol 287(6):H2535–2544. doi:10.1152/ajpheart.00889.2003
56. Goldman D, Bateman RM, Ellis CG (2006) Effect of decreased O2 supply on skeletal muscle oxygenation and O2 consumption during sepsis: role of heterogeneous capillary spacing and blood flow. Am J Physiol Heart Circ Physiol 290(6):H2277–2285. doi:10.1152/ajpheart.00547.2005
57. Goldman D, Popel AS (2000) A computational study of the effect of capillary network anastomoses and tortuosity on oxygen transport. J Theor Biol 206(2):181–194. doi:10.1006/jtbi.2000.2113
58. Gray LH, Conger AD, Ebert M, Hornsey S, Scott OCA (1953) The concentration of oxygen dissolved in tissues at the time of irradiation as a factor in radiotherapy. Br J Radiol 26(312):638–648.doi:10.1259/0007-1285-26-312-638. PMID: 13106296
59. Griffon-Etienne G, Boucher Y, Brekken C, Suit HD, Jain RK (1999) Taxane-induced apoptosis decompresses blood vessels and lowers interstitial fluid pressure in solid tumors: clinical implications. Cancer Res 59(15):3776–3782. http://cancerres.aacrjournals.org/content/59/15/3776.abstract
60. Grosenick D, Wabnitz H, Moesta KT, Mucke J, Schlag PM, Rinneberg H (2005) Timedomain scanning optical mammography: II. Optical properties and tissue parameters of 87 carcinomas. Phys Med Biol 50(11):2451–2468. doi:10.1088/0031-9155/50/11/002
61. Guibert R, Fonta C, Plouraboue F (2010) Cerebral blood flow modeling in primate cortex. J Cereb Blood Flow Metab 30(11):1860–1873. doi:10.1038/jcbfm.2010.105
62. Hanahan D,Weinberg RA (2011) Hallmarks of cancer: the next generation. Cell 144(5):646–674. doi:10.1016/j.cell.2011.02.013
63. Harris AL (2002) Hypoxia–a key regulatory factor in tumour growth. Nat Rev Cancer 2(1):38–47 doi:10.1038/nrc704
64. Heldin CH, Rubin K, Pietras K, Ostman A (2004) High interstitial fluid pressure – an obstacle in cancer therapy. Nat Rev Cancer 4(10):806–813. doi:10.1038/nrc1456
65. Hellums JD, Nair PK, Huang NS, Ohshima N (1996) Simulation of intraluminal gas transport processes in the microcirculation. Ann Biomed Eng 24(1):1–24. doi:10.1007/bf02770991
66. Höckel M, Vaupel P (2001) Tumor hypoxia: definitions and current clinical, biologic, and molecular aspects. J Natl Cancer Inst 93(4):266–276. doi:10.1093/jnci/93.4.266

67. Holash J, Maisonpierre PC, Compton D, Boland P, Alexander CR, Zagzag D, Yancopoulos GD, Wiegand SJ (1999) Vessel cooption, regression, and growth in tumors mediated by angiopoietins and vegf. Science 284(5422):1994–1998. doi:10.1126/science.284.5422.1994
68. Holash J, Wiegand S, Yancopoulos G (1999) New model of tumor angiogenesis: dynamic balance between vessel regression and growth mediated by angiopoietins and vegf. Oncogene 18(38):5356–5362. doi:10.1038/sj.onc.1203035
69. Hopcroft J, Tarjan R (1973) Algorithm 447: efficient algorithms for graph manipulation. Commun ACM 16(6):372–378. doi:10.1145/362248.362272
70. Hsu R, Secomb TW (1989) A green's function method for analysis of oxygen delivery to tissue by microvascular networks. Math Biosci 96(1):61–78. doi:10.1016/0025-5564(89)90083-7
71. Hubbard M, Byrne H (2013) Multiphase modelling of vascular tumour growth in two spatial dimensions. J Theor Biol 316(0):70–89. http://dx.doi.org/10.1016/j.jtbi.2012.09.031
72. Ito M, Lammertsma AA, Wise RJ, Bernardi S, Frackowiak RS, Heather JD, McKenzie CG, Thomas DG, Jones T (1982) Measurement of regional cerebral blood flow and oxygen utilisation in patients with cerebral tumours using 15O and positron emission tomography: analytical techniques and preliminary results. Neuroradiology 23(2):63–74. doi:10.1007/bf00367239
73. Jain RK (1987) Transport of molecules in the tumor interstitium: a review. Cancer Res 47(12):3039–3051. http://cancerres.aacrjournals.org/content/47/12/3039.abstract
74. Jain RK (1987) Transport of molecules in the tumor interstitium: a review. Cancer Res 47(12):3039–3051. http://cancerres.aacrjournals.org/content/47/12/3039.abstract
75. Jain RK (1988) Determinants of tumor blood flow: a review. Cancer Res 48(10):2641–2658. http://cancerres.aacrjournals.org/content/48/10/2641.abstract
76. Jain RK (1999) Transport of molecules, particles, and cells in solid tumors. Annu Rev Biomed Eng 1(1):241–263. doi:10.1146/annurev.bioeng.1.1.241
77. Jain RK (2005) Normalization of tumor vasculature: an emerging concept in antiangiogenic therapy. Science 307(5706):58–62. doi:10.1126/science.1104819
78. Jain RK (2013) Normalizing tumor microenvironment to treat cancer: bench to bedside to biomarkers. J Clin Oncol 31(17):2205–2218. doi:10.1200/jco.2012.46.3653
79. Jain RK (2014) An indirect way to tame cancer. Sci Am 310(2):46–53. doi:10.1038/scientificamerican0214-46
80. Jain RK (2015) Antiangiogenesis strategies revisited: from starving tumors to alleviating hypoxia. Cancer Cell 26(5):605–622. doi:10.1016/j.ccell.2014.10.006
81. Jain RK, Baxter LT (1988) Mechanisms of heterogeneous distribution of monoclonal antibodies and other macromolecules in tumors: significance of elevated interstitial pressure. Cancer Res 48(24 Pt 1):7022–7032. http://cancerres.aacrjournals.org/content/48/24_Part_1/7022.abstract
82. Jain RK, Stylianopoulos T (2010) Delivering nanomedicine to solid tumors. Nat Rev Clin Oncol 7(11):653–664. doi:10.1038/nrclinonc.2010.139
83. Jain RK, Tong RT, Munn LL (2007) Effect of vascular normalization by antiangiogenic therapy on interstitial hypertension, peritumor edema, and lymphatic metastasis: insights from a mathematical model. Cancer Res 67(6):2729–2735. doi:10.1158/0008-5472.can-06-4102
84. Kallinowski F, Schlenger KH, Kloes M, Stohrer M, Vaupel P (1989) Tumor blood flow: the principal modulator of oxidative and glycolytic metabolism, and of the metabolic micromilieu of human tumor xenografts in vivo. Int J Cancer 44(2):266–272. doi:10.1002/ijc.2910440214
85. Karakashev SV, Reginato MJ (2015) Progress toward overcoming hypoxia-induced resistance to solid tumor therapy. Cancer Manag Res 7:253–264. doi:10.2147/cmar.s58285
86. Kelly CJ, Brady M (2006) A model to simulate tumour oxygenation and dynamic [18F]-Fmiso PET data. Phys Med Biol 51(22):5859–5873. doi:10.1088/0031-9155/51/22/009
87. Krogh A (1919) The number and distribution of capillaries in muscles with calculations of the oxygen pressure head necessary for supplying the tissue. J Physiol (Lond) 52(6):409–415. doi:10.1113/jphysiol.1919.sp001839
88. Lagerlöf JH, Kindblom J, Bernhardt P (2014) The impact of including spatially longitudinal heterogeneities of vessel oxygen content and vascular fraction in 3D tumor oxygenation models on predicted radiation sensitivity. Med Phys 41(4):044,101. doi:10.1118/1.4866887
89. Lagerlöf JH, Kindblom J, Cortez E, Pietras K, Bernhardt P (2013) Image-based 3D modeling study of the influence of vessel density and blood hemoglobin concentration on tumor oxygenation and response to irradiation. Med Phys 40(2):024,101. doi:10.1118/1.4773886
90. Lee D, Rieger H, Bartha K (2006) Flow correlated percolation during vascular remodeling in growing tumors. Phys Rev Lett 96(5):058,104-1-058,104-4. doi:10.1103/PhysRevLett.96.058104
91. Lee J, Beighley P, Ritman E, Smith N (2007) Automatic segmentation of 3D micro-CT coronary vascular images. Med Image Anal 11(6):630–647. doi:10.1016/j.media.2007.06.012
92. Leenders KL, Beaney RP, Brooks DJ, Lammertsma AA, Heather JD, McKenzie CG (1985) Dexamethasone treatment of brain tumor patients: effects on regional cerebral blood flow, blood volume, and oxygen utilization. Neurology 35(11):1610–1616. doi:10.1212/wnl.35.11.1610

93. LeVeque RJ (2007) Finite difference methods for ordinary and partial differential equations – steady-state and time-dependent problems. SIAM. doi:10.1137/1.9780898717839
94. Logsdon EA, Finley SD, Popel AS, Gabhann FM (2014) A systems biology view of blood vessel growth and remodelling. J Cell Mol Med 18(8):1491–1508. doi:10.1111/jcmm.12164
95. Lowengrub JS, Frieboes HB, Jin F, Chuang YL, Li X, Macklin P, Wise SM, Cristini V (2010) Nonlinear modelling of cancer: bridging the gap between cells and tumours. Nonlinearity 23(1):R1–R9. doi:10.1088/0951-7715/23/1/r01
96. Macklin P, Lowengrub J (2007) Nonlinear simulation of the effect of microenvironment on tumor growth. J Theor Biol 247(3):677–704. doi:10.1016/j.jtbi.2006.12.004
97. Macklin P, McDougall S, Anderson AR, Chaplain MA, Cristini V, Lowengrub J (2009) Multiscale modelling and nonlinear simulation of vascular tumour growth. J Math Biol 58(4-5):765–798. doi:10.1007/s00285-008-0216-9
98. Mandelbrot BB (1983) The fractal geometry of nature, vol 51. Freeman, San Francisco. doi:10.1119/1.13295
99. Mankoff DA, Dunnwald LK, Gralow JR, Ellis GK, Charlop A, Lawton TJ, Schubert EK, Tseng J, Livingston RB (2002) Blood flow and metabolism in locally advanced breast cancer: relationship to response to therapy. J Nucl Med 43(4):500-509. http://jnm.snmjournals.org/content/43/4/500.abstract
100. Marieb E, Hoehn K (2013) Human anatomy & physiology. Pearson, San Francisco
101. McDonald DM, Choyke PL (2003) Imaging of angiogenesis: from microscope to clinic. Nat Med 9(6):713–725. doi:10.1038/nm0603-713
102. Minchinton AI, Tannock IF (2006) Drug penetration in solid tumours. Nat Rev Cancer 6(8):583–592. doi:10.1038/nrc1893
103. Mönnich D, Troost EG, Kaanders JH, Oyen WJ, Alber M, Thorwarth D (2011) Modelling and simulation of [18F]fluoromisonidazole dynamics based on histology-derived microvessel maps. Phys Med Biol 56(7):2045–2057. doi:10.1088/0031-9155/56/7/009
104. Moschandreou TE, Ellis CG, Goldman D (2011) Influence of tissue metabolism and capillary oxygen supply on arteriolar oxygen transport: a computational model. Math Biosci 232(1):1–10. doi:10.1016/j.mbs.2011.03.010
105. Murray C (1926) The physiological principle of minimum work: the vascular system and the cost of blood volume. Proc Natl Acad Sci USA 12:207–214. doi:10.1073/pnas.12.3.207
106. Nair PK, Hellums JD, Olson JS (1989) Prediction of oxygen transport rates in blood flowing in large capillaries. Microvasc Res 38(3):269–285. doi:10.1016/0026-2862(89)90005-8
107. Nair PK, Huang NS, Hellums JD, Olson JS (1990) A simple model for prediction of oxygen transport rates by flowing blood in large capillaries. Microvasc Res 39(2):203–211. doi:10.1016/0026-2862(90)90070-8
108. Nehls V, Herrmann R, H´uhnken M (1998) Guided migration as a novel mechanism of capillary network remodeling is regulated by fibroblast growth factor. Histochem Cell Biol 109(4):319–329. doi:10.1007/s004180050232
109. Netti PA, Roberge S, Boucher Y, Baxter LT, Jain RK (1996) Effect of transvascular fluid exchange on pressure-flow relationship in tumors: a proposed mechanism for tumor blood flow heterogeneity. Microvasc Res 52(1):27–46. doi:10.1006/mvre.1996.0041
110. Nico B, Benagiano V, Mangieri D, Maruotti N, Vacca A, Ribatti D (2008) Evaluation of microvascular density in tumors: pro and contra. Histol Histopathol 23(5):601–607
111. Owen MR, Alarcon T, Maini PK, Byrne HM (2009) Angiogenesis and vascular remodelling in normal and cancerous tissues. J Math Biol 58(4–5):689–721. doi:10.1007/s00285-008-0213-z
112. Peirce SM, Van Gieson EJ, Skalak TC (2004) Multicellular simulation predicts microvascular patterning and in silico tissue assembly. FASEB J 18(6):731–733. doi:10.1096/fj.03-0933fje
113. Pennacchietti S, Michieli P, Galluzzo M, Mazzone M, Giordano S, Comoglio PM (2003) Hypoxia promotes invasive growth by transcriptional activation of the met protooncogene. Cancer Cell 3(4):347–361. doi:10.1016/s1535-6108(03)00085-0
114. Penta R, Ambrosi D (2015) The role of the microvascular tortuosity in tumor transport phenomena. J Theor Biol 364(0):80–97 (2015). doi:10.1016/j.jtbi.2014.08.007
115. Perfahl H, Byrne HM, Chen T, Estrella V, Alarcon T, Lapin A, Gatenby RA, Gillies RJ, Lloyd MC, Maini PK, Reuss M, Owen MR (2011) Multiscale modelling of vascular tumour growth in 3D: the roles of domain size and boundary conditions. PLoS ONE 6(4):e14,790. doi:10.1371/journal.pone.0014790
116. Peskin CS (2002) The immersed boundary method. Acta Numer 11:479–517. doi:10.1017/S0962492902000077
117. Preziosi L, Tosin A (2009) Multiphase modelling of tumour growth and extracellular matrix interaction: mathematical tools and applications. J Math Biol 58(4–5):625–656. doi:10.1007/s00285-008-0218-7
118. Pries A, Secomb T, Gaehtgens P, Gross J (1990) Blood flow in microvascular networks. Experiments and simulation. Circ Res 67(4):826–834. doi:10.1161/01.res.67.4.826
119. Pries AR, Reglin B, Secomb TW (2005) Remodeling of blood vessels: responses of diameter and wall thickness to hemodynamic and metabolic stimuli. Hypertension 46(4):725–731. doi:10.1161/01.hyp.0000184428.16429.be
120. Pries AR, Secomb TW, Gessner T, Sperandio MB, Gross JF, Gaehtgens P (1994) Resistance to blood flow in microvessels in vivo. Circ Res 75(5):904–915. doi:10.1161/01.res.75.5.904

121. Primeau AJ, Rendon A, Hedley D, Lilge L, Tannock IF (2005) The distribution of the anticancer drug Doxorubicin in relation to blood vessels in solid tumors. Clin Cancer Res 11(24 Pt 1):8782-8788. doi:10.1158/1078-0432.ccr-05-1664
122. Raica M, Cimpean AM, Ribatti D (2009) Angiogenesis in pre-malignant conditions. Eur J Cancer 45(11):1924–1934. doi:10.1016/j.ejca.2009.04.007
123. Rieger H, Thome C, Sadjadi Z (2015) Meniscus arrest dominated imbibition front roughening in porous media with elongated pores. J Phys Conf Ser 638(1):012,007. doi:10.1088/1742-6596/638/1/012007
124. Rieger H, Welter M (2015) Integrative models of vascular remodeling during tumor growth. Wiley Interdiscip Rev Syst Biol Med 7(3):113–129. doi:10.1002/wsbm.1295
125. Roose T, Chapman S, Maini P (2007) Mathematical models of avascular tumor growth. SIAM Rev 49(2):179–208. doi:10.1137/S0036144504446291
126. Safaeian N, David T (2013) A computational model of oxygen transport in the cerebrocapillary levels for normal and pathologic brain function. J Cereb Blood Flow Metab 33(10):1633–1641. doi:10.1038/jcbfm.2013.119
127. Safaeian N, Sellier M, David T (2011) A computational model of hemodynamic parameters in cortical capillary networks. J Theor Biol 271(1):145–156. doi:10.1016/j.jtbi.2010.11.038
128. Safaeian N (2012) Computational modelling of capillaries in neuro-vascular coupling. University of Canterbury. doi:10092/8038
129. Sahani DV, Kalva SP, Hamberg LM, Hahn PF, Willet CG, Saini S, Mueller PR, Lee TY (2005) Assessing tumor perfusion and treatment response in rectal cancer with multisection ct: initial observations. Radiology 234(3):785–792. doi:10.1148/radiol.2343040286
130. Sainson RC, Harris AL (2007) Anti-Dll4 therapy: can we block tumour growth by increasing angiogenesis? Trends Mol Med 13(9):389–395. doi:10.1016/j.molmed.2007.07.002
131. Sanga S, Sinek JP, Frieboes HB, Ferrari M, Fruehauf JP, Cristini V (2006) Mathematical modeling of cancer progression and response to chemotherapy. Expert Rev Anticancer Ther 10(10):1361–1376. doi:10.1586/14737140.6.10.1361
132. Schreiner W (1993) Computer generation of complex arterial tree models. J Biomed Eng 15(2):148–150. doi:10.1016/0141-5425(93)90046-2
133. Schreiner W, Buxbaum P (1993) Computer-optimization of vascular trees. IEEE Trans Biomed Eng 40(5):482–491. doi:10.1109/10.243413
134. Scianna M, Bell CG, Preziosi L (2013) A review of mathematical models for the formation of vascular networks. J Theor Biol 333:174–209. doi:10.1016/j.jtbi.2013.04.037
135. Sciume G, Shelton S, Gray W, Miller C, Hussain F, Ferrari M, Decuzzi P, Schrefler B (2013) A multiphase model for three-dimensional tumor growth. New J Phys 15(1):015,005. doi:10.1088/1367-2630/15/1/015005
136. Secomb TW, Hsu R, Park EYH, Dewhirst MW (2004) Green's function methods for analysis of oxygen delivery to tissue by microvascular networks. Ann Biomed Eng 32(11):1519–1529. doi:10.1114/b:abme.0000049036.08817.44
137. Sefidgar M, Soltani M, Raahemifar K, Bazmara H, Nayinian SM, Bazargan M (2014) Effect of tumor shape, size, and tissue transport properties on drug delivery to solid tumors. J Biol Eng 8(1):1–13. doi:10.1186/1754-1611-8-12
138. Sefidgar M, Soltani M, Raahemifar K, Sadeghi M, Bazmara H, Bazargan M, Naeenian MM (2015) Numerical modeling of drug delivery in a dynamic solid tumor microvasculature. Microvasc Res 99(0):43–56. doi:10.1016/j.mvr.2015.02.007
139. Sethian JA, Smereka P (2003) Level set methods for fluid interfaces. Ann Rev Fluid Mech 35(1):341–372. doi:10.1146/annurev.fluid.35.101101.161105
140. Shieh AC, Swartz MA (2011) Regulation of tumor invasion by interstitial fluid flow. Phys Biol 8(1):015,012. doi:10.1088/1478-3975/8/1/015012
141. Sinek JP, Sanga S, Zheng X, Frieboes HB, Ferrari M, Cristini V (2009) Predicting drug pharmacokinetics and effect in vascularized tumors using computer simulation. J Math Biol 58(4–5):485–510. doi:10.1007/s00285-008-0214-y
142. Skeldon AC, Chaffey G, Lloyd DJ, Mohan V, Bradley DA, Nisbet A (2012) Modelling and detecting tumour oxygenation levels. PLoS ONE 7(6):e38,597. doi:10.1371/journal.pone.0038597
143. Song JW, Munn LL (2011) Fluid forces control endothelial sprouting. Proc Natl Acad Sci USA 108(37):15,342–15,347. doi:10.1073/pnas.1105316108
144. Spinelli L, Torricelli A, Pifferi A, Taroni P, Danesini G, Cubeddu R (2005) Characterization of female breast lesions from multi-wavelength time-resolved optical mammography. Phys Med Biol 50(11):2489–2502. doi:10.1088/0031-9155/50/11/004
145. McDougall SR, Anderson A, Chaplain MAJ (2006) Mathematical modelling of dynamic adaptive tumour-induced angiogenesis: clinical implications and therapeutic targeting strategies. J Theor Biol 241(3):564–589. doi:10.1016/j.jtbi.2005.12.022
146. McDougall SR, Anderson A, Chaplain MAJ, Sherratt J (2002) Mathematical modelling of flow through vascular networks: implications for tumor-induced angiogenesis and chemotherapy strategies. Bull Math Biol 64(4):673–702. doi:10.1006/bulm.2002.0293
147. Stamatelos SK, Kim E, Pathak AP, Popel AS (2014) A bioimage informatics based reconstruction of breast tumor microvasculature with computational blood flow predictions. Microvasc Res 91:8–21. doi:10.1016/j.mvr.2013.12.003
148. Stamper IJ, Byrne HM, Owen MR, Maini PK (2007) Modelling the role of angiogenesis and vasculo-

genesis in solid tumour growth. Bull Math Biol 69(8):2737–2772. doi:10.1007/s11538-007-9253-6
149. Stauffer D, Aharony A (1994) Introduction to percolation theory. CRC press. doi: 10.4324/9780203211595
150. Stylianopoulos T, Martin JD, Chauhan VP, Jain SR, Diop-Frimpong B, Bardeesy N, Smith BL, Ferrone CR, Hornicek FJ, Boucher Y, Munn LL, Jain RK (2012) Causes, consequences, and remedies for growth-induced solid stress in murine and human tumors. Proc Natl Acad Sci USA 109(38):15,101–15,108. doi:10.1073/pnas.1213353109
151. Swanson KR, Rockne RC, Claridge J, Chaplain MA, Alvord EC, Anderson AR (2011) Quantifying the role of angiogenesis in malignant progression of gliomas: in silico modeling integrates imaging and histology. Cancer Res 71(24):7366–7375. doi:10.1158/0008 5472.can-11-1399
152. Swartz MA, Lund AW (2012) Lymphatic and interstitial flow in the tumour microenvironment: linking mechanobiology with immunity. Nat Rev Cancer 12(3):210–219. doi:10.1038/nrc3186
153. Tammela T, Alitalo K (2010) Lymphangiogenesis: molecular mechanisms and future promise. Cell 140(4):460–476. doi:10.1016/j.cell.2010.01.045
154. Taroni P (2012) Diffuse optical imaging and spectroscopy of the breast: a brief outline of history and perspectives. Photochem Photobiol Sci 11(2):241–250. doi:10.1039/c1pp05230f
155. Taroni P, Torricelli A, Spinelli L, Pifferi A, Arpaia F, Danesini G, Cubeddu R (2005) Timeresolved optical mammography between 637 and 985 nm: clinical study on the detection and identification of breast lesions. Phys Med Biol 50(11):2469–2488. doi:10.1088/0031-9155/50/11/003
156. Thomas H, Cormen Charles E, Leiserson RLR, Stein C (2009) Introduction to algorithms, 3rd edn., The MIT Press
157. Tong RT, Boucher Y, Kozin SV, Winkler F, HicklinDJ, Jain RK (2004) Vascular normalization by vascular endothelial growth factor receptor 2 blockade induces a pressure gradient across the vasculature and improves drug penetration in tumors. Cancer Res 64(11):3731–3736. doi:10.1158/0008 5472.CAN-04-0074
158. Tracqui P (2009) Biophysical models of tumour growth. Rep Prog Phys 72(5):056,701. doi:10.1088/0034-4885/72/5/056701
159. Tsai AG, Johnson PC, Intaglietta M (2003) Oxygen gradients in the microcirculation. Physiol Rev 83(3):933–963. doi:10.1152/physrev.00034.2002
160. Tsoukias NM, Goldman D, Vadapalli A, Pittman RN, Popel AS (2007) A computational model of oxygen delivery by hemoglobin-based oxygen carriers in three-dimensional microvascular networks. J Theor Biol 248(4):657–674. doi:10.1016/j.jtbi.2007.06.012
161. Vaupel P, Kallinowski F, Okunieff P (1989) Blood flow, oxygen and nutrient supply, and metabolic microenvironment of human tumors: a review. Cancer Res 49(23):6449–6465. http://cancerres.aacrjournals.org/content/49/23/6449.abstract
162. Verdier C, Etienne J, Duperray A, Preziosi L (2009) Review: rheological properties of biological materials. Comptes Rendus Physique 10(8): 790–811. http://dx.doi.org/10.1016/j.crhy.2009.10.003. Complex and biofluidsFluides complexes et biologiques
163. Volokh KY (2006) Stresses in growing soft tissues. Acta Biomater 2(5):493–504. doi:10.1016/j.actbio.2006.04.002
164. Weislo R, Dzwinel W, Yuen DA, Dudek AZ (2009) A 3-D model of tumor progression based on complex automata driven by particle dynamics. J Mol Model 15(12):1517–1539. doi:10.1007/s00894-009-0511-4
165. Welter M, Bartha K, Rieger H (2008) Emergent vascular network inhomogeneities and resulting blood flow patterns in a growing tumor. J Theor Biol 250(2):257–280. http://dx.doi.org/10.1016/j.jtbi.2007.09.031
166. Welter M, Bartha K, Rieger H (2009) Vascular remodelling of an arterio-venous blood vessel network during solid tumour growth. J Theor Biol 259(3):405–422. http://dx.doi.org/10.1016/j.jtbi.2009.04.005
167. Welter M, Fredrich T, Rinneberg H, Rieger H (2015, Submitted) Relation between tumor oxygenation, vascular remodeling, and blood flow: a computational model with applications to breast cancer. PLOS Comput Biol. http://www.uni saarland.de/fak7/rieger/Paper/welter2015_submitted.pdf
168. Welter M, Rieger H (2010) Physical determinants of vascular network remodeling during tumor growth. Eur Phys J E Soft Matter 33(2):149–163. doi:10.1140/epje/i2010-10611-6
169. Welter M, Rieger H (2013) Interstitial fluid flow and drug delivery in vascularized tumors: a computational model. PLoS ONE 8(8):e70,395. doi:10.1371/journal.pone.0070395
170. Wilson CB, Lammertsma AA, McKenzie CG, Sikora K, Jones T (1992) Measurements of blood flow and exchanging water space in breast tumors using positron emission tomography: a rapid and noninvasive dynamic method. Cancer Res 52(6):1592–1597. http://cancerres.aacrjournals.org/content/52/6/1592.abstract
171. Wise SM, Lowengrub JS, Frieboes HB, Cristini V (2008) Three-dimensional multispecies nonlinear tumor growth-I model and numerical method. J Theor Biol 253(3):524–543 (2008). doi:10.1016/j.jtbi.2008.03.027
172. Wu J, Long Q, Xu S, Padhani AR (2009) Study of tumor blood perfusion and its variation due to vascular normalization by anti-angiogenic therapy based on 3d angiogenic microvasculature. J Biomech 42(6), 712–721. http://dx.doi.org/10.1016/j.jbiomech.2009.01.009
173. Wu J, Xu S, Long Q, Collins MW, Konig CS, Zhao G, Jiang Y, Padhani AR (2008) Coupled model-

ing of blood perfusion in intravascular, interstitial spaces in tumor microvasculature. J Biomech 41(5): 996–1004 (2008). doi:10.1016/j.jbiomech.2007.12.008

174. Wu M, Frieboes HB, McDougall SR, Chaplain MA, Cristini V, Lowengrub J (2013) The effect of interstitial pressure on tumor growth: coupling with the blood and lymphatic vascular systems. J Theor Biol 320(0):131–151. doi:10.1016/j.jtbi.2012.11.031
175. Yuan F, Salehi HA, Boucher Y, Vasthare US, Tuma RF, Jain RK (1994) Vascular permeability and microcirculation of gliomas and mammary carcinomas transplanted in rat and mouse cranial windows. Cancer Res 54(17):4564–4568
176. Zhao J, Salmon H, Sarntinoranont M (2007) Effect of heterogeneous vasculature on interstitial transport within a solid tumor. Microvasc Res 73(3):224–236. doi:10.1016/j.mvr.2006.12.003
177. Zheng JH, Chen CT, Au JL, Wientjes MG (2001) Time- and concentration-dependent penetration of doxorubicin in prostate tumors. AAPS PharmSci 3(2):E15. doi:10.1208/ps030215
178. Zheng X, Wise SM, Cristini V (2005) Nonlinear simulation of tumor necrosis, neovascularization and tissue invasion via an adaptive finite-element/level-set method. Bull Math Biol 67(2):211–259. doi:10.1016/j.bulm.2004.08.001

Cell-ECM Interactions in Tumor Invasion

4

Xiuxiu He, Byoungkoo Lee, and Yi Jiang

Abstract

The cancer cells obtain their invasion potential not only by genetic mutations, but also by changing their cellular biophysical and biomechanical features and adapting to the surrounding microenvironments. The extracellular matrix, as a crucial component of the tumor microenvironment, provides the mechanical support for the tissue, mediates the cell-microenvironment interactions, and plays a key role in cancer cell invasion. The biomechanics of the extracellular matrix, particularly collagen, have been extensively studied in the biomechanics community. Cell migration has also enjoyed much attention from both the experimental and modeling efforts. However, the detailed mechanistic understanding of tumor cell-ECM interactions, especially during cancer invasion, has been unclear. This chapter reviews the recent advances in the studies of ECM biomechanics, cell migration, and cell-ECM interactions in the context of cancer invasion.

Keywords

Extracellular matrix • Cell-ECM interactions • Cell migration • Mathematical models • Collagen • Mechanotransduction • Cancer invasion

4.1 Introduction

The tumor microenvironment is created by proliferating tumor cells and dominated by tumor-induced interactions [112]. It has been well accepted that the tumor microenvironment plays a significant role in disease progression, but the precise function of each constituent remains unclear. The tissue microenvironment of a developing tumor can be broken down into three categories: the biological, the chemical, and the biophysical/biomechanical. The biological environment is comprised of the cellular constituents that surround the malignant cancer cells. A variety of infiltrating immune cells [112], cancer-associated fibroblasts [100], and

X. He • B. Lee • Y. Jiang (✉)
Department of Mathematics and Statistics, Georgia State University, Atlanta, GA 30303, USA
e-mail: xhe9@student.gsu.edu; blee37@gsu.edu; yjiang12@gsu.edu

K.A. Rejniak (eds.), *Systems Biology of Tumor Microenvironment*, Advances in Experimental Medicine and Biology 936, DOI 10.1007/978-3-319-42023-3_4

angiogenic endothelial cells [110] perform critical functions in sustaining cell proliferation, evading growth suppressors, promoting survival, activating invasion and metastasis, as well as reprogramming energy metabolism. The chemical environment refers to the abnormal distribution of oxygen, nutrients, wastes, and cytokines, as well as many growth factors and inhibitors. For example, excess growth of the tumor cells leads to a hypoxic environment [55], elevated oxidative stress [22], and consequently, the accumulation of lactic acid due to anaerobic metabolism [44] and up-regulation growth factor production (e.g., VEGF). The biophysical and biomechanical aspect of the tumor is both the physical and geometrical constraints from the tissue structure, and the mechanical interactions between the tumor and surrounding environment, most importantly the extracellular matrix (ECM). This category of the microenvironment has only recently begun to receive an increased level of attention, including studies on the hydrostatic stress from interstitial fluid [12], substrate topography [61, 66, 82], and the biomechanics of the extracellular matrix [71]. In this chapter, we focus on the emerging concepts in the contribution of ECM heterogeneity and remodeling to tumor growth and invasion.

It has been appreciated for some time that the extracellular matrix (ECM) plays an important role in all the stages of cancer development. In breast cancer, dense breast tissue on mammography is associated with increased collagen content. Women with more than 75 % dense regions have been shown to have an increased risk of breast cancer by up to five fold in comparison to women with less than 5 % density [13, 80]. Breast density is common, heritable, and has been postulated to account for up to one third of breast cancers [80]. In mouse models of breast cancer, it has been shown that increasing either the density or the crosslinking of collagen promotes invasiveness and, to a lesser extent, the formation of breast cancer [74, 87, 90], confirming the critical role of extracellular matrix (ECM) in assisting tumor progression.

Moving beyond correlative data and understanding the underlying mechanisms is difficult using traditional experimentation alone, since ECM affects many aspects of both host and tumor cell behavior, such as migration, differentiation, invasion, and proliferation. Furthermore, the properties of ECM itself are also complex, with diverse topographies and mechanical properties possibly depending on density, alignment, polymerization, and crosslinking. Because tumor invasion and growth are emergent outcomes of the complex interactions between cells and ECM, computational and mathematical models are becoming necessary tools to help dissect this complexity. Here, we will review the recent advances in the understanding of how cell-ECM interactions help to regulate cancer invasion, focusing on the biomechanical effects.

4.2 ECM in Cancer Invasion

The ECM, a fibrous macromolecular network outside cells, plays a crucial role in tissue environments, providing mechanical structures [94] as well as promoting cell phenotype change [75]. Through direct or indirect means, the ECM regulates almost all cellular behavior and is indispensable for developmental processes [78]. Recent experimental evidence has suggested that cancer cells interact with ECM fibers during invasion, condensing [106], remodeling [106], and aligning [89] fibers. Using *in vitro* mouse breast cancer models, Provenzano and coworkers discovered three tumor-associated collagen signatures (TACS): TACS-1 with dense collagen near the tumor, TACS-2 with stretched collagen fibers encasing the tumor, and TACS-3 with aligned collagen fibers normal to the tumor boundary. Despite the fact that the mechanisms are still unclear, it has been well accepted that breast tumors are associated with dense breast tissue, notably dense collagen [89]. At the early stage of cancer, collagen fibers condense near the tumor, interacted with growing cancer cells (Fig. 4.1A). As cancer progresses, cancer cells invade outward. Migrating cells supposedly pull on the surrounding ECM fibers, producing stretched fibers, but the mechanism for the aligned fibers normal to the tumor boundary

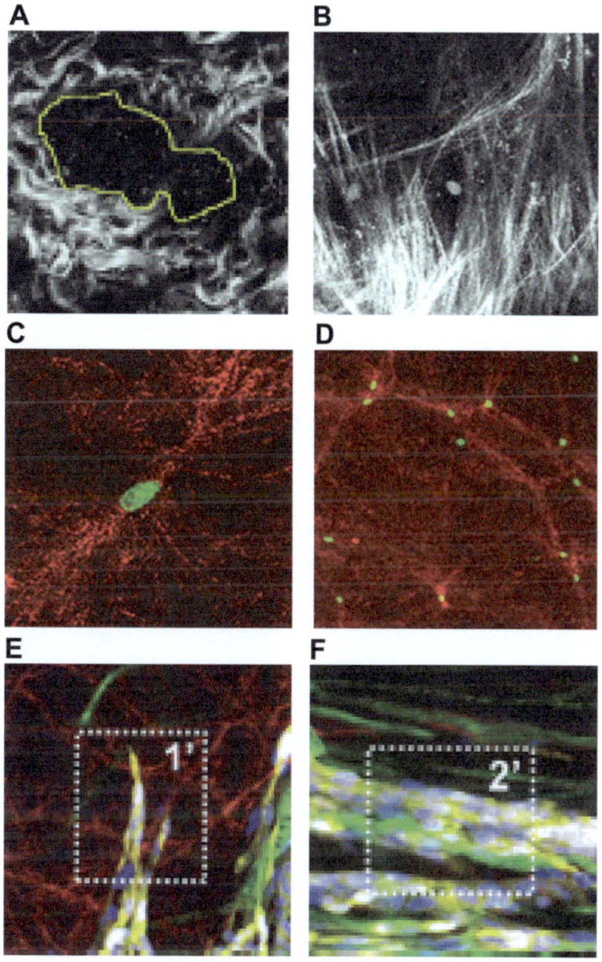

Fig. 4.1 The ECM in cancer invasion. Multiphoton microscopy images of mouse breast tumor: (**a**) dense collagen, and (**b**) aligned collagen fibers (From Provenzano et al. [89] with permission). *Yellow outline* in (**a**) is a tumor boundary. Single (**c**) and multiple (**d**) U87 glioblastoma cells modified collagen fiber structures 10 h after gel polymerization (From Vader et al. [106] with permission), cell nuclei are green, and collagen fibers are *red*. Multiphoton intravital microscopy images of a tip cell of invasion into a mouse dermis (**e**) and the multicellular core (**f**) (From Alexander et al. [3] with permission)

is still unclear (Fig. 4.1B) [89]. A recent review summarizes remodeled ECM as an anomaly that deregulates behavior of stromal cells, facilitates tumor angiogenesis and inflammation, and leads to a tumorigenic microenvironment [79].

The remodeling of collagen fibers by invasive cancer cells has also been observed *in vitro*, where collagen fibers condense near single glioblastoma cells (Fig. 4.1C) and aligned fiber tracks appear between multiple migrating glioblastoma cells (Fig. 4.1D) [106]. Cancer invasion *in vivo* is more complicated. Using intravital microscopy, Alexander et al. [3] observed melanoma cells invading into the mouse dermis. They suggested that heterogeneous connective tissue, in particular the porous 3D ECM network, provides a guidance or track for invasive cancer cells (Fig. 4.1E). They also observed that, in addition to individual migrating cells, cancer cells often invade collectively as a multicellular unit with cell-cell junctions retained [37]. This suggests that the leader cell searches for a pore space in the ECM fiber network and squeezes itself through the space, whereas the following cells collectively invade using the track (Fig. 4.1F) [3].

Both *in vitro* and *in vivo* evidence shows substantial ECM remodeling associated with proliferating and invading cancer cells. However, because of the complexity of the microenvironment, many other factors could potentially contribute to ECM remodeling, including tumor associated fibroblasts [100] that can produce or degrade the ECM. In order to understand how mechan-

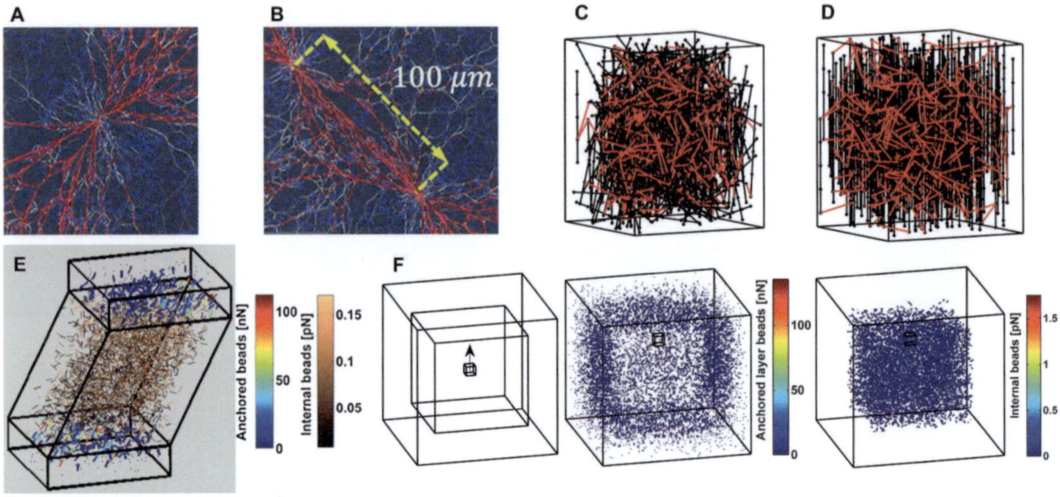

Fig. 4.2 Computational models of ECM. A two-dimensional ECM fiber model showing configurations upon anisotropic contraction from (**a**) a single cell and (**b**) two cells (From Abhilash et al. [1] with permission). A three-dimensional elastic bead-spring fiber network model for random (**c**), and pre-aligned structure (**d**) *Black lines* are fibers, and red lines are crosslinkers. The residual stress distribution for a random fiber network upon a shear strain (**e**) and a local box displacement at the center of the fiber network, mimicking a local deformation imposed by a migrating cell (**f**) (From Lee et al. [71] with permission)

ical cell-ECM interactions contribute to the remodeling, theoretical and computational models have been developed to further investigate the mechanical properties of ECM fibers and their interactions with cells. A two-dimensional (2D) discrete fiber network model using the finite element method simulated ECM fiber remodeling by contractile force from a single cell and between two cells [1]. Anisotropic contractile forces produce ECM fiber patterns (Fig. 4.2a, b) resembling the experimental observations [1]. More recently, a three-dimensional (3D) elastic fiber network model using a bead-and-spring fiber representation with elastic crosslinking simulated tensile and shear tests for random and aligned fiber networks [71] (Fig. 4.2c, d). Their simulations show that aligned fiber network structure is stiffer than the random network, while both structures showed nonlinear strain-stiffening. The stress-strain curve of a random fiber network illustrates how the matrix responds to external strain. Upon small strain, the network first responds with minimal stress, like a fluid. As the external strain increases, the stress increases slowly until the strain reaches about 10 %, when the fibers start to align. Between 10 and 30 % strain, the stress of the fiber network increases linearly, indicating that the network behaves like an elastic material. At 30 % strain, the fiber alignment reaches 70 % [93], after which the fibers will be stretched to show a much stiffer bulk response. The residual stress distributions, depicted as force vectors, after a shear test and a local displacement showed the nonaffine deformation of the network and the accumulation of stress at the boundary of displacement (Fig. 4.2e, f). Feng et al. [32] explored the role of fiber alignment in a fiber network using a Landau-type theory for the nonlinear elasticity with the order parameter taking into account the kinematic order of fibers. Comparing the theory and simulation of a disordered lattice model, they concluded that the nonlinear elastic behavior of biopolymer gels arises from strain-induced fiber alignment, suggesting that it explained contact guidance of cell motility.

4.3 Cell-ECM Interactions

Regulation of cell motility and invasiveness in the ECM is complex. On one hand, deposition of fibrillar collagen appears to promote tumor cell motility by providing one-dimensional or two-dimensional "tracks" for cell movement [31, 38]. Further crosslinking of collagen fibrils by enzymes such as lysyl oxidase may increase the alignment and rigidity of those tracks, aiding cell invasiveness [2, 74]. Increasing collagen density may also inhibit cell migration and require proteolytic activity to allow tumor cell migration [113, 115]. On the other hand, cellular machinery that recognizes not only the biochemical diversity of the ECM, but also its physical and topographical characteristics, such as rigidity, dimensionality and ligand spacing is critical for the response of cells to ECM.

It has been increasing clear that the cellular response to environmental signaling goes far beyond the ability of chemically sensing specific ECM ligands and encompass a wide range of physical cues and the adhesive interface [47]. More attempts on understanding cell migration during tumor invasion are focusing on the interplay of multiscale mechanotransduction, which is comprised of how the cell sense and react to internally generated and externally applied signals [45]. Current understanding of the biomechanics of cell-matrix interactions is based primarily on *in vitro* studies of the cell leading edge of migration (focal adhesion and membrane remodeling), and intra-cellular cytoskeletal activities (actin protrusion, actomyosin contraction, and cell motility signaling pathway). All these elements need to work in concert to regulate cell migration speed, directionality, and cell migration plasticity. We discuss these elements of cell motility below.

4.3.1 Focal Adhesion

Focal adhesions are integrin-based structures that mediate strong cell-substrate adhesion and transmit information between the extracellular matrix and the cytoplasm [46, 48]. During the formation of focal adhesion, a subset of adhesion components with actin nucleates the nascent adhesion, which is stabilized by its association with integrin to form stable focal adhesion assembly [70, 72, 98]. Increasing the strength and longevity of integrin binding and integrin clustering is a crucial step in this adhesion process. Active integrin complexes promote recruitment of cytoskeletal components, activate signaling molecules, and enhance adhesive force [23]. In particular, integrin activation regulates microtubule dynamics and helps to stabilize microtubules at the cell cortex [14]. Integrins connect the ECM to the cytoskeleton and provide cells with mechanical anchorages and signaling platforms. At the molecular level, force-induced strengthening of cell adhesion [4, 19, 42] has been explained in terms of recruitment of integrins and cytoskeletal proteins [92] and/or ligand-integrin catch bonds [39]. Furthermore, cyclic mechanical reinforcement [68] is found to be a more effective regulatory mechanism than the catch bond, as it prolongs the bond lifetime for fibronectin and integrin-$\alpha_5\beta_1$ [28]. While the short-lived integrin-ligand bonds may allow the cell to rapidly explore its environment, the long-lived integrin-ligand bonds are critical to adhesion maturation and downstream signaling, which takes tens of seconds to minutes [42]. The mechanically reinforced ligand-integrin bonds enable nascent adhesion to be stabilized by myosin-generated contractile forces.

Live-cell microscopy studies reveal four main stages in the "life cycle" of integrin-mediated adhesions, including nascent adhesions, focal complexes, focal adhesions, and fibrillar adhesions [116]. Nascent adhesions are submicron-sized, and are barely visible by means of ordinary fluorescence microscopy. The process of generating focal complexes is on a timescale of seconds and involves only a small number of integrin that triggers actin polymerization [119]. Measurements of mechanical tension across vinculin, a protein that connects integrins to actin filaments, also showed that vinculin recruitment to focal adhesions and force transmission to vinculin are regulated separately [49]. The subsequent strengthening of adhesions through myosin pulling is believed to lead to the recruitment of additional adhesive proteins, which promotes the growth of larger focal complexes. The growth processes depend on actomyosin-based stress fibers and also require the stress fibers to serve as physical

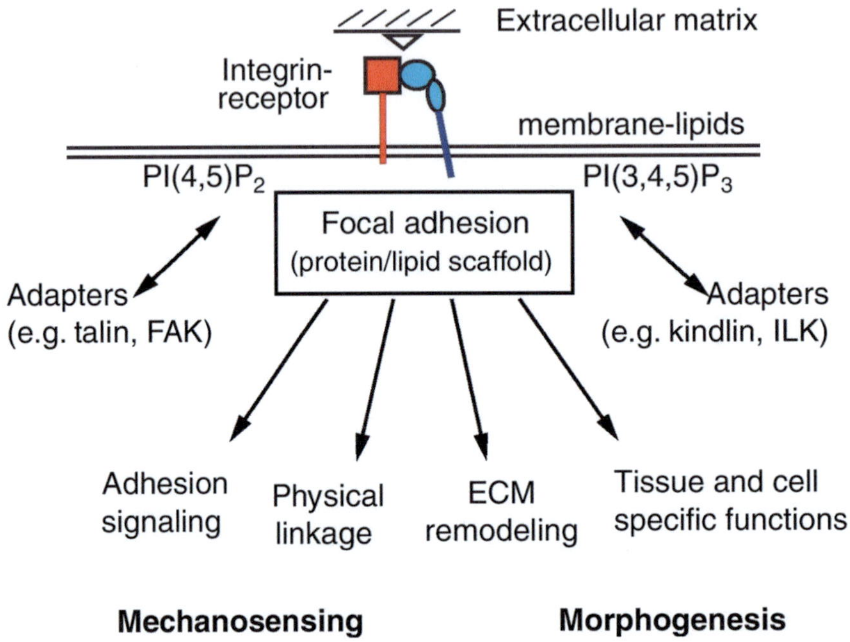

Fig. 4.3 Molecular architecture of cell-ECM interactions centered around focal adhesion (From Wehrle-Haller [109] with permission)

contractile anchors [83]. The transformation of one form of adhesion into another is tightly regulated by the cellular signaling system and is also mediated by cues from ECM and intracellular structures (Fig. 4.3).

Both ECM rigidity and ligand spacing have been found to influence focal adhesion, stress fiber assembly, cell spreading, cell migration speed, and adhesive forces [60]. Adhesive area is also found to strongly modulate adhesion strength, integrin binding, and vinculin and talin recruitment [43]. Interestingly, cells cannot integrate signals from integrin-ligand complexes spaced more than 58 nm from each other, as demonstrated in experiments of cells sitting on fibronectin nano-islands within non-adhesive background [16]. The minimal area of integrin-fibronectin clusters required for stable focal adhesion assembly and force transmission is not a predetermined value; it arises dynamically from the interaction between pathways controlling adhesive force, cytoskeletal tension, and the structural linkage that transmits these forces [23, 81].

4.3.2 Intracellular Mechanical Structures

The intracellular mechanical structures that play a key role in cell migration include actin microfilaments, intermediate filaments, lamin, nucleoskeleton and cytoskeleton linker, microtubules, and cell nucleus. The latter adds an additional layer of mechanical stability because of its significant stiffness [25, 50] and the possibility to physically divide the cytoplasm into forward and rear compartments [88]. These structures can be altered during cancer progression [9]; e.g., cell nucleus deformation can be a function of malignancy [27, 40].

Actin microfilaments provide the largest contribution to cell body stiffness when probed at adhesion sites [10, 11, 42, 58]. They are organized into different structures, including actin bundles and stress fibers. They stabilize cell architecture, including the formation of lamellipodia and filopodia, which play important roles in cell motility [57, 108]. Actin participates as an internal stabilizer and a dynamic mechanical structure in cells for migration and mechanosensing [9]. The inherent elastic features and the myosin-mediated contractility of actin fibers [10, 42, 59] and the linkage to ECM via focal adhesion [42, 57] together regulate cell-ECM interaction.

As the load–bearing element in the cell [59], the microtubule network provides internal structural support while contributing to the polarization and initiation of cell migration [64, 101]. The microtubules allowing the cell to polarize in response to ECM cues contribute to spatial organization and participate in initiating cell migration [9]. Large scale disruption of the microtubules has dramatic mechanical consequences on cell stiffness [10]. During cell migration, the microtubule depolymerization and the inhibition of the microtubule-associated molecular motors can effectively impair cell motility [63]. Focal adhesion is found to be necessary for the microtubule depolymerization and the microtubules are also required for focal adhesion disassembly and regulation [52, 62].

Intermediate filaments are the most diverse family of cytoskeletal components that exist as associated effectors of the cytoskeletal framework through connections with actin and the microtubules. The overexpression of intermediate filament proteins during transformation process is notably connected to carcinomas [21]. In most epithelia cells, intermediate filaments span the cell cortex and wind around the nucleus to form an interconnected network that provides a continuous link between focal adhesions, cell-cell adhesions and the nucleus through the linker of nucleoskeleton and cytoskeleton complex [26, 77].

The cell nucleus is the largest and the stiffest organelle with the ability to affect cell migration through nucleocytoskeletal connections [9]. The nucleoskeleton links directly to the cytoplasmic cytoskeleton through the linkers that connects the lamin network in the nucleus to actin and intermediate filaments [20]. Functionally, the nucleus sustains global deformation and changes in its sub-nuclear spatial organization when the cell is subjected to mechanical stress, indicating that the nucleus is also a mechanosensitive element participating in cell-ECM interactions. The features of the nucleus regulate cell migration, but the exact mechanisms are not clear. However, it has been found that the nucleus physically divides the cytoplasm into forward and rear pressure compartments when a human fibroblast migrates through a 3D ECM [88]. This finding suggests that the nucleus can act as a piston to increase the hydrostatic pressure between the nucleus and the leading edge of the cell in order to drive lamellipodia-independent 3D cell migration.

4.3.3 Cell Membrane Remodeling and Mechanotransduction Signaling Network

Cell membrane tension together with the pressure generated by intra-cellular structures and the focal adhesion strength are the forces that define the movement of cell membrane. The contact angle between substrate and membrane has been found to correlate with the load on actin polymerization and cell protrusion rate [41]. This result emphasizes the fundamental importance of membrane configuration for cellular force balance and the subcellular scale biophysical dynamics. In a model trying to explain the cell morphology of slime mold *Dictyostelium* during its chemotactic migration, an analogy was drawn between the interaction of pushing microtubules with the surface tension of the plasma membrane and the Marangoni effect that generates the tear drops of wine covered glass (Fig. 4.4a) [97]. The protrusion results from a dynamic balance between the outward push from cytoskeletal fibers and membrane surface tension, while the wine tear pattern arises from the balance between gravity and the surface tension gradient between alcohol and water. Furthermore, the recent biochemical understanding of reaction-diffusion inside the cell has led to the similar results of cell morphology. For chemotactic cell migration, the local excitation global inhibition (LEGI) model [97] (Fig. 4.4) and its variations were proposed to explain the signaling responses of cells exposed to gradients of chemoattractant [86]. The response to a stimulus is mediated through the balance between a fast, local excitation and a slower, global inhibi-

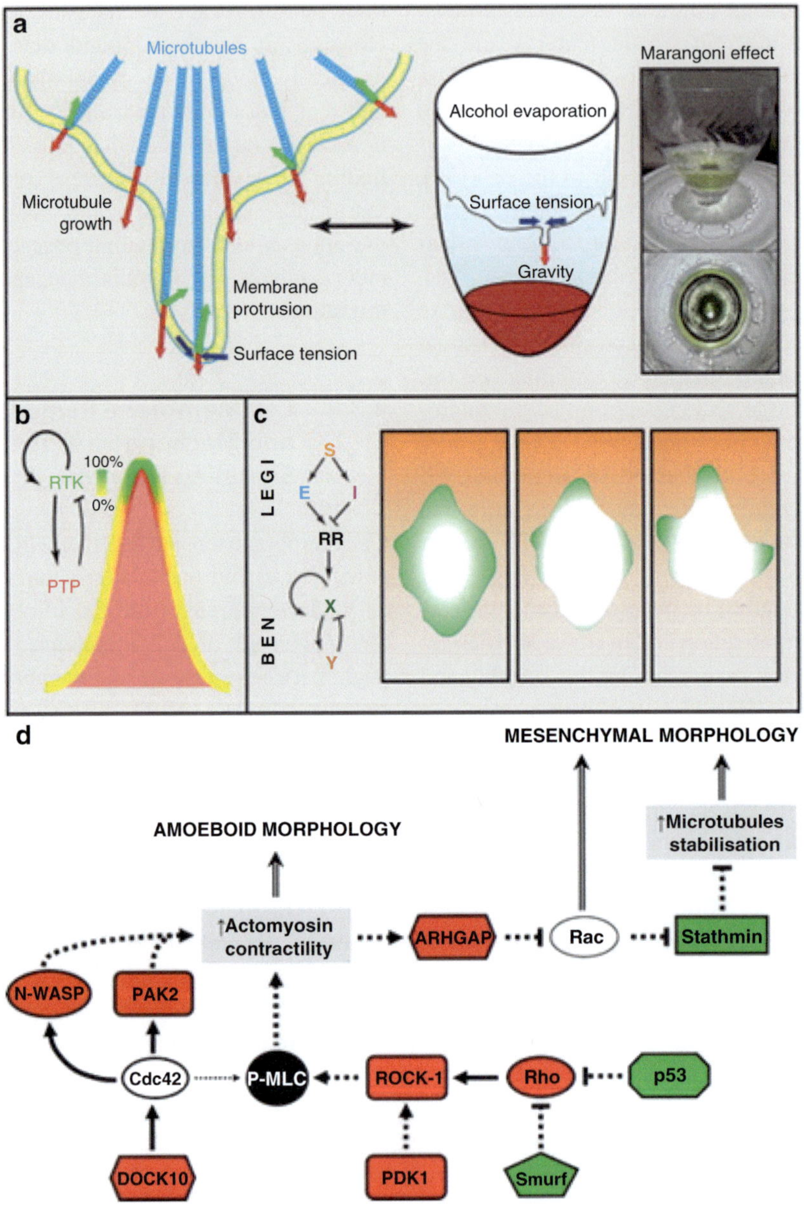

Fig. 4.4 Signaling, cytoskeletal dynamics, and cell shape. (**a**) The interaction of pushing microtubules (\textit{red arrows}) with the surface tension (*blue arrows*) of the plasma membrane (*left*) resembles the balance of gravitational pull (*red arrow*) and alcohol-dependent surface tension (*blue arrows*) along the edge of a wine-covered glass (*right*). (**b**) Activator-inhibitor system of an autocatalytically activated RTK. The high curvature at the tip of a protrusion facilitates initial RTK activation by effectively exposing the receptors to more extracellular space. The faster diffusing phosphatase limits spreading of autocatalytic activation by lateral inhibition. (**c**) (*Left*) The LEGI-BEN model: in the local excitation global inhibition (LEGI) model, a stimulus (S) turns on excitation (E) and inhibition (I) processes that act in parallel on a response regulator RR, which activates the biased excitable network (BEN), consisting of autocatalytic activity (X) that activates its own inhibitor (Y). (*Right*) Activity of X at different times after initial exposure to an extracellular chemotactic gradient (From Schmick and Bastiaens [97] with permission). (**d**) Interactions among the components of signaling pathways involved in the MAT/AMT transitions of cells in a 3D environment. The inhibition of the activity of the proteins highlighted in red was shown to trigger amoeboid to mesenchymal transitions. Inactivation of the proteins depicted in *green* induces a conversion from the mesenchymal to the amoeboid mode of invasiveness (From Pankova et al. [85] with permission)

tion process, both of which are controlled by receptor occupancy [30, 69, 73]. When stimulated by uniform concentration of chemoattractant, the faster local excitation rises with receptor occupancy, leading to an increase in the response. As the slower inhibition rises, the response subsides, ensuring perfect adaptation. When in a gradient, local excitation mirrors receptor occupancy, and hence, chemoattractant concentration gradients. The inhibition process integrates the global signal, leading to an inhibitory signal that is equivalent to the average level of receptor occupancy in the cell. This model has satisfactorily explained Ras, PTEN and PI3K activation during amoeboid *Dictyostelium* cell migration [118]. Thus far the LEGI models seem most promising in providing a plausible mechanism for chemotactic migration, possibly applicable to more generic cell migration as well.

Many studies of the molecular mechanisms of cell motility signaling pathways have focused on the Rho family of small GTPases that regulate the cytoskeleton-dependent processes. The complexity of the interactions among the Rho family of proteins, their regulators, and effectors (Fig. 4.4d) is challenging to both experimental and mathematical studies. Integrating the cell motility pathway to mechanotransduction network alone is difficult. Moreover, the spatiotemporal reaction-diffusion dynamics of the signaling molecules in the cytosol and on cell membrane are thought to be the key determinants of cell migration plasticity [85].

4.3.4 Cell Migration Modes

Cell migration plasticity refers to the cell's ability to switch between different cell migration modes. The migration modes were originally classified based on the cell morphology alone, but have since been extended to describe the multi-scale properties of cell migration, including cell shape, cell migration speed, and organization of intracellular structures. The main categories are individual (amoeboid and mesenchymal), and collective (as cohesive multicellular units) migration [34]. The amoeboid migration refers to the movement of round or ellipsoid cells that lack mature focal adhesions and stress fibers [38, 85] with blebby membrane dynamics and faster migration speed. Mesenchymal migration is characterized by a spindle-like elongated cell shape, actin-rich filopodia and more focalized cell–matrix interactions; mesenchymal movement resembles the migration of a fibroblast [103], whereby the cell entangles with the ECM [38].

Individual cell migration modes differ depending on cell type, developmental stage, local environment, and disease state [111]. Cell migration mode can be dynamically changed by the strength of adhesion, physical confinement (e.g., squeezed between two surfaces), contractility, and chemical cues [76, 95]. With low adhesion and strong confinement, slow mesenchymal cells can switch to fast amoeboid migration, suggesting that no specific genetic alteration is necessary for tumor cells to escape the primary tumors [29]. *In vitro* evidence shows that, in confined 3D ECM, the intrinsic fluctuation in cortical contraction is sufficient to trigger the switch from embryonic progenitor cells to prototypic amoeboid migration mode [95]. Cancer cell migration persistence and local membrane protrusion persistence have been measured *in vivo* [102] (Fig. 4.5a). However, because of the lack of local measurements of ECM dynamics concurrently, cellular and subcellular imaging has not been able to offer comprehensive understanding of the cell-ECM interaction. Recent advances in combination of live cell imaging, molecular manipulation and force measurement have revealed multiscale cell migration details with extraordinary precision that allowed for mechanistic mathematical modeling. An ideal 2D experimental cell migration model has been the fish epithelial keratocytes for investigating cell shape determination [6, 65]. Individual keratocytes maintain nearly constant shape, speed, and direction over many cell lengths of migration, with considerable heterogeneity within a population of keratocyte [65] (Fig. 4.5b). Several mathematical models have been developed to take advantage of such data and explained the detailed intracellular signaling and mechanical interactions leading to the specific cell shape during migration (Fig. 4.5c, d).

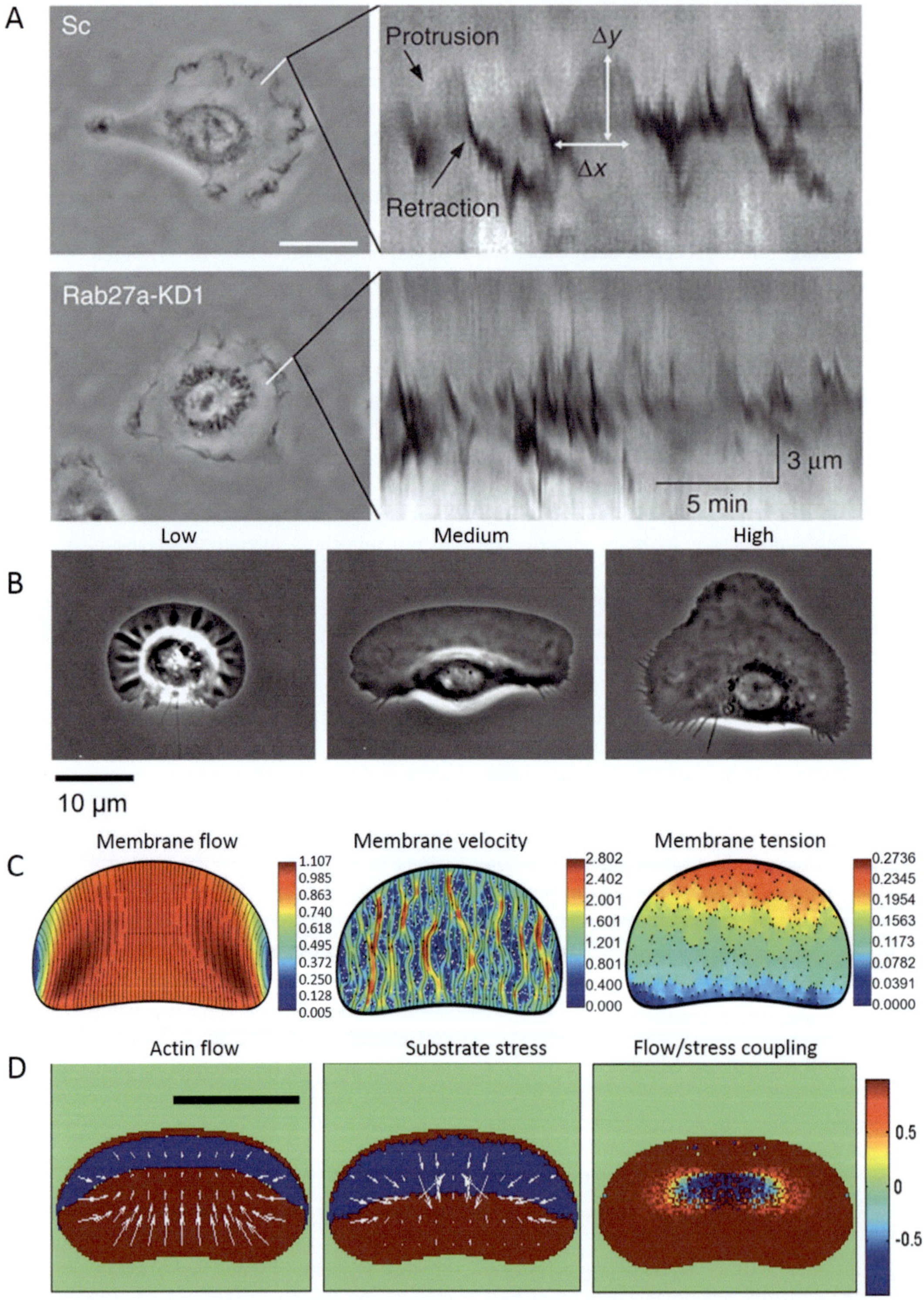

Fig. 4.5 Cell shape and membrane dynamics during migration. (**a**) Kymographs of an HT1080 cell on tissue culture-treated dishes showing membrane protrusion dynamics (From Sung et al. [102] with permission). (**b**) Phase contrast images of keratocytes crawling at low (*left*), intermediate (*center*), and high (*right*) adhesion strengths (From Barnhart et al. [6] with permission). (**c**) Simulations of membrane dynamics for keratocyte migration: membrane flow, velocity and tension (From Fogelson and Mogilner [33] with permission). (**d**) Steady-state maps of actin flow and substrate stress for keratocyte migration (From Shao et al. [99] with permission)

Collective cell migration plays a crucial role in many biological processes, including embryonic development, wound healing, as well as cancer invasion and metastasis [35, 36]. During collectively cell migration, the enhanced migration is led by a subset of "leader cells" that extend filopodia at the leading edge of the cell cluster [105]. The "invasion-competent" malignant cells induced the collective invasion of otherwise "invasion incompetent" epithelial cells, and that these two cell types consistently exhibited distinct leader and follower roles during invasion. Analysis of extracellular matrix (ECM) microarchitecture revealed that malignant cell invasion was accompanied by extensive ECM remodeling including matrix alignment and proteolytic track making [15].

Physical characteristics of ECM strongly modulate cell migration by outside-in signaling from microenvironment, while morphological properties of cell and intracellular dynamics feedback to ECM by inside-out signaling [91]. Current knowledge about the focal adhesion, cell migration, mechano-signaling, and cytoskeletal function is derived primarily from studies on planar 2D tissue culture substrates. The 2D substrate may induce artificial polarity between the basal and apical surfaces of the normally nonpolar cells, e.g., fibroblastic cells [24]. It also may exclude ECM-dependent regulators of 3D cell migration, including ECM porosity, ECM compliance, collagen fiber size, and collagen concentration [114]. The microarchitecture of 3D scaffolds has been found to influence cell migration behavior via junction interactions [54]. The pore size of collagen-glycosaminoglycan scaffolds influences the fibroblast migration: the migration speed decreases as pore size increases across a range from 90 to 150 μm [84]. Importantly, ECM density, stiffness and alignment also contributes to cell migration speed and persistence differently; ECM density and stiffness influences cell speed, but ECM alignment does not change cell speed; instead, alignment increases cell migration persistence [93].

4.4 2D Cell Migration Models

Computational and mathematical modeling has benefited from the availability of new data with combination of live-cell imaging, molecular manipulation and force measurement. To date, most models in cell-ECM interactions focus on cell shape and cell motility. These models have treated implicit or explicit focal-adhesion, motility related diffusion-reaction of molecules, cytoskeletal dynamics, intracellular flows, and cell morphology related protrusion and contraction.

A rule-based model was developed for cell migration, in which the underlying mechanochemical events are incorporated implicitly using rules describing the evolution of cell shapes and regulatory signals [96]. The main rules are about the local/global feedbacks and deterministic/stochastic signaling regulations. A cell is modeled using a collection of perimeter points and a center. The perimeter points can move according to the balance between protrusion signal and retraction signal. The local protrusion signal propagates and decays, with a stochastic positive feedback loop that accounts for both "local stimulation" and generation of random noises. Focal adhesion is a probabilistic event with a fixed average halftime. This simple model was capable of generating the dynamic shapes and persistence of amoeboid cells migration without the chemoattractants.

Using the keratocyte migration as a model, a whole series of mathematical models explained the keratocyte cell shape [5, 6, 32, 94, 99]. Actin fibers polymerize pushing on the cell membrane from within, generating membrane tension that rapidly equilibrates. The membrane tension in turn exerts a constant force on the actin network. The spatiotemporal dynamics of adhesion-dependent actin polymerization, retrograde flow, myosin distribution, and traction forces together provide a more complete understanding of distribution of cell motility molecules and cell shape [33, 56] (Fig. 4.5c).

As the cell morphology adapts to the local forces from focal adhesion, actin flow, and myosin activities, the macromolecular distribution inside the cell presents a moving boundary reaction-diffusion problem for modeling. Because of the computational complexity, few models have integrated or implemented this problem. A recent model used the phase-field method to integrate the adhesion dynamics with the dynamics of the actin filaments modeled as a viscous network, and to solve for the moving boundary with membrane tension. The model included a reaction-diffusion model for the actin-myosin machinery and discrete adhesion sites that can be in a "gripping" or "slipping" mode. This model suggested the pattern of the actin flow inside the cell, the cell velocity, and the cell morphology are determined by the integration of actin polymerization, myosin contraction, adhesion forces, and membrane forces (Fig. 4.5d) [99].

The interaction between migrating cells and the ECM has also become a focal point of modeling in the past decade. In the context of angiogenesis, Bauer et al. developed a 2D model based on the cellular Potts model to study the effects of ECM topography on the collective migration morphology of endothelial cells [7]. They varied the density and alignment of the matrix fibers to simulate different tissue environments and to explore the possibility of manipulating the extracellular matrix to achieve pro- and anti-angiogenic effects. The ECM in this model only provided contact guidance, without mechanical interactions with the cells. Van Oers et al. coupled a 2D cellular Potts model with a finite element model for the ECM substrate, to simulate the mechanical interaction between cells and the ECM [107]. They showed that the resulting matrix strain could in turn mediate the interaction between cells and promote collective migration (Fig. 4.6a). The effect of ECM geometry on cell migration mode determination was studied by Tozluoglu et al. [104], using a 2D hybrid agent-based/finite element model of cell blebbing migration. The model integrated actin-polymerization-based protrusion, actomyosin contractility, and membrane blebbing due to actin-plasma membrane linkage, cell-ECM adhesion and varied matrix geometries (Fig. 4.6d–f) [104]. Actomyosin cortex and cell membrane were agents, with local levels of actin cortex density, myosin concentration, cortex-membrane linker proteins recorded at each agent. The model predicted the optimal migration strategies with different matrix geometries.

4.5 3D Cell-ECM Model

Most cells encounter a 3D matrix environment during processes such as wound healing or cancer metastasis. Increasing evidence from literature suggests that 2D ECM models are inherently limited in their scope to capture the ability of cells to form adhesions in three dimensions. Therefore, it is important that we use 3D systems to study cell-matrix interactions to gain more physiologically-relevant insights. The 3D matrix structure, focal adhesion, cell signaling, and cell morphology are more complex. But with the advance of imaging tools, such as multi-photon microscopy for imaging the ECM, and lattice light-sheet microscopy to image both cell and ECM with very high spatial and temporal resolutions [17], the hope is high for a more complete understanding of 3D cell-ECM interactions in the near future.

A phenomenological 3D model of single cell migration through cell-ECM interaction is developed taking into account the ECM deposition, cell protrusion, adhesion detachment and MMP activities [18] [53]. In this model, cells can degrade, deposit, or pull local fibers, depending on the fiber density around each cell. The cells can also move within the 3D matrix. The model produced results consistent with the current understanding: in low density environments, cells deposit more collagen to increase fibril fraction; in higher density environments, the less invasive cell line reduced the fibril fraction as compared to the highly invasive phenotype. Riching et al. showed another 3D cell-ECM interaction model (in the supporting materials in [93]). In this model, the cells migrate in 2D but interact with a 3D ECM environment they are embeded in. The cells send out protrusion vectors around their perimeters,

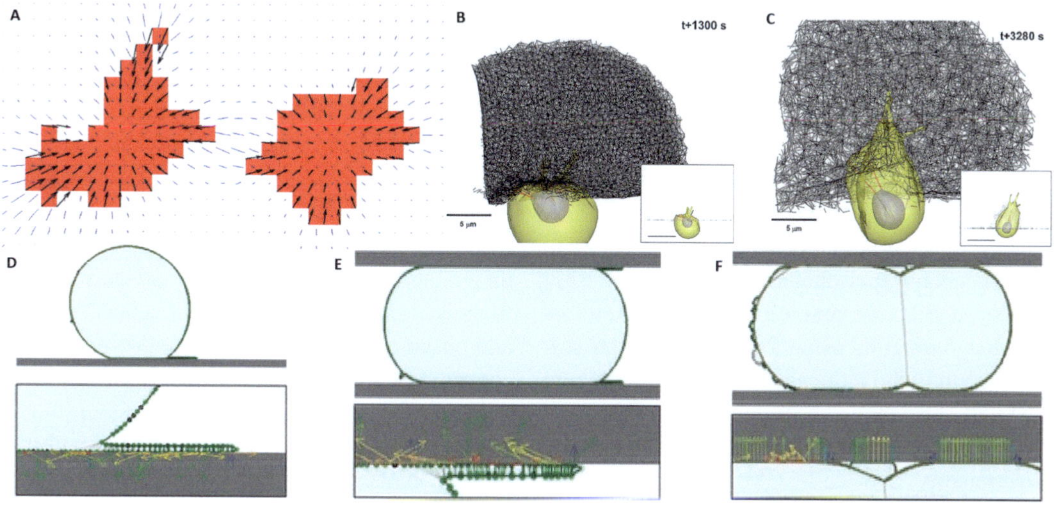

Fig. 4.6 Simulated cell-ECM interaction. (**a**) Traction forces (*black arrow*) and resulting matrix strains (*blue line* segments) generated in the hybrid cellular Potts and finite element model (From van Oers et al. [107] with permission). Cell invasion into ECM fiber network with pore sizes of (**b**) 0.5 and (**c**) 1.5 μm (From Kim et al. [67] with permission). Simulations of cell moving through different matrix geometries show different optimal migration strategies. (**d**) Cell crawling on a surface. (**e**) Actin-protrusion-based solution within confined continuous environments. (**f**) Blebbing-driven solution for cells with 50 % more overall contractility (From Tozluoglu [104] with permission)

the magnitudes of these protrusion vectors are determined by its interaction with the local matrix stiffness, alignment and ligand density. Mechanics was only considered implicitly through the coefficient of matrix rigidity. The simple cell-ECM model was able to qualitatively agree with experiments in concluding that matrix alignment does not change cell speed but increases cell migration persistence [93].

Borau et al. [8] developed a probabilistic, cell voxel-based finite element for 3D cell-ECM interactions in a microfluidic environment. A cell is a collection of voxels, where stress, chemical concentration and fluid flow surrounding the cell drives cell migration by adding and removing voxels. Cell contains cortex, cytoplasm and nucleus. The nucleus is an elastic material that only plays a passive role during cell migration. The cortex and cytoplasm contractility depends on the mechanosensing of ECM stiffness, which is modeled implicitly. It provides a methodology for testing and designing experiments in microfluidic systems.

Kim et al. [67] reported a more biomechanically realistic cell-ECM interaction model, which accounted for intracellular mechanics of cellular and nuclear membranes, contractile actin stress fibers, focal adhesion dynamics, structural mechanics of ECM fiber networks, and reaction-diffusion mass transfers of seven biochemical concentrations associated with chemotaxis, proteolysis, haptotaxis, and degradation in ECM. Simulations of cell invasion into fiber networks with various pore sizes, such as 0.5 μm pore size (Fig. 4.6b) and 1.5 μm pore size (Fig. 4.6c), show that filopodia invaded more deeply in the large pore ECM fibers [67]. The results were successfully compared with experiments of 3D HUVEC migration for ECMs with different pore sizes and stiffness.

4.6 Modeling Collective Behavior of Cell Migration

Comparing to single cell modeling, less effort has been directed towards understanding how clusters of cells migrate collectively through microenvironments. A few models have been developed to explore how the cell-cell and cell-

ECM interactions influence collective behavior of migrating cells. Models at this scale are usually complicated, but computationally less expensive than single cell level, because most cell details have been coarse-grained.

Most multi-cellular models contain three parts: single agent with simple properties to represent a cell, cell-cell adhesion, and heterogeneous environment. Guven et al. [51] developed a coarse-grained stochastic model of *Dictyostelium* cells using 2D self-propelled soft disks to study the influence of signal relay. Wynn et al. [117] developed an agent-based cell model that treats point-like cells with biased migration directionality and cell-ECM interactions, and modeled the leader-follower dynamic patterns of collective migration in neural crest cells. In this model the ECM is a passive substrate that can be degraded by cells to form tracks of less resistance. Zaritsky et al. [120] proposed a new analytical framework to explicitly detect and quantify cell clusters that move coordinately in a monolayer, and reported the finding of waves of coordinated migration in wound healing experiments. They explained the wave by Met activation with hepatocyte growth factor or scatter factor. The data and model suggested that collective migration emerges from spatial and temporal accumulation and directionality, which can be a basic cellular mechanism for long-term cell guidance during collective cell migration.

4.7 Summary

Cancer cell invasion into ECM is the first step of metastasis, the main difficulty in treating cancer. Biomechanical experiments and simulations of the ECM, cell, and interactions between the cell and ECM are necessary to better understand the invasion behavior of cancer cells. Recent technologies in microscopy, biomechanical rheology, image processing, and 2D and 3D computational modeling shed light on cancer invasion. We highlighted recent studies on tumor microenvironment, especially ECM, cell, and their interaction.

The importance of the mechanics of ECM and cell-ECM interactions in regulating and contributing to cancer invasion has been increasingly accepted. Moreover, it is necessary to combine these understandings into a unified framework of cancer invasion. The biophysical and biomechanical aspects of the microenvironment should be integrated with the biological and the biochemical aspects, to form a comprehensive description of the tumor microenvironment. The integrated understanding of the cell, ECM, and their interactions is required to better predict cancer invasion and possibly develop new tools to prevent or stop cancer. The strong interplay between cancer cell biology and the mechanical microenvironment suggests new possibilities of regulation and manipulation of cell behavior to alter the outcome of cancer.

References

1. Abhilash AS, Baker BM, Trappmann B, Chen CS, Shenoy VB (2014) Remodeling of fibrous extracellular matrices by contractile cells: predictions from discrete fiber network simulations. Biophys J 107(8):1829–1840. doi:10.1016/j.bpj.2014.08.029
2. Akiri G, Sabo E, Dafni H, Vadasz Z, Kartvelishvily Y, Gan N, Kessler O, Cohen T, Resnick M, Neeman M, Neufeld G (2003) Lysyl oxidase-related protein-1 promotes tumor fibrosis and tumor progression in vivo. Cancer Res 63(7):1657–1666
3. Alexander S, Weigelin B, Winkler F, Friedl P (2013) Preclinical intravital microscopy of the tumour-stroma interface: invasion, metastasis, and therapy response. Curr Opin Cell Biol 25(5):659–671. doi:10.1016/j.ceb.2013.07.001
4. Balaban NQ, Schwarz US, Riveline D, Goichberg P, Tzur G, Sabanay I, Mahalu D, Safran S, Bershadsky A, Addadi L, Geiger B (2001) Force and focal adhesion assembly: a close relationship studied using elastic micropatterned substrates. Nat Cell Biol 3(5):466–472
5. Barnhart E, Lee K-C, Allen GM, Theriot JA, Mogilner A (2015) Balance between cell—substrate adhesion and myosin contraction determines the frequency of motility initiation in fish keratocytes. Proc Natl Acad Sci 112(16):5045–5050
6. Barnhart EL, Lee KC, Keren K, Mogilner A, Theriot JA (2011) An adhesion-dependent switch between mechanisms that determine motile cell shape. PLoS Biol 9(5):e1001059. doi:10.1371/journal.pbio.1001059
7. Bauer AL, Jackson TL, Jiang Y (2009) Topography of extracellular matrix mediates vascular morphogenesis and migration speeds in angiogenesis. PLoS Comput Biol 5(7):e1000445. doi:10.1371/journal.pcbi.1000445

8. Borau C, Polacheck WJ, Kamm RD, García-Aznar JM (2014) Probabilistic Voxel-Fe model for single cell motility in 3D. In Silico Cell Tissue Sci 1(1):2
9. Bordeleau F, Ta A, Ca R-K (2014) Physical biology in cancer. 5. The rocky road of metastasis: the role of cytoskeletal mechanics in cell migratory response to 3D matrix topography. Am Physiol Cell Physiol 306(2):C110–C120
10. Bordeleau F, Bessard J, Marceau N, Sheng Y (2011) Measuring integrated cellular mechanical stress response at focal adhesions by optical tweezers. J Biomed Opt 16(9):095005
11. Bordeleau F, Bessard J, Sheng Y, Marceau N (2008) Keratin contribution to cellular mechanical stress response at focal adhesions as assayed by laser tweezers. Biochem Cell Biol 86(4):352–359
12. Boucher Y, Salehi H, Witwer B, Harsh GR, Jain RK (1997) Interstitial fluid pressure in intracranial tumours in patients and in rodents. Br J Cancer 75(6):829–836
13. Boyd NF, Guo H, Martin LJ, Sun L, Stone J, Fishell E, Jong RA, Hislop G, Chiarelli A, Minkin S, Yaffe MJ (2007) Mammographic density and the risk and detection of breast cancer. N Engl J Med 356(3):227–236. doi:10.1056/NEJMoa062790
14. Byron A, Ja A, Humphries JD, Jacquemet G, Koper EJ, Warwood S, Choi CK, Stroud MJ, Chen CS, Knight D, Humphries MJ (2015) A proteomic approach reveals integrin activation state-dependent control of microtubule cortical targeting. Nat Commun 6(6135):1–14
15. Carey SP, Starchenko A, McGregor AL, Reinhart-King CA (2013) Leading malignant cells initiate collective epithelial cell invasion in a three-dimensional heterotypic tumor spheroid model. Clin Exp Metastasis 30(5):615–630
16. Ea C-A, Micoulet A, Blümmel J, Auernheimer J, Kessler H, Spatz JP (2006) Lateral spacing of integrin ligands influences cell spreading and focal adhesion assembly. Eur J Cell Biol 85(3-4):219–224
17. Chen B-C, Legant WR, Wang K, Shao L, Milkie DE, Davidson MW, Janetopoulos C, Wu XS, Hammer JA, Liu Z, English BP, Mimori-Kiyosue Y, Romero DP, Ritter AT, Lippincott-Schwartz J, Fritz-Laylin L, Mullins RD, Mitchell DM, Bembenek JN, Reymann A C, Bohme R, Grill SW, Wang JT, Seydoux G, Tulu US, Kiehart DP, Betzig E (2014) Lattice light-sheet microscopy: imaging molecules to embryos at high spatiotemporal resolution. Science 346(6208):1257998
18. Chisholm RH, Hughes BD, Landman KA, Zaman MH (2013) Analytic study of three-dimensional single cell migration with and without proteolytic enzymes. Cell Mol Bioeng 6(2):239–249. doi:10.1007/s12195-012-0261-8
19. Choquet D, Felsenfeld DP, Sheetz MP (1997) Extracellular matrix rigidity causes strengthening of integrin-cytoskeleton linkages. Cell 88(1):39–48
20. Chung BM, Rotty JD, Pa C (2013) Networking galore: intermediate filaments and cell migration. Curr Opin Cell Biol 25(5):600–612
21. Condeelis J, Segall JE (2003) Intravital imaging of cell movement in tumours. Nat Rev Cancer 3(12):921–930. doi:10.1038/nrc1231
22. Cook JA, Gius D, Wink DA, Krishna MC, Russo A, Mitchell JB (2004) Oxidative stress, redox, and the tumor microenvironment. Semin Radiat Oncol 14(3):259–266
23. Coyer SR, Singh A, Dumbauld DW, Calderwood DA, Craig SW, Delamarche E, Garcia AJ (2012) Nanopatterning reveals an ECM area threshold for focal adhesion assembly and force transmission that is regulated by integrin activation and cytoskeleton tension. J Cell Sci 125(21):5110–5123. doi:10.1242/jcs.108035
24. Cukierman E, Pankov R, Stevens DR, Yamada KM (2001) Taking cell-matrix adhesions to the third dimension. Science 294(5547):1708–1712. doi:10.1126/science.1064829
25. Dahl KN, Engler AJ, Pajerowski JD, Discher DE (2005) Power-law rheology of isolated nuclei with deformation mapping of nuclear substructures. Biophys J 89(4):2855–2864. doi:10.1529/biophysj.105.062554
26. Dahl KN, Ribeiro AJ, Lammerding J (2008) Nuclear shape, mechanics, and mechanotransduction. Circ Res 102(11):1307–1318. doi:10.1161/CIRCRESAHA.108.173989
27. Davidson PM, Sliz J, Isermann P, Denais C, Lammerding J (2015) Design of a microfluidic device to quantify dynamic intra-nuclear deformation during cell migration through confining environments. Integr Biol 7(12):1534–1546. doi:10.1039/c5ib00200a
28. Dembo M, Torney DC, Saxman K, Hammer D (1988) The reaction-limited kinetics of membrane-to surface adhesion and detachment. Proc R Soc Lond B Biol Sci 234(1274):55–83
29. Diaz-Cano SJ (2012) Tumor heterogeneity: mechanisms and bases for a reliable application of molecular marker design. Int J Mol Sci 13(2):1951–2011. doi:10.3390/ijms13021951
30. Dujon B, Sherman D, Fischer G, Durrens P, Casaregola S, Lafontaine I, De Montigny J, Marck C, Neuveglise C, Talla E, Goffard N, Frangeul L, Aigle M, Anthouard V, Babour A, Barbe V, Barnay S, Blanchin S, Beckerich JM, Beyne E, Bleykasten C, Boisrame A, Boyer J, Cattolico L, Confanioleri F, De Daruvar A, Despons L, Fabre E, Fairhead C, Ferry-Dumazet H, Groppi A, Hantraye F, Hennequin C, Jauniaux N, Joyet P, Kachouri R, Kerrest A, Koszul R, Lemaire M, Lesur I, Ma L, Muller H, Nicaud JM, Nikolski M, Oztas S, Ozier-Kalogeropoulos O, Pellenz S, Potier S, Richard GF, Straub ML, Suleau A, Swennen D, Tekaia F, Wesolowski-Louvel M, Westhof E, Wirth B, Zeniou-Meyer M, Zivanovic I, Bolotin-

Fukuhara M, Thierry A, Bouchier C, Caudron B, Scarpelli C, Gaillardin C, Weissenbach J, Wincker P, Souciet JL (2004) Genome evolution in yeasts. Nature 430(6995):35–44. doi:10.1038/nature02579
31. Even-Ram S, Yamada KM (2005) Cell migration in 3D matrix. Curr Opin Cell Biol 17(5):524–532. doi:10.1016/j.ceb.2005.08.015
32. Feng J, Levine H, Mao X, Sander LM (2015) Alignment and nonlinear elasticity in biopolymer gels. Phys Rev E Stat Nonlin Soft Matter Phys 91(4):042710. doi:10.1103/PhysRevE.91.042710
33. Fogelson B, Mogilner A (2014) Computational estimates of membrane flow and tension gradient in motile cells. PLoS One 9(1):e84524. doi:10.1371/journal.pone.0084524
34. Friedl P (2004) Prespecification and plasticity: shifting mechanisms of cell migration. Curr Opin Cell Biol 16(1):14–23. doi:10.1016/j.ceb.2003.11.001
35. Friedl P, Gilmour D (2009) Collective cell migration in morphogenesis, regeneration and cancer. Nat Rev Mol Cell Biol 10(7):445–457. doi:10.1038/nrm2720
36. Friedl P, Hegerfeldt Y, Tusch M (2004) Collective cell migration in morphogenesis and cancer. Int J Dev Biol 48(5-6):441–449
37. Friedl P, Locker J, Sahai E, Segall JE (2012) Classifying collective cancer cell invasion. Nat Cell Biol 14(8):777–783. doi:10.1038/ncb2548
38. Friedl P, Wolf K (2010) Plasticity of cell migration: a multiscale tuning model. J Cell Biol 188(1):11–19
39. Friedland JC, Lee MH, Boettiger D (2009) Mechanically activated integrin switch controls $\alpha5\beta1$ function. Science (New York) 323(5914):642–644
40. Fu Y, Chin LK, Bourouina T, Liu AQ, VanDongen AM (2012) Nuclear deformation during breast cancer cell transmigration. Lab Chip 12(19):3774–3778. doi:10.1039/c2lc40477j
41. Gabella C, Bertseva E, Bottier C, Piacentini N, Bornert A, Jeney S, Forro L, Sbalzarini IF, Meister JJ, Verkhovsky AB (2014) Contact angle at the leading edge controls cell protrusion rate. Curr Biol 24(10):1126–1132
42. Galbraith CG, Yamada KM, Sheetz MP (2002) The relationship between force and focal complex development. J Cell Biol 159(4):695–705. doi:10.1083/jcb.200204153
43. Gallant ND, García AJ (2005) Cell adhesion strengthening and focal adhesion assembly on micropatterned substrates. Mol Biol Cell 16(9):4329–4340
44. Gatenby RA, Gawlinski ET (2003) The glycolytic phenotype in carcinogenesis and tumor invasion: insights through mathematical models. Cancer Res 63(14):3847–3854
45. Geiger B, Bershadsky A (2002) Exploring the neighborhood: adhesion-coupled cell mechanosensors. Cell 110(2):139–142
46. Geiger B, Bershadsky A, Pankov R, Yamada KM (2001) Transmembrane crosstalk between the extracellular matrix–cytoskeleton crosstalk. Nat Rev Mol Cell Biol 2(11):793–805. doi:10.1038/35099066
47. Geiger B, Spatz JP, Bershadsky AD (2009) Environmental sensing through focal adhesions. Nat Rev Mol Cell Biol 10(1):21–33. doi:10.1038/nrm2593
48. Geiger B, Yamada KM (2011) Molecular architecture and function of matrix adhesions. Cold Spring Harb Perspect Biol 3(5):1–21. doi:10.1101/cshperspect.a005033
49. Grashoff C, Hoffman BD, Brenner MD, Zhou R, Parsons M, Yang MT, McLean MA, Sligar SG, Chen CS, Ha T, Schwartz MA (2010) Measuring mechanical tension across vinculin reveals regulation of focal adhesion dynamics. Nature 466(7303):263–266. doi:10.1038/nature09198
50. Guilak F, Tedrow JR, Burgkart R (2000) Viscoelastic properties of the cell nucleus. Biochem Biophys Res Commun 269(3):781–786. doi:10.1006/bbrc.2000.2360
51. Guven C, Rericha E, Ott E, Losert W (2013) Modeling and measuring signal relay in noisy directed migration of cell groups. PLoS Comput Biol 9(5):e1003041. doi:10.1371/journal.pcbi.1003041
52. Hamadi A, Bouali M, Dontenwill M, Stoeckel H, Takeda K, Ronde P (2005) Regulation of focal adhesion dynamics and disassembly by phosphorylation of FAK at tyrosine 397. J Cell Sci 118(19):4415–4425. doi:10.1242/jcs.02565
53. Harjanto D, Zaman MH (2013) Modeling extracellular matrix reorganization in 3D environments. PLoS One 8(1):e52509. doi:10.1371/journal.pone.0052509
54. Harley BA, Kim HD, Zaman MH, Yannas IV, Lauffenburger DA, Gibson LJ (2008) Microarchitecture of three-dimensional scaffolds influences cell migration behavior via junction interactions. Biophys J 95(8):4013–4024. doi:10.1529/biophysj.107.122598
55. Hockel M, Vaupel P (2001) Tumor hypoxia: definitions and current clinical, biologic, and molecular aspects. J Natl Cancer Inst 93(4):266–276
56. Holmes WR, Edelstein-Keshet L (2012) A comparison of computational models for eukaryotic cell shape and motility. PLoS Comput Biol 8(12):e1002793. doi:10.1371/journal.pcbi.1002793
57. Huveneers S, Danen EH (2009) Adhesion signaling – crosstalk between integrins, Src and Rho. J Cell Sci 122(Pt 8):1059–1069. doi:10.1242/jcs.039446
58. Icard-Arcizet D, Cardoso O, Richert A, Henon S (2008) Cell stiffening in response to external stress is correlated to actin recruitment. Biophys J 94(7):2906–2913. doi:10.1529/biophysj.107.118265
59. Ingber DE (2003) Tensegrity I. Cell structure and hierarchical systems biology. J Cell Sci 116(Pt 7):1157–1173
60. Jiang GY, Giannone G, Critchley DR, Fukumoto E, Sheetz MP (2003) Two-piconewton slip bond between fibronectin and the cytoskeleton depends

on talin. Nature 424(6946):334–337. doi:10.1038/nature01805
61. Kaltenbrunner M, White MS, Glowacki ED, Sekitani T, Someya T, Sariciftci NS, Bauer S (2012) Ultrathin and lightweight organic solar cells with high flexibility. Nat Commun 3:770. doi:10.1038/ncomms1772
62. Kaverina I, Rottner K, Small JV (1998) Targeting, capture, and stabilization of microtubules at early focal adhesions. J Cell Biol 142(1):181–190. doi:10.1083/jcb.142.1.181
63. Kaverina I, Straube A (2011) Regulation of cell migration by dynamic microtubules. Semin Cell Dev Biol 22(9):968–974. doi:10.1016/j.semcdb.2011.09.017
64. Kaverina I, Straube A (2011) Regulation of cell migration by dynamic microtubules. Semin Cell Dev Biol 22(9):968–974
65. Keren K, Pincus Z, Allen GM, Barnhart EL, Marriott G, Mogilner A, Theriot JA (2008) Mechanism of shape determination in motile cells. Nature 453(7194):475–480. doi:10.1038/nature06952
66. Kim DH, Han K, Gupta K, Kwon KW, Suh KY, Levchenko A (2009) Mechanosensitivity of fibroblast cell shape and movement to anisotropic substratum topography gradients. Biomaterials 30(29):5433–5444. doi:10.1016/j.biomaterials.2009.06.042
67. Kim MC, Whisler J, Silberberg YR, Kamm RD, Asada HH (2015) Cell invasion dynamics into a three dimensional extracellular matrix fibre network. PLoS Comput Biol 11(10):e1004535. doi:10.1371/journal.pcbi.1004535
68. Kong F, Li Z, Parks WM, Dumbauld DW, Garcia AJ, Mould AP, Humphries MJ, Zhu C (2013) Cyclic mechanical reinforcement of integrin-ligand interactions. Mol Cell 49(6):1060–1068. doi:10.1016/j.molcel.2013.01.015
69. Krishnan J, Iglesias PA (2003) Analysis of the signal transduction properties of a module of spatial sensing in eukaryotic chemotaxis. Bull Math Biol 65(1):95–128. doi:10.1006/bulm.2002.0323
70. Kuo JC (2013) Mechanotransduction at focal adhesions: integrating cytoskeletal mechanics in migrating cells. J Cell Mol Med 17(6):704–712. doi:10.1111/jcmm.12054
71. Lee B, Zhou X, Riching K, Eliceiri KW, Keely PJ, Guelcher SA, Weaver AM, Jiang Y (2014) A three-dimensional computational model of collagen network mechanics. PLoS One 9(11):e111896. doi:10.1371/journal.pone.0111896
72. Legate KR, Wickstrom SA, Fassler R (2009) Genetic and cell biological analysis of integrin outside-in signaling. Genes Dev 23(4):397–418. doi:10.1101/gad.1758709
73. Levchenko A, Iglesias PA (2002) Models of eukaryotic gradient sensing: application to chemotaxis of amoebae and neutrophils. Biophys J 82(1 Pt 1):50–63. doi:10.1016/S0006-3495(02)75373-3
74. Levental KR, Yu H, Kass L, Lakins JN, Egeblad M, Erler JT, Fong SF, Csiszar K, Giaccia A, Weninger W, Yamauchi M, Gasser DL, Weaver VM (2009) Matrix crosslinking forces tumor progression by enhancing integrin signaling. Cell 139(5):891–906. doi:10.1016/j.cell.2009.10.027
75. Lin CQ, Bissell MJ (1993) Multi-faceted regulation of cell differentiation by extracellular matrix. FASEB J 7(9):737–743
76. Liu YJ, Le Berre M, Lautenschlaeger F, Maiuri P, Callan-Jones A, Heuze M, Takaki T, Voituriez R, Piel M (2015) Confinement and low adhesion induce fast amoeboid migration of slow mesenchymal cells. Cell 160(4):659–672. doi:10.1016/j.cell.2015.01.007
77. Lombardi ML, Lammerding J (2011) Keeping the LINC: the importance of nucleocytoskeletal coupling in intracellular force transmission and cellular function. Biochem Soc Trans 39(6):1729–1734. doi:10.1042/BST20110686
78. Lu P, Takai K, Weaver VM, Werb Z (2011) Extracellular matrix degradation and remodeling in development and disease. Cold Spring Harb Perspect Biol 3(12):a005058. doi:10.1101/cshperspect.a005058
79. Lu P, Weaver VM, Werb Z (2012) The extracellular matrix: a dynamic niche in cancer progression. J Cell Biol 196(4):395–406. doi:10.1083/jcb.201102147
80. McCormack VA, dos Santos SI (2006) Breast density and parenchymal patterns as markers of breast cancer risk: a meta-analysis. Cancer Epidemiol Biomark Prev 15(6):1159–1169. doi:10.1158/1055-9965.EPI-06-0034
81. Michael KE, Dumbauld DW, Burns KL, Hanks SK, Garcia AJ (2009) Focal adhesion kinase modulates cell adhesion strengthening via integrin activation. Mol Biol Cell 20(9):2508–2519. doi:10.1091/mbc.E08-01-0076
82. Mrksich M, Chen CS, Xia Y, Dike LE, Ingber DE, Whitesides GM (1996) Controlling cell attachment on contoured surfaces with self-assembled monolayers of alkanethiolates on gold. Proc Natl Acad Sci U S A 93(20):10775–10778
83. Oakes PW, Beckham Y, Stricker J, Gardel ML (2012) Tension is required but not sufficient for focal adhesion maturation without a stress fiber template. Dev Cell 196:3
84. Oryan A, Moshiri A, Sharifi P (2012) Advances in injured tendon engineering with emphasis on the role of collagen implants. Hard Tissue 1(2):12
85. Panková K, Rösel D, Novotný M, Brábek J (2010) The molecular mechanisms of transition between mesenchymal and amoeboid invasiveness in tumor cells. Cell Mol Life Sci: CMLS 67(1):63–71
86. Parent CA, Devreotes PN (1999) A cell's sense of direction. Science 284(5415):765–770. doi:10.1126/science.284.5415.765
87. Paszek MJ, Zahir N, Johnson KR, Lakins JN, Rozenberg GI, Gefen A, Reinhart-King CA, Mar-

gulies SS, Dembo M, Boettiger D, Hammer DA, Weaver VM (2005) Tensional homeostasis and the malignant phenotype. Cancer Cell 8(3):241–254. doi:10.1016/j.ccr.2005.08.010
88. Petrie RJ, Koo H, Yamada KM (2014) Generation of compartmentalized pressure by a nuclear piston governs cell motility in a 3D matrix. Science 345(6200):1062–1065. doi:10.1126/science.1256965
89. Provenzano PP, Eliceiri KW, Campbell JM, Inman DR, White JG, Keely PJ (2006) Collagen reorganization at the tumor-stromal interface facilitates local invasion. BMC Med 4(1):38. doi:10.1186/1741-7015-4-38
90. Provenzano PP, Inman DR, Eliceiri KW, Knittel JG, Yan L, Rueden CT, White JG, Keely PJ (2008) Collagen density promotes mammary tumor initiation and progression. BMC Med 6:11. doi:10.1186/1741-7015-6-11
91. Provenzano PP, Keely PJ (2011) Mechanical signaling through the cytoskeleton regulates cell proliferation by coordinated focal adhesion and Rho GTPase signaling. J Cell Sci 124(8):1195–1205. doi:10.1242/jcs.067009
92. Puklin-Faucher E, Sheetz MP (2009) The mechanical integrin cycle. J Cell Sci 122(Pt 2):179–186. doi:10.1242/jcs.042127
93. Riching KM, Cox BL, Salick MR, Pehlke C, Riching AS, Ponik SM, Bass BR, Crone WC, Jiang Y, Weaver AM, Eliceiri KW, Keely PJ (2014) 3D collagen alignment limits protrusions to enhance breast cancer cell persistence. Biophys J 107(11):2546–2558. doi:10.1016/j.bpj.2014.10.035
94. Rosso F, Giordano A, Barbarisi M, Barbarisi A (2004) From cell-ECM interactions to tissue engineering. J Cell Physiol 199(2):174–180. doi:10.1002/jcp.10471
95. Ruprecht V, Wieser S, Callan-Jones A, Smutny M, Morita H, Sako K, Barone V, Ritsch-Marte M, Sixt M, Voituriez R, Heisenberg CP (2015) Cortical contractility triggers a stochastic switch to fast amoeboid cell motility. Cell 160(4):673–685. doi:10.1016/j.cell.2015.01.008
96. Satulovsky J, Lui R, Wang YL (2008) Exploring the control circuit of cell migration by mathematical modeling. Biophys J 94(9):3671–3683. doi:10.1529/biophysj.107.117002
97. Schmick M, Bastiaens PI (2014) The interdependence of membrane shape and cellular signal processing. Cell 156(6):1132–1138. doi:10.1016/j.cell.2014.02.007
98. Serrels B, Serrels A, Brunton VG, Holt M, McLean GW, Gray CH, Jones GE, Frame MC (2007) Focal adhesion kinase controls actin assembly via a FERM-mediated interaction with the Arp2/3 complex. Nat Cell Biol 9(9):1046–1056. doi:10.1038/ncb1626
99. Shao D, Levine H, Rappel WJ (2012) Coupling actin flow, adhesion, and morphology in a computational cell motility model. Proc Natl Acad Sci U S A 109(18):6851–6856. doi:10.1073/pnas.1203252109
100. Spaeth EL, Dembinski JL, Sasser AK, Watson K, Klopp A, Hall B, Andreeff M, Marini F (2009) Mesenchymal stem cell transition to tumor-associated fibroblasts contributes to fibrovascular network expansion and tumor progression. PLoS One 4(4):e4992. doi:10.1371/journal.pone.0004992
101. Stamenović D (2005) Microtubules may harden or soften cells, depending of the extent of cell distension. J Biomech 38(8):1728–1732
102. Sung BH, Ketova T, Hoshino D, Zijlstra A, Weaver AM (2015) Directional cell movement through tissues is controlled by exosome secretion. Nat Commun 13(6):2546–2558
103. Takahashi R, Nagayama S, Furu M, Kajita Y, Jin Y, Kato T, Imoto S, Sakai Y, Toguchida J (2014) AFAP1L1, a novel associating partner with vinculin, modulates cellular morphology and motility, and promotes the progression of colorectal cancers. Cancer Med 3(4):759–774. doi:10.1002/cam4.237
104. Tozluoglu M, Tournier AL, Jenkins RP, Hooper S, Bates PA, Sahai E (2013) Matrix geometry determines optimal cancer cell migration strategy and modulates response to interventions. Nat Cell Biol 15(7):751–762. doi:10.1038/ncb2775
105. Tse JM, Cheng G, Tyrrell JA, Wilcox-Adelman SA, Boucher Y, Jain RK, Munn LL (2012) Mechanical compression drives cancer cells toward invasive phenotype. Proc Natl Acad Sci U S A 109(3):911–916. doi:10.1073/pnas.1118910109
106. Vader D, Kabla A, Weitz D, Mahadevan L (2009) Strain-induced alignment in collagen gels. PLoS One 4(6):e5902. doi:10.1371/journal.pone.0005902
107. van Oers RF, Rens EG, LaValley DJ, Reinhart-King CA, Merks RM (2014) Mechanical cell-matrix feedback explains pairwise and collective endothelial cell behavior in vitro. PLoS Comput Biol 10(8):e1003774. doi:10.1371/journal.pcbi.1003774
108. Vicente-Manzanares M, Horwitz A (2011) Cell migration: an overview. Springer, Berlin
109. Wehrle-Haller B (2012) Structure and function of focal adhesions. Curr Opin Cell Biol 24(1):116–124. doi:10.1016/j.ceb.2011.11.001
110. Weidner N, Folkman J, Pozza F, Bevilacqua P, Allred EN, Moore DH, Meli S, Gasparini G (1992) Tumor angiogenesis: a new significant and independent prognostic indicator in early-stage breast carcinoma. J Natl Cancer Inst 84(24):1875–1887
111. Welch MD (2015) Cell migration, freshly squeezed. Cell 160(4):581–582. doi:10.1016/j.cell.2015.01.053
112. Whiteside TL (2008) The tumor microenvironment and its role in promoting tumor growth. Oncogene 27(45):5904–5912. doi:10.1038/onc.2008.271
113. Wolf K, Mazo I, Leung H, Engelke K, von Andrian UH, Deryugina EI, Strongin AY, Brocker EB, Friedl P (2003) Compensation mechanism in tumor cell

migration: mesenchymal-amoeboid transition after blocking of pericellular proteolysis. J Cell Biol 160(2):267–277. doi:10.1083/jcb.200209006
114. Wolf K, Te Lindert M, Krause M, Alexander S, Te Riet J, Willis AL, Hoffman RM, Figdor CG, Weiss SJ, Friedl P (2013) Physical limits of cell migration: control by ECM space and nuclear deformation and tuning by proteolysis and traction force. J Cell Biol 201(7):1069–1084. doi:10.1083/jcb.201210152
115. Wolf K, Wu YI, Liu Y, Geiger J, Tam E, Overall C, Stack MS, Friedl P (2007) Multi-step pericellular proteolysis controls the transition from individual to collective cancer cell invasion. Nat Cell Biol 9(8):893–904. doi:10.1038/ncb1616
116. Wolfenson H, Lavelin I, Geiger B (2013) Dynamic regulation of the structure and functions of integrin adhesions. Dev Cell 24(5):447–458. doi:10.1016/j.devcel.2013.02.012
117. Wynn ML, Rupp P, Trainor PA, Schnell S, Kulesa PM (2013) Follow-the-leader cell migration requires biased cell-cell contact and local microenvironmental signals. Phys Biol 10(3):035003. doi:10.1088/1478-3975/10/3/035003
118. Xiong Y, Huang CH, Iglesias PA, Devreotes PN (2010) Cells navigate with a local-excitation, global-inhibition-biased excitable network. Proc Natl Acad Sci U S A 107(40):17079–17086. doi:10.1073/pnas.1011271107
119. Yu CH, Law JBK, Suryana M, Low HY, Sheetz MP (2011) Early integrin binding to Arg-Gly-Asp peptide activates actin polymerization and contractile movement that stimulates outward translocation. Proc Natl Acad Sci U S A 108(51):20585–20590. doi:10.1073/pnas.1109485108
120. Zaritsky A, Kaplan D, Hecht I, Natan S, Wolf L, Gov NS, Ben-Jacob E, Tsarfaty I (2014) Propagating waves of directionality and coordination orchestrate collective cell migration. PLoS Comput Biol 10(7):e1003747. doi:10.1371/journal.pcbi.1003747

Circulating Tumor Cells: When a Solid Tumor Meets a Fluid Microenvironment

5

Katarzyna A. Rejniak

Abstract

Solid tumor dissemination from the primary site to the sites of metastasis involves tumor cell transport through the blood or lymph circulation systems. Once the tumor cells enter the bloodstream, they encounter a new hostile microenvironment. The cells must withstand hemodynamic forces and overcome the effects of fluid shear. The cells are exposed to immunological signaling insults from leukocytes, to collisions with erythrocytes, and to interactions with platelets or macrophages. Finally, the cells need to attach to the blood vessel walls and extravasate to the surrounding stroma to form tumor metastases. Although only a small fraction of invasive cells is able to complete the metastatic process, most cancer-related deaths are the result of tumor metastasis. Thus, investigating the intracellular properties of circulating tumor cells and the extracellular conditions that allow the tumor cells to survive and thrive in this microenvironment is of vital interest. In this chapter, we discuss the intravascular microenvironment that the circulating tumor cells must endure. We summarize the current experimental and computational literature on tumor cells in the circulation system. We also illustrate various aspects of the intravascular transport of circulating tumor cells using a mathematical model based on immersed boundary principles.

Keywords

Circulating tumor cells • Metastatic cascade • Tumor microemboli • Cell deformation • Computational modeling • Immersed boundary method

K.A. Rejniak (✉)
Integrated Mathematical Oncology Department, Center of Excellence in Cancer Imaging and Technology, H. Lee Moffitt Cancer Center & Research Institute, Tampa, FL, USA

Department of Oncologic Sciences, College of Medicine, University of South Florida, Tampa, FL, USA
e-mail: kasia@rejniak.net

5.1 Introduction

Metastasis to distant organs is the major cause of mortality for patients with many types of cancer. Metastasis accounts for nearly 90 % of

K.A. Rejniak (eds.), *Systems Biology of Tumor Microenvironment*, Advances in Experimental Medicine and Biology 936, DOI 10.1007/978-3-319-42023-3_5

cancer-related deaths, although no more than 0.01 % of circulating tumor cells (CTCs) is able to successfully form secondary lesions [67]. In order to metastasize, the cancer cells must detach from the primary tumor mass, migrate through the stromal tissue surrounding the primary tumor, enter the blood or lymph circulation system, travel with the blood flow, and, finally, extravasate into the new site, invade the local matrix, and colonize the target organ [39, 43, 71]. The process of CTC intravenous transport and the physical properties that allow CTCs to survive under the physiological blood flow is still poorly understood. One of the hypotheses is called the "adhesion cascade" (Fig. 5.1) and was formulated and observed in the context of circulating leukocytes (reviewed in [44]). During this process, a single tumor cell is expected to switch among various locomotion strategies, from floating with the bloodstream to rolling on the endothelial wall, to tumor cell arrest and crawling, and finally, to tumor cell transmigration through the endothelial layer [44, 71].

Tumor cell dissemination from primary to metastatic sites has not only a biological meaning but also important clinical implications. With the detection of metastasis in a patient, the clinical strategy for therapy changes from localized to systemic. However, how chemotherapeutic drugs directed at primary tumors affect the distant metastases is not known. Neither is it known how intravenously administered drugs affect CTCs. Therefore, investigating how to prevent CTCs from metastatic spread is of vital interest to cancer biology. A deeper understanding of what factors allow CTCs to survive in the circulation system could be used to design novel therapeutic treatments. Moreover, understanding and characterizing CTCs is the first step toward utilizing them as biopsy material and directly as a biomarker of disease progression or as a measure of patient response to an anti-cancer treatment.

Although CTCs can be found in the blood of patients with various types of cancers, these cells are quite rare—one in a million in comparison with leukocytes [31] or one in a billion in comparison with blood cells [74]. In recent years, several approaches were undertaken to design highly specialized technical devices capable of capturing and isolating CTCs from patients' peripheral blood [2, 9, 10, 18, 28]. These methods allowed the cancer biology community to gain enormous knowledge about the biological and physical properties of CTCs and to correlate the presence of CTCs with patients' clinical outcomes [17, 45, 61, 69].

One such method, a high-definition CTC assay (HD-CTC), was developed by Kuhn and colleagues to capture CTCs from patients' blood without using surface protein-based enrichment. This method allows not only for identification of significant numbers of CTCs in patients with different types of cancers but also for imaging of the captured cells. Significant heterogeneity has been observed between tumor cells of various origins, as well as within cells of the same type of cancer [30, 32, 33]. The patients' CTCs varied in size, shape, expression of specific proteins, and the cytoplasm to nucleus ratio. For example, the prostate cancer cell diameter ranged from 8 to 16 μm, and the breast cancer cells reached 9–19 μm in diameter [30, 51]. Several blood

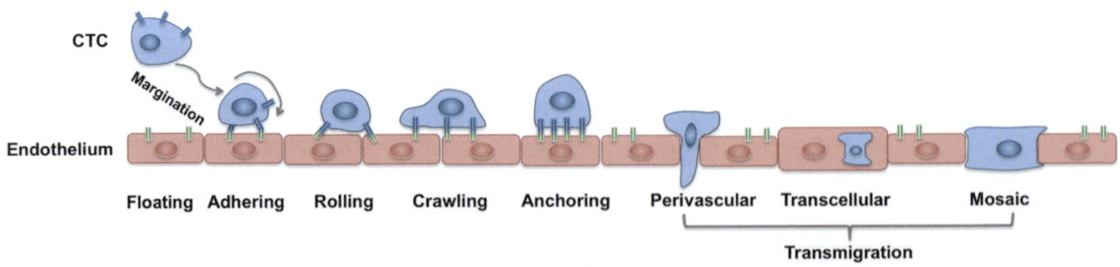

Fig. 5.1 Schematics of the CTC adhesive cascade. A cell floating with the blood flow need to reach an endothelial wall via margination process; once near the endothelial wall, it needs to adhere to the endothelium, undergo the transitions from rolling to crawling migration before anchoring to the endothelium and transmigrate the endothelial wall using one of the following ways: perivascular migration, transcellular migration or a mosaic process

samples contained also CTCs' clusters (emboli) of various sizes and shapes [7, 32, 34]. These cell aggregates were found in blood from patients with breast, lung, pancreatic, and prostate cancers which indicates that the emboli may not be a sporadic case and that cell aggregates may present certain advantages for CTC survival in blood flow.

Recent advances in real-time fluorescent microscopy imaging allow for a visual account of CTCs in the circulation system in animal models (Fig. 5.2). Experiments conducted by the Hoffman group showed both individual cells and multicellular emboli in the blood and lymph circulation systems in mouse skin [15, 16, 72, 73]. In particular, the researchers showed that injected human fibrosarcoma cells acquire different shapes depending on the size of the vessel. The CTCs in microvessels were relatively circular and nondeformed, with the minor and major axes equal to 19.3 and 23.3 μm on average, respectively (Fig. 5.2a). However, in capillaries (typically 3–8 μm in diameter), the cells were elongated reaching four times their normal size (7.8 × 92.6 μm on average), and the length of the nuclei increased 1.6 times reaching 7.8 × 27.6 μm on average. These cells occupied the full diameter of the vessel (Fig. 5.2b). The same group also observed occasional small masses of tumor cells and larger multicellular emboli traveling through the large venules [15, 16].

Once in the circulation system, CTCs are exposed to various microenvironmental factors that are novel for cells arising from a solid tumor mass. In order to survive, the cells must withstand hemodynamic forces and overcome the effects of fluid shear [41, 51, 71]. Blood flow velocities in the circulation system can range from 0.03 to 40 cm/s depending on the vessel size [36], and the hemodynamic shear forces in the bloodstream can reach 0.5–4.0 dyn/s^2 in the venous circulation and 4.0–30.0 dyn/s^2 in the arterial circulation [41, 68]. The pattern of the blood flow and its shear stress, as well as the movement of other cells in the circulation system can affect CTC trajectory. Moreover, CTCs may also be exposed to collisions with red blood cells or adhesion to leukocytes, platelets, and microphages [42, 51, 66]. How individual CTCs and CTC emboli can withstand these intercellular interactions and persist in the fluid flow is not yet fully understood, and many questions are still unanswered. Mathematical modeling based on physical principles and available experimental and clinical data can provide an invaluable tool for reconstructing the intravascular microenvironment and for testing what properties of CTCs allow them to survive in the circulation system.

In this chapter, we present a mathematical model of circulating tumor cells in the microvessel that utilizes fluid-structure interaction principles. We also discuss CTCs' biophysical properties, including cell deformability and dynam-

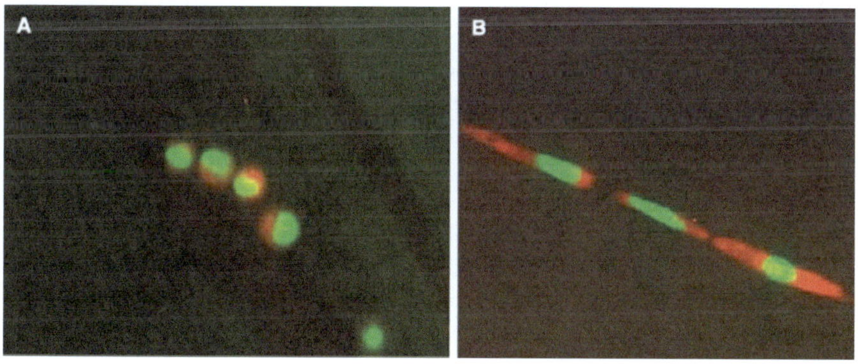

Fig. 5.2 Circulating tumor cells in the vessel in the mouse skin. An image of tumor HT-1080 sarcoma cells: (**a**) nondeformed cells within a microvessel; (**b**) deformed cells occupying the full diameter of the vessel. Cells are labeled with *green* fluorescent protein in the nucleus, and *red* fluorescent protein in the cytoplasm. Image acquired using the Olympus OV100 system at ×100 magnification. (Reproduced from Yamauchi et al. Cancer Research 2005 Fig. 3, with permission [73])

ical changes in cell structure (both locally and globally) that are necessary for cell survival in the blood flow, and switches in cell locomotion strategies. This will allow us to identify combinations of parameters that can be manipulated experimentally in order to disrupt some steps of the CTC adhesive cascade.

5.2 The IBCell Mathematical Model of Circulating Tumor Cells

To model interactions between the bodies of circulating tumor cells and the surrounding blood plasma flow, fluid-structure-interactions methods are a natural modeling choice. One such modeling framework is the Immersed Boundary Method developed by Charles Peskin to model blood flow in the heart [47–50]. We adapted this computational framework to design a model of a deformable eukaryotic cell, *IBCell* (an Immersed Boundary model of a Cell [55, 56]). This model includes various cellular processes that the cells can experience, such as cell growth, division, death, adhesion, and migration. Subsequently, we applied the *IBCell* framework to model circulating tumor cells in the blood flow [57].

We present here the details of this two-dimensional (2D) model of deformable circulating tumor cells traveling through a microvessel. Note, that a 3D version of this model can be based on the same principles, as described in [57]. Here, we consider a microvessel large enough that the cells are able to preserve a circular shape as shown in Fig. 5.2a. During the simulations, cell deformation is a result of the blood flow, not the occlusion within the vessel. The cells are exposed to both hemodynamic forces exerted by the blood plasma flow and adhesive-repulsive forces between either other tumor cells or the tumor cell attaching to or migrating along the endothelial cells of the microvessel. All components of this model are described below and are presented in Fig. 5.3.

- Blood plasma flow—this model takes into account the fluid phase of the blood (the plasma) only. No other cells, such as red blood cells, leukocytes or platelets, are included. The blood plasma is modeled as a viscous incompressible Newtonian fluid governed by Navier-Stokes equations. The fluid flow inside the vessel is laminar of a parabolic profile with zero velocity at the microvessel walls and a maximum velocity of 0.6 mm/s in the center of the microvessel. However, in the presence of obstacles, such as deformable tumor cells, the plasma flow profile may be distorted. In these simulations, we use a small 30-μm-wide and 75-μm-long vessel.
- Endothelium structure—the endothelium is modeled as a mesh of short and relatively stiff linear springs that form a uniform rigid wall. For simplicity, no individual endothelial cells are included in the model. While we are aware that the blood flow can alter the biophysical state of the endothelium (for example, some membrane receptors can be expressed differently under the flow), for simplicity, we neglect any changes that the plasma flow may have on the endothelium.
- Circulating tumor cell structure—in this model, we consider a circular cell with a diameter of 10 μm, a 4-μm-wide nucleus, and a 2-μm-wide cortex band. No other cell cytoskeleton elements and intracellular organelles are included in the model, but the whole cell structure is interpenetrated by viscous incompressible cytoplasm. Cell shape and stiffness are modulated by the cell actin cortex and the cell nuclear envelope. Both are modeled as dense networks of linear Hookean springs. The model allows modifications in spring stiffness globally (i.e., for all springs forming the structure) and locally (i.e., for an individual spring in the structure), and separately for the cell nucleus and the cortex. This enables testing the relative role of the nucleus and the cortex in preserving the overall cell shape under the blood flow.
- Circulating tumor cells-endothelium interactions—all interactions between circulating tumor cells and the endothelium are modeled via the pseudo-receptors located on

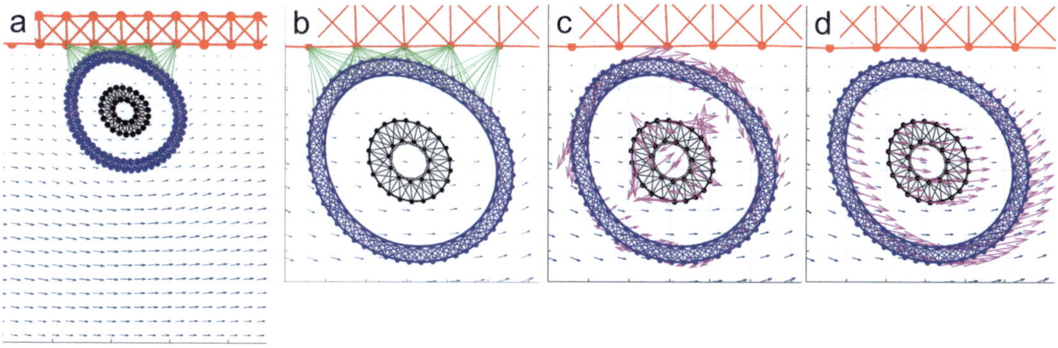

Fig. 5.3 Schematics of the CTC and endothelium structure. A schematic representation of the 2D model of the tumor cell in circulation. (**a**) The blood vessel (*red top* wall) is interpenetrated by a blood flow (*grey* arrows representing a velocity field); the CTC nucleus (*black*) is surrounded by the cell cortex (*blue*) both composed from cross-linked Hookean springs. (**b**) The cell near the endothelial wall develops adhesive connections (*green* links). (**c**) As a result of cell deformation, the springs are stretched and exert restoring forces (*magenta* arrows). (**d**) The cell boundary points are moved on a local fluid velocity (*magenta* arrows). This mathematical model utilizes the computational frameworks of the Immersed Boundary method

$$(1)\quad \rho\left(\frac{\partial \mathbf{u}(\mathbf{x},t)}{\partial t} + (\mathbf{u}(\mathbf{x},t)\cdot\nabla)\mathbf{u}(\mathbf{x},t)\right) = -\nabla p(\mathbf{x},t) + \mu\Delta\mathbf{u}(\mathbf{x},t) + \mathbf{f}^*(\mathbf{x},t),$$

$$(2)\quad \rho\nabla\cdot\mathbf{u}(\mathbf{x},t) = 0,$$

$$(3)\quad \mathbf{f}(\mathbf{x},t) = \textstyle\int_{\Gamma_t\cup\Gamma_e} \mathbf{F}(l,t)\,\delta(\mathbf{x}-\mathbf{X}(l,t))\,dl,$$

$$(4)\quad \frac{\partial \mathbf{X}(l,t)}{\partial t} = \mathbf{u}(\mathbf{x},t) = \textstyle\int_{\Omega} \mathbf{u}(\mathbf{x},t)\,\delta(\mathbf{x}-\mathbf{X}(l,t))\,d\mathbf{x},$$

$$(5)\quad \mathbf{F}(l,t) = \mathcal{S}\,\frac{\|\mathbf{X}(l,t)-\mathbf{X}^*(l,t)\|-\mathcal{L}}{\|\mathbf{X}(l,t)-\mathbf{X}^*(l,t)\|}\,(\mathbf{X}(l,t)-\mathbf{X}^*(l,t)), \quad \text{if}\ \ \|\mathbf{X}(l,t)-\mathbf{X}^*(l,t)\| \le \mathcal{L}_{max}.$$

Fig. 5.4 *IBCell* model equations. Equations Eqs. 1–5 define the mathematical frameworks of the Immersed Boundary method

the tumor and the endothelial cell membranes. The adhesive links emerge when these receptors are in close proximity. These adhesive links, are modeled as short linear Hookean springs, and can be dynamically assembled and disassembled based on the distance between the receptors, as well as the adhesive spring stiffness. For simplicity, we assume that the receptor-ligand binding is always effective when the distance between the receptors is small enough.

The mathematical framework of the immersed boundary method is given in Fig. 5.4. In this method, Equation 1 is the Navier–Stokes equation of a viscous incompressible fluid defined on the Cartesian grid $\boldsymbol{x} = (\boldsymbol{x_1},\boldsymbol{x_2})$, where p is the fluid pressure, μ is the fluid viscosity, ρ is the fluid density, $\boldsymbol{u}$ is the fluid velocity, and $\boldsymbol{f}$ is the external force density. Equation 2 is the law of mass balance. Interactions between the fluid and the material points $\boldsymbol{X(l,t)}$ on the tumor cell membrane and the endothelial wall boundaries (l is an index along either the boundaries of the tumor cells Γ_t or the boundaries of the endothelial walls Γ_e) are defined in Equations 3–4. Here, the force density $\boldsymbol{F(l,t)}$ acting on the cell and wall boundaries is applied to the fluid using the 2D Dirac delta function δ, while all material boundary points $\boldsymbol{X(l,t)}$ are carried along with the fluid. The boundary forces $\boldsymbol{F}(l,t)$ arise from the elastic properties of the tumor cell membranes, from the rigid properties of the endothelial walls and from tumor cell-endothelial cell adhesion and are all represented by the short linear Hookean springs in Equation 5, where S is the spring stiffness, L is the spring resting length and $\boldsymbol{X}^*(l,t)$ is the adjacent, opposite, or neighboring point for the elastic, rigid, or adhesive forces, respectively. These equations are solved using the finite difference methods with a discrete approximation of the Dirac delta function and are described in detail in [55, 58].

The equations of the immersed boundary method are solved in the following steps:

1. Define the structure of the endothelium and the circulating tumor cells.
2. Define the forces that determine the cells and the endothelium structure; define the adhesive forces between the cells and between the cells and the endothelium.
3. Spread the forces to the underlying fluid grid.
4. Solve the Navier-Stokes equations on the Cartesian fluid grid for the fluid velocities.
5. Interpolate the fluid velocities to the boundary points of the cells and the endothelium.
6. Move the immersed boundary points based on the computed velocities.
7. Repeat steps 2–6.

In the following sections, we discuss several stages of the intravascular transport and, whenever possible, illustrate them with simulations from the *IBCell* model. In particular, we discuss the relative stiffness of the CTC cortex versus the stiffness of the CTC nuclear envelope that controls cell deformation under blood flow (Sect. 5.3), collisions with other cells (Sect. 5.4), how CTC cytoskeletal properties need to be modified in order to attach and roll on the endothelium (Sect. 5.5), how CTC cytoskeleton reorganization and cell membrane receptor redistribution enable CTC anchorage and migration on the endothelium (Sect. 5.6), and CTC transmigration through the endothelium (Sect. 5.7). Finally, we discuss how the formation of cell emboli changes the dynamics of the cells in the circulation system (Sect. 5.8).

5.3 Circulating Tumor Cell Survival in the Blood Flow

Laboratory experiments with CTCs in the mouse circulation system showed that the cells are able to undergo high levels of deformability (Fig. 5.2 and [16, 73]). The overall cell shape and the shape of the cell nucleus were able to adjust to the size of the small capillaries. Similar deformability was also observed in human tumor cells transplanted into the vasculature of a transparent zebrafish [59, 60] and in modeled nucleus-free red blood cells [54] and neutrophils [22].

We previously used the *IBCell* model to investigate how changes in the stiffness of the cell cortex and the cell nucleus influence the survival of the whole cell when it is exposed to the same pattern of blood flow [57]. The results are summarized in Fig. 5.5 in the form of a parameter space graph. We varied the stiffness constant of the individual fibers forming the cell cortex and, independently, the stiffness constant of the fibers forming the cell nuclear envelope. Following the values reported in the literature, the range was 1–10^6 g/cm $\cdot$ s^2 for each fiber [29, 35, 46, 58]. The final cell shapes were recorded after the cell reached two thirds of the length of a straight vessel (*i.e.*, at 50 μm).

All simulations started with a cell with a perfectly circular cortex and nucleus. The cell was initially located in the middle of the vessel. The cases in which both the cortex and the nucleus were very stiff resulted in no significant deformation during the whole simulation, and the cells retained their initial circular shapes (Fig. 5.5(iv)). However, when both the cell cortex and the cell nucleus were very soft, the cells did not survive the flow and underwent clasmatosis—cytoplasmic fragmentation. For the relatively soft cortex (fiber stiffness of 10–100 g/cm $\cdot$ s^2) the cells acquired a concave shape with the central inward deformation (a parachute-like shape) at the location of the highest blood

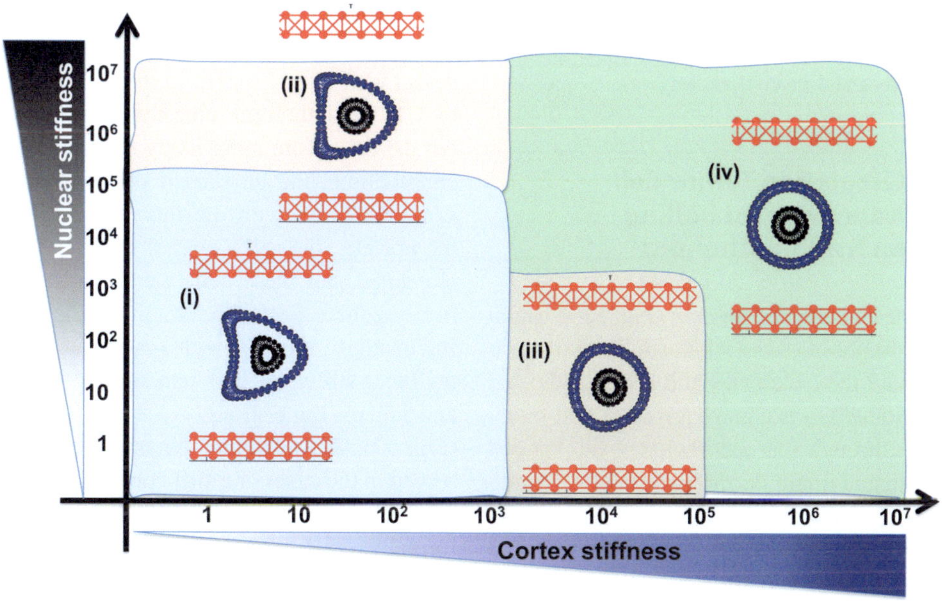

Fig. 5.5 Deformation of a CTC under a steady blood flow. A parameter space of final cellular morphologies when the stiffness of the cell cortex (*blue*) and the cell nuclear envelope (*black*) is varied by 7 orders of magnitude. Insets (*i–iv*)—the representative deformable (*i–ii*) and non-deformable (*iii–iv*) cell morphologies are shown within each parameter range

flow (Fig. 5.5(i)). For the same cortex stiffness, these invaginations became smaller when the nuclear envelope became stiffer (Fig. 5.5(ii)). Interestingly, a stiff nucleus with a softer (but not too soft) cortex prevented significant cell deformations (Fig. 5.5(iii)). Thus, actin cortex stiffness is crucial for cell survival during passive transport of the cell by the blood flow, and the combination of a soft cortex and a soft nuclear envelope may lead to cell damage by the bloodstream.

5.4 Circulating Tumor Cell Collision Dynamics

An interesting question is, how do CTCs approach the endothelial vessel walls? Is the CTCs' movement a result of collisions with other cells? Or do the CTCs need to be pushed by the flow created by other cells? The phenomenon of the cells' translocation to the periphery of the vessel wall is known as margination. This process has been computationally modeled for large colonies of red blood cells [75], for interactions between red blood cells and platelets [8, 12], and for interactions between red and white blood cells [11, 63]. These results indicated that red blood cell-induced drift effectively leads to margination of leukocytes and platelets in both small and large vessel channels. The duration of the margination process depends on the vessel size, but once margination is achieved, the process is sustained for a long time. Moreover, the size of the cell-free layer depends on the cell deformability and the aggregation strength. Recently published mathematical models [27, 64] were drawn on previous studies of leukocyte-red blood cell interactions and tested whether margination processes can be also observed for CTCs. In particular, a mathematical model discussed in [27] showed that when three different CTC membrane elasticities were considered the cells with rigid bodies were directed toward the vessel walls the quickest. The softest CTCs showed only small fluctuations along the initial trajectory. How such properties of the CTCs in the circulation system can be used for therapeutic purposes should be investigated in a quantitative and predictive way.

5.5 Circulating Tumor Cell Adhesion and Rolling on the Endothelium

In the adhesive cascade process (Fig. 5.1), which has been proposed as a mechanism for extravasation for both the leukocytes and CTCs, adhesion to the endothelium is a prerequisite of cell transmigration through the endothelial wall. Various receptor-ligand bond formations (via a family of selectin adhesion molecules) and hydrodynamic interactions have previously been modeled for leukocytes-endothelium adhesion [19, 23, 26, 65]. Recent experiments in microfluidic devices have also reported either selectin-mediated [40] or platelet-enhanced [37] adhesive capture of tumor cells. Moreover, cell rolling along the endothelium that was observed in *in vitro* experiments and modeled *in silico* in the case of leukocytes [3, 19, 25, 38] has also been postulated to work in the case of circulating tumor cells [37, 67].

The model we used in [57] does not include detailed receptor-ligand kinetics. The adhesion process between CTCs and the endothelium was simplified to have high affinity and to assemble easily if both the receptor and the ligand were in close proximity to one another. Instead, we focused again on the cortex and nucleus structures and their deformation when the CTCs adhere to the endothelium under the blood flow. Using the same range of stiffness values as in Sect. 5.3, *IBCell* produced the parameter space shown in Fig. 5.6. We, again, started with a perfectly circular cell (the cell cortex and the cell nucleus), but this time, the cell was located near the bottom wall of the vessel. This allowed the cell to adhere to the endothelium. Moreover, this location also resulted in the cell being exposed to nonsymmetrical blood shear stress with increasing blood velocity far from the endothelium.

The simulated cases revealed that cells with a soft cortex underwent quite significant deformation due to the flow pushing on one side of the cells while the opposite side was attached to the vessel wall (Fig. 5.6a(i)). In these cases, the area of CTC-endothelium adhesion was quite large, but the cells were more likely to detach from the endothelium and be carried out with the blood flow. However, with the increasing stiffness of the nucleus, the cell's overall shape became less elongated, but these cells were also not able to firmly adhere to the endothelium (Fig. 5.6a(ii)). Interestingly, the cells with a moderately soft cortex but a stiffer nucleus remained less deformed and adhered quite strongly to the endothelial cells (Fig. 5.6a(iii)). Similar dynamics was also observed in the cases of a stiff cortex (Fig. 5.6a(iv)). The simulations showed that although the deformable cells with a softer cortex had a larger area of adhesion to the endothelium, they were also more responsive to blood shear stress, as their area of contact with higher-velocity flow could increase with time due to their deformability. The stiffer cells had a smaller area of adhesion to the endothelium, but as the cells' shape was not deforming, they were able to maintain this area of adhesion over time, and were less likely to detach.

We stained a fixed part of the cell cortex in each cell to show its translocation along the cell perimeter (Figs. 5.6b, c). When both the cortex and the nucleus were relatively stiff, the cells showed a rolling motion. These simulations indicated that the transition in the CTC transport phase from floating with the blood flow to attaching and rolling on the endothelial wall requires stiffening of the whole CTC cortex.

5.6 Circulating Tumor Cell Crawling Dynamics

One of the steps of the leukocyte adhesive cascade is cell crawling—a slow migration along the endothelium before the cells can completely stop and start the extravasation process [44, 70]. Such intraluminal cell crawling has been observed in time-lapse video-microscopy within the postcapillary venules [53]. It has also been shown that leukocyte crawling can trigger specific signaling

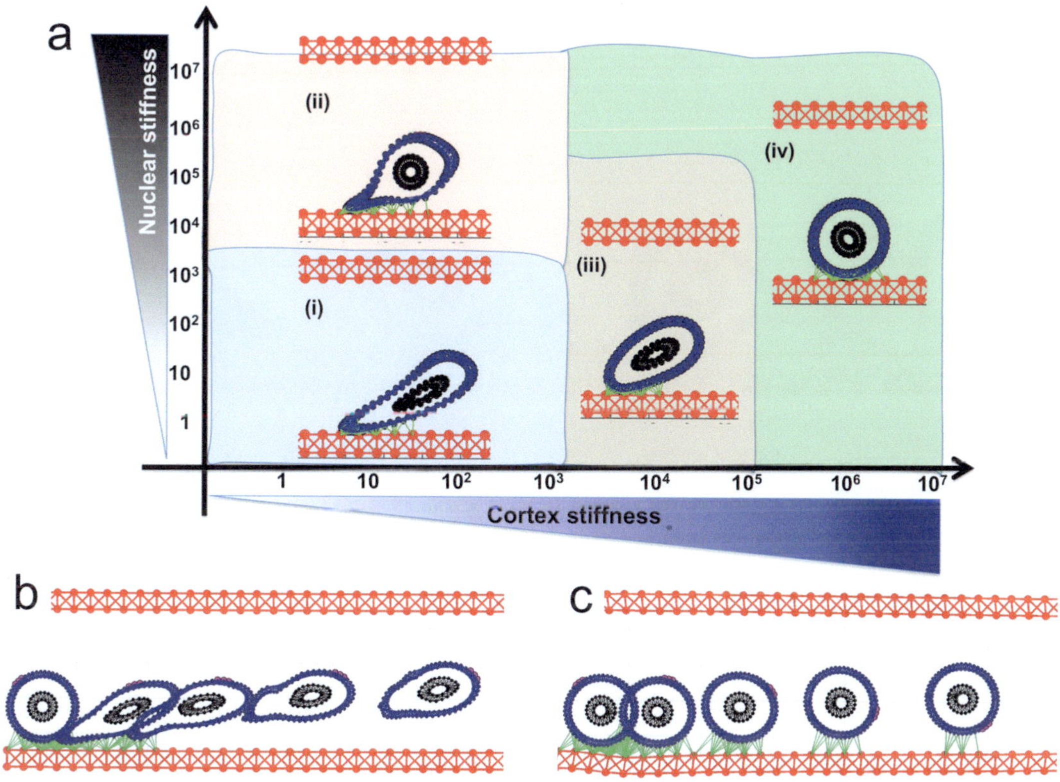

Fig. 5.6 CTC attachment to endothelium under a steady blood flow. (**a**) A parameter space of final cellular morphologies when the stiffness of the cell cortex (*blue*) and the cell nuclear envelope (*black*) is varied by 7 orders of magnitude. Insets (*i–iv*)—the representative cell morphologies for each category; the adhesive links between CTC and the endothelium shown in *green*. The time course image (**b** and **c**) shows the same cell at five different time point overimposed on the same image: (**b**) shows cell detachment from the endothelium, and (**c**) cell rolling on the endothelium. A fixed small part of the cell cortex is stained in magenta to illustrate the rolling effect

between white blood cells and endothelial cells that can induce the transient weakening of endothelial cell junctions that may play an important role in enabling cell transmigration [44].

We used the *IBCell* model to reproduce the cell crawling movement under the blood flow and to test the mechanical properties of the migrating cell. Based on the results described in Sect. 5.5, we chosen a cell configuration for which the cell was able to attach to the endothelium and withstand the hemodynamic forces of the flowing blood (Fig. 5.6c). In the case we considered here, the cell had a relatively stiff nucleus but a softer cortex. Therefore, we tested whether further softening of the cell cortex around the cell's focal adhesions would result in better cell migration. The softer and stiffer parts of the cell cortex were stained differently, as shown in Fig. 5.7. Following the example in Fig. 5.6, we also stained a small part of the cortex opposite the cell-endothelium adhesions, to indicate their translocation along the cell perimeter. These simulations revealed that in order for the circulating tumor cell to progress from rolling to anchoring and to crawling, the stiffness of the cell's cytoskeleton has to be dynamically altered between the part in adhesion to the endothelium and the opposite side (Fig. 5.7). Cell rolling required a quite stiff actin cortex that could be pushed by the blood flow without extensive deformation. Upon transition to anchoring, the CTC cytoskeleton must become more flexible along the contact surface. Cell crawling required that the cortex fibers were softened around the cell focal adhesions with

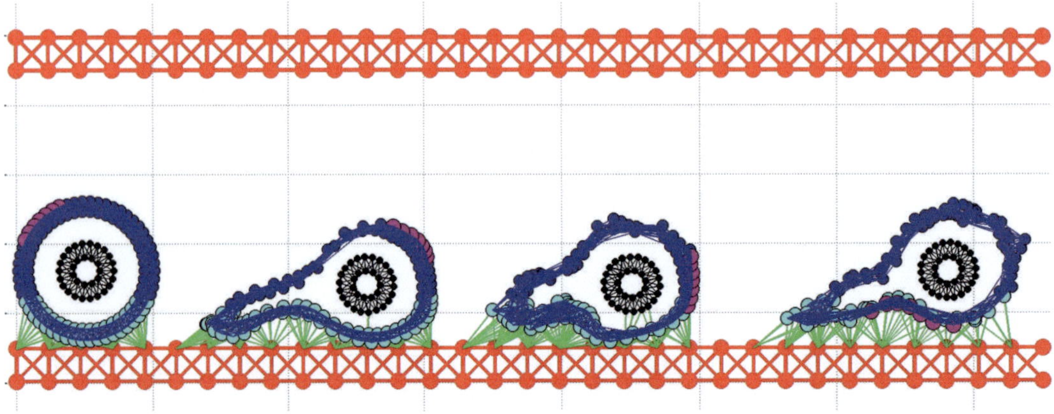

Fig. 5.7 CTC migrating on the endothelium. A time course shows the same cell at different time point overimposed on the same image. To illustrate cell crawling a fixed small part of the cell cortex is stained in magenta. Cell cortex is stained differently depending on its local stiffness: soft (*cyan*) close to cell focal adhesion with the endothelium, and stiff (*blue*) far from the adhesive contacts (*green*)

the endothelium, and the cortex fibers became stiffer immediately after the focal adhesions were broken. This local softening of the cell cytoskeleton along the contact area with the endothelium may be potentially modulated by signals from the endothelial cells in contact.

5.7 Circulating Tumor Cell Anchorage to the Endothelium and Transmigration

The final step in the adhesive cascade is cell transmigration through the endothelium and extravasation to the surrounding stromal tissue. This is an area of very active research with in *in vitro* microfluidic devices and *in vivo* animal models [5, 6, 20, 24]. It has been shown experimentally using the laminar flow chamber that cancer cells can utilize three distinct mechanisms of tumor cell-endothelial cell interactions in order to initialize the transmigration process [67]. Tumor cells can use paracellular transmigration that takes place at the junction of three adjacent endothelial cells. The second mechanism allowed the tumor cell to form a mosaic with endothelial cells in which the tumor cell became inserted within the endothelial layer, sometimes for as long as 24 h. The third observed transmigration mechanism was penetration through the endothelial cell. This transcellular migration process did not occur on the endothelial cell junctions, but the tumor cell was engulfed in a large vacuole and transported through the endothelial cell [67]. The complex process of CTC transmigration through the endothelium has not yet been explored with mathematical modeling.

5.8 The Dynamics of Circulating Tumor Cell Microemboli

Although tumor emboli have been observed clinically in the blood of patients with different tumors [1, 4, 7], how such emboli can arise is not yet well understood, especially given how rare CTCs in the circulation system are. The observed CTC clusters numbered from two to several cells in clinical samples [4, 7] and even up to 50 cells in animal experiments [15, 16]. It has also been observed that CTC emboli can contain other cells, such as platelets, that may guard the tumor cells from immune elimination and promote their arrest at the endothelium [14, 42].

With the *IBCell* model, we considered microemboli that consisted only of CTCs. The simulations showed that such CTC clusters were able to survive the fluid flow shear stress longer than individual CTCs (Fig. 5.8). Their collective abil-

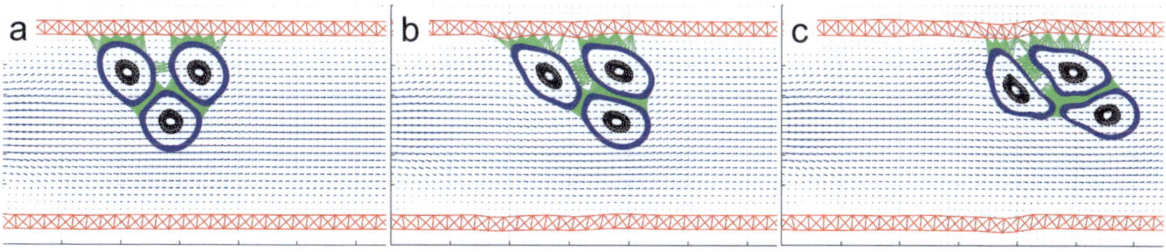

Fig. 5.8 CTC microemboli in the blood circulation system. A time course (**a–c**) shows the same CTC cluster at different time points with cellular adhesions and adhesion to the endothelium indicated as *green* links. The cluster structure allows individual cells to withstand the blood shear stress and survive in the blood flow (showed as *grey* vector field)

ity to adhere to the endothelium was also superior to the adhesion capabilities of individual CTCs. However, further, more systematic investigations are needed to determine the trade-offs between the size of the CTC cluster, the probability of surviving intravenous transport, and the chances of creating vessel occlusion because of the embolus size. The latter was investigated in [52] by reconstructing the size and shape of emboli observed in patients' blood. This work showed that small CTC clusters tumbled during the flow and were able to reach the vessel walls, even in the absence of other cells. However, large CTC clusters were capable of vascular occlusion, slowing the blood's velocity and increasing the levels of local generation of coagulation factors.

5.9 Summary

The fluid environment of the blood to which circulating tumor cells are exposed presents a formidable challenge to cells of epithelial origin. Although experimental studies showed that dysplastic and metastatic cells exhibit a different degree of mechanical stiffness and cell deformability [13, 21, 62], the precise mechanisms and the extent of such deformations in circulating tumor cells in the blood flow are not known. Most of the currently used methods for capturing CTCs require cell attachment or anchorage to the substrate. This may result in changes in the cytoskeletal properties of the cell in a way that subsequent measurements will not reflect the cell properties in the circulation system. Numerous *in vivo* studies showed that metastatic tumor cells are quite deformable, and the cell cytoplasm and the cell nucleus can undergo strong compression and shape deformation in small capillaries [16, 59, 60, 72, 73].

Computational simulations can help determine which properties of CTCs must be altered so that they can successfully travel through the blood or lymph circulation systems. This knowledge can lead to designing the next generation of therapeutic interventions that will prevent CTCs from completing the adhesive cascade or from forming multicellular emboli that can result in vessel occlusions. However, such models need to be based on physical principles and need to be able to formulate and test various hypotheses of CTC biomechanics and mechanotransduction. Quantitative integration of physics-based computational models with *in vitro* experimentation and *ex vivo* CTC analysis will lead to advancements from the biomechanical and therapeutic perspectives.

Acknowledgment This work was initiated through a Moffitt-PSOC Pilot Project from the Physical Sciences-Oncology Program Phase I at the National Institute of Health U54-CA-143970.

References

1. Aceto N, Bardia A, Miyamoto DT, Donaldson MC, Wittner BS, Spencer JA, Yu M, Pely A, Engstrom A, Zhu H, Brannigan BW, Kapur R, Stott SL, Shioda T, Ramaswamy S, Ting DT, Lin CP, Toner M, Haber DA, Maheswaran S (2014) Circulating tumor cell clusters are oligoclonal precursors of breast cancer metastasis. Cell 158(5):1110–1122. doi:10.1016/j.cell.2014.07.013

2. Alunni-Fabbroni M, Sandri MT (2010) Circulating tumour cells in clinical practice: methods of detection and possible characterization. Methods 50(4):289–297. doi:10.1016/j.ymeth.2010.01.027
3. Bose S, Das SK, Karp JM, Karnik R (2010) A semianalytical model to study the effect of cortical tension on cell rolling. Biophys J 99(12):3870–3879. doi:10.1016/j.bpj.2010.10.038
4. Carlsson A, Nair VS, Luttgen MS, Keu KV, Horng G, Vasanawala M, Kolatkar A, Jamali M, Iagaru AH, Kuschner W, Loo BW Jr, Shrager JB, Bethel K, Hoh CK, Bazhenova L, Nieva J, Kuhn P, Gambhir SS (2014) Circulating tumor microemboli diagnostics for patients with non-small-cell lung cancer. J Thorac Oncol 9(8):1111–1119. doi:10.1097/JTO.0000000000000235
5. Chen MB, Lamar JM, Li R, Hynes RO, Kamm RD (2016) Elucidation of the roles of tumor integrin ss1 in the extravasation stage of the metastasis cascade. Cancer Res 76:2513–2524. doi:10.1158/0008-5472.CAN-15-1325
6. Chen MB, Whisler JA, Jeon JS, Kamm RD (2013) Mechanisms of tumor cell extravasation in an in vitro microvascular network platform. Integr Biol (Camb) 5(10):1262–1271. doi:10.1039/c3ib40149a
7. Cho EH, Wendel M, Luttgen M, Yoshioka C, Marrinucci D, Lazar D, Schram E, Nieva J, Bazhenova L, Morgan A, Ko AH, Korn WM, Kolatkar A, Bethel K, Kuhn P (2012) Characterization of circulating tumor cell aggregates identified in patients with epithelial tumors. Phys Biol 9(1):016001. doi:10.1088/1478-3975/9/1/016001
8. Crowl L, Fogelson AL (2011) Analysis of mechanisms for platelet near-wall excess under arterial blood flow conditions. J Fluid Mech 676:348–375
9. Diamond E, Lee GY, Akhtar NH, Kirby BJ, Giannakakou P, Tagawa ST, Nanus DM (2012) Isolation and characterization of circulating tumor cells in prostate cancer. Front Oncol 2:131. doi:10.3389/fonc.2012.00131
10. Farace F, Massard C, Vimond N, Drusch F, Jacques N, Billiot F, Laplanche A, Chauchereau A, Lacroix L, Planchard D, Le Moulec S, Andre F, Fizazi K, Soria JC, Vielh P (2011) A direct comparison of cell search and ISET for circulating tumour-cell detection in patients with metastatic carcinomas. Br J Cancer 105(6):847–853. doi:10.1038/bjc.2011.294
11. Fedosov DA, Gompper G (2014) White blood cell margination in microcirculation. Soft Matter 10(17):2961–2970. doi:10.1039/c3sm52860j
12. Fitzgibbon S, Spann AP, Qi QM, Shaqfeh ES (2015) In vitro measurement of particle margination in the microchannel flow: effect of varying hematocrit. Biophys J 108(10):2601–2608. doi:10.1016/j.bpj.2015.04.013
13. Fuhrmann A, Staunton JR, Nandakumar V, Banyai N, Davies PC, Ros R (2011) AFM stiffness nanotomography of normal, metaplastic and dysplastic human esophageal cells. Phys Biol 8(1):015007. doi:10.1088/1478-3975/8/1/015007
14. Gay LJ, Felding-Habermann B (2011) Contribution of platelets to tumour metastasis. Nat Rev Cancer 11(2):123–134. doi:10.1038/nrc3004
15. Hayashi K, Jiang P, Yamauchi K, Yamamoto N, Tsuchiya H, Tomita K, Moossa AR, Bouvet M, Hoffman RM (2007) Real-time imaging of tumor-cell shedding and trafficking in lymphatic channels. Cancer Res 67(17):8223–8228. doi:10.1158/0008-5472.CAN-07-1237
16. Hoffman RM (2008) Imaging in mice with fluorescent proteins: from macro to subcellular. Sensors 8:1157–1173
17. Hou JM, Krebs M, Ward T, Morris K, Sloane R, Blackhall F, Dive C (2010) Circulating tumor cells, enumeration and beyond. Cancers (Basel) 2(2):1236–1250. doi:10.3390/cancers2021236
18. Hsieh HB, Marrinucci D, Bethel K, Curry DN, Humphrey M, Krivacic RT, Kroener J, Kroener L, Ladanyi A, Lazarus N, Kuhn P, Bruce RH, Nieva J (2006) High speed detection of circulating tumor cells. Biosens Bioelectron 21(10):1893–1899. doi:10.1016/j.bios.2005.12.024
19. Jadhav S, Eggleton CD, Konstantopoulos K (2005) A 3-D computational model predicts that cell deformation affects selectin-mediated leukocyte rolling. Biophys J 88(1):96–104. doi:10.1529/biophysj.104.051029
20. Kelley LC, Lohmer LL, Hagedorn EJ, Sherwood DR (2014) Traversing the basement membrane in vivo: a diversity of strategies. J Cell Biol 204(3):291–302. doi:10.1083/jcb.201311112
21. Ketene AN, Schmelz EM, Roberts PC, Agah M (2012) The effects of cancer progression on the viscoelasticity of ovarian cell cytoskeleton structures. Nanomedicine 8(1):93–102. doi:10.1016/j.nano.2011.05.012
22. Khismatullin DB, Truskey GA (2005) Three-dimensional numerical simualtion of receptor-mediated leukocyte adhesion to surfaces: effects of cell deformability and viscoelasticity. Phys Fluids 17:031505
23. Khismatullin DB, Truskey GA (2012) Leukocyte rolling on P-selectin: a three-dimensional numerical study of the effect of cytoplasmic viscosity. Biophys J 102(8):1757–1766. doi:10.1016/j.bpj.2012.03.018
24. Kim Y, Williams KC, Gavin CT, Jardine E, Chambers AF, Leong HS (2016) Quantification of cancer cell extravasation in vivo. Nat Protoc 11(5):937–948. doi:10.1038/nprot.2016.050
25. King MR, Hammer DA (2001) Multiparticle adhesive dynamics: hydrodynamic recruitment of rolling leukocytes. Proc Natl Acad Sci U S A 98(26):14919–14924. doi:10.1073/pnas.261272498
26. King MR, Hammer DA (2001) Multiparticle adhesive dynamics. Interactions between stably rolling cells. Biophys J 81(2):799–813. doi:10.1016/S0006-3495(01)75742-6
27. King MR, Phillips KG, Mitrugno A, Lee TR, de Guillebon AM, Chandrasekaran S, McGuire MJ, Carr RT, Baker-Groberg SM, Rigg RA, Kolatkar A,

Luttgen M, Bethel K, Kuhn P, Decuzzi P, McCarty OJ (2015) A physical sciences network characterization of circulating tumor cell aggregate transport. Am J Physiol Cell Physiol 308(10):C792–C802. doi:10.1152/ajpcell.00346.2014

28. Kirby BJ, Jodari M, Loftus MS, Gakhar G, Pratt ED, Chanel-Vos C, Gleghorn JP, Santana SM, Liu H, Smith JP, Navarro VN, Tagawa ST, Bander NH, Nanus DM, Giannakakou P (2012) Functional characterization of circulating tumor cells with a prostate-cancer-specific microfluidic device. PLoS One 7(4):e35976. doi:10.1371/journal.pone.0035976
29. Laurent VM, Planus E, Fodil R, Isabey D (2003) Mechanical assessment by magnetocytometry of the cytosolic and cortical cytoskeletal compartments in adherent epithelial cells. Biorheology 40(1–3):235–240
30. Lazar DC, Cho EH, Luttgen MS, Metzner TJ, Uson ML, Torrey M, Gross ME, Kuhn P (2012) Cytometric comparisons between circulating tumor cells from prostate cancer patients and the prostate-tumor-derived LNCaP cell line. Phys Biol 9(1):016002. doi:10.1088/1478-3975/9/1/016002
31. Maheswaran S, Haber DA (2010) Circulating tumor cells: a window into cancer biology and metastasis. Curr Opin Genet Dev 20(1):96–99. doi:10.1016/j.gde.2009.12.002
32. Marrinucci D, Bethel K, Kolatkar A, Luttgen MS, Malchiodi M, Baehring F, Voigt K, Lazar D, Nieva J, Bazhenova L, Ko AH, Korn WM, Schram E, Coward M, Yang X, Metzner T, Lamy R, Honnatti M, Yoshioka C, Kunken J, Petrova Y, Sok D, Nelson D, Kuhn P (2012) Fluid biopsy in patients with metastatic prostate, pancreatic and breast cancers. Phys Biol 9(1):016003. doi:10.1088/1478-3975/9/1/016003
33. Marrinucci D, Bethel K, Lazar D, Fisher J, Huynh E, Clark P, Bruce R, Nieva J, Kuhn P (2010) Cytomorphology of circulating colorectal tumor cells: a small case series. J Oncol 2010:861341. doi:10.1155/2010/861341
34. Marrinucci D, Bethel K, Luttgen M, Bruce RH, Nieva J, Kuhn P (2009) Circulating tumor cells from well-differentiated lung adenocarcinoma retain cytomorphologic features of primary tumor type. Arch Pathol Lab Med 133(9):1468–1471. doi:10.1043/1543-2165-133.9.1468
35. Mathur AB, Truskey GA, Reichert WM (2000) Atomic force and total internal reflection fluorescence microscopy for the study of force transmission in endothelial cells. Biophys J 78(4):1725–1735
36. McCarty OJ, Ku D, Sugimoto M, King MR, Cosemans JM, Neeves KB, Subcommittee on Biorheology (2016) Dimensional analysis and scaling relevant to flow models of thrombus formation: communication from the SSC of the ISTH. J Thromb Haemost 14(3):619–622. doi:10.1111/jth.13241
37. McCarty OJ, Mousa SA, Bray PF, Konstantopoulos K (2000) Immobilized platelets support human colon carcinoma cell tethering, rolling, and firm adhesion under dynamic flow conditions. Blood 96(5):1789–1797
38. Migliorini C, Qian Y, Chen H, Brown EB, Jain RK, Munn LL (2002) Red blood cells augment leukocyte rolling in a virtual blood vessel. Biophys J 83(4):1834–1841. doi:10.1016/S0006-3495(02)73948-9
39. Mina LA, Sledge GW Jr (2011) Rethinking the metastatic cascade as a therapeutic target. Nat Rev Clin Oncol 8(6):325–332. doi:10.1038/nrclinonc.2011.59
40. Mitchell MJ, Castellanos CA, King MR (2015) Immobilized surfactant-nanotube complexes support selectin-mediated capture of viable circulating tumor cells in the absence of capture antibodies. J Biomed Mater Res A 103(10):3407–3418. doi:10.1002/jbm.a.35445
41. Mitchell MJ, King MR (2013) Computational and experimental models of cancer cell response to fluid shear stress. Front Oncol 3:44. doi:10.3389/fonc.2013.00044
42. Mitrugno A, Tormoen GW, Kuhn P, McCarty OJ (2016) The prothrombotic activity of cancer cells in the circulation. Blood Rev 30(1):11–19. doi:10.1016/j.blre.2015.07.001
43. Nguyen DX, Bos PD, Massague J (2009) Metastasis: from dissemination to organ-specific colonization. Nat Rev Cancer 9(4):274–284. doi:10.1038/nrc2622
44. Nourshargh S, Hordijk PL, Sixt M (2010) Breaching multiple barriers: leukocyte motility through venular walls and the interstitium. Nat Rev Mol Cell Biol 11(5):366–378. doi:10.1038/nrm2889
45. Oakman C, Pestrin M, Bessi S, Galardi F, Di Leo A (2010) Significance of micrometastases: circulating tumor cells and disseminated tumor cells in early breast cancer. Cancers (Basel) 2(2):1221–1235. doi:10.3390/cancers2021221
46. Park S, Koch D, Cardenas R, Kas J, Shih CK (2005) Cell motility and local viscoelasticity of fibroblasts. Biophys J 89(6):4330–4342. doi:10.1529/biophysj.104.053462
47. Peskin CS (1972) Flow patterns around heart valves: a numerical method. J Comput Phys 10:252–271
48. Peskin CS (1977) Numerical analysis of blood flow in the heart. J Comput Phys 25:220–252
49. Peskin CS (2002) The immersed boundary method. Acta Numer 11:479–517
50. Peskin CS, McQueen DM (1995) A general method for the computer simulation of biological systems interacting with fluids. Symp Soc Exp Biol 49:265–276
51. Phillips KG, Kuhn P, McCarty OJ (2014) Physical biology in cancer. 2. The physical biology of circulating tumor cells. Am J Physiol Cell Physiol 306(2):C80–C88. doi:10.1152/ajpcell.00294.2013
52. Phillips KG, Lee AM, Tormoen GW, Rigg RA, Kolatkar A, Luttgen M, Bethel K, Bazhenova L, Kuhn P, Newton P, McCarty OJ (2015) The thrombotic potential of circulating tumor microemboli: computational modeling of circulating tumor cell-induced coagulation. Am J Physiol Cell Physiol 308(3):C229–C236. doi:10.1152/ajpcell.00315.2014

53. Phillipson M, Heit B, Colarusso P, Liu L, Ballantyne CM, Kubes P (2006) Intraluminal crawling of neutrophils to emigration sites: a molecularly distinct process from adhesion in the recruitment cascade. J Exp Med 203(12):2569–2575. doi:10.1084/jem.20060925
54. Pozrikidis C (2005) Numerical simulation of cell motion in tube flow. Ann Biomed Eng 33(2): 165–178
55. Rejniak KA (2007) An immersed boundary framework for modelling the growth of individual cells: an application to the early tumour development. J Theor Biol 247(1):186–204. doi:10.1016/j.jtbi.2007.02.019
56. Rejniak KA (2007) Modelling the development of complex tissues using individual viscoelastic cells. In: Anderson A, Chaplain MC, Rejniak KA (eds) Single-cell-based models in biology and medicine. Birkhauser-Verlag, Boston
57. Rejniak KA (2012) Investigating dynamical deformations of tumor cells in circulation: predictions from a theoretical model. Front Oncol 2:111. doi:10.3389/fonc.2012.00111
58. Rejniak KA, Dillon RH (2007) A single cell based model of the Ductal tumor microarchitecture. Comput Math Methods Med 8:51–69
59. Stoletov K, Kato H, Zardouzian E, Kelber J, Yang J, Shattil S, Klemke R (2010) Visualizing extravasation dynamics of metastatic tumor cells. J Cell Sci 123(Pt 13):2332–2341. doi:10.1242/jcs.069443
60. Stoletov K, Montel V, Lester RD, Gonias SL, Klemke R (2007) High-resolution imaging of the dynamic tumor cell vascular interface in transparent zebrafish. Proc Natl Acad Sci U S A 104(44):17406–17411. doi:10.1073/pnas.0703446104
61. Swaby RF, Cristofanilli M (2011) Circulating tumor cells in breast cancer: a tool whose time has come of age. BMC Med 9:43. doi:10.1186/1741-7015-9-43
62. Swaminathan V, Mythreye K, O'Brien ET, Berchuck A, Blobe GC, Superfine R (2011) Mechanical stiffness grades metastatic potential in patient tumor cells and in cancer cell lines. Cancer Res 71(15):5075–5080. doi:10.1158/0008-5472.CAN-11-0247
63. Takeishi N, Imai Y, Nakaaki K, Yamaguchi T, Ishikawa T (2014) Leukocyte margination at arteriole shear rate. Physiol Rep 2(6):e12037. doi:10.14814/phy2.12037
64. Takeishi N, Imai Y, Yamaguchi T, Ishikawa T (2015) Flow of a circulating tumor cell and red blood cells in microvessels. Phys Rev E Stat Nonlin Soft Matter Phys 92(6–1):063011. doi:10.1103/PhysRevE.92.063011
65. Tang J, Hunt CA (2010) Identifying the rules of engagement enabling leukocyte rolling, activation, and adhesion. PLoS Comput Biol 6(2):e1000681. doi:10.1371/journal.pcbi.1000681
66. Tormoen GW, Haley KM, Levine RL, McCarty OJ (2012) Do circulating tumor cells play a role in coagulation and thrombosis? Front Oncol 2:115. doi:10.3389/fonc.2012.00115
67. Tremblay PL, Huot J, Auger FA (2008) Mechanisms by which E-selectin regulates diapedesis of colon cancer cells under flow conditions. Cancer Res 68(13):5167–5176. doi:10.1158/0008-5472.CAN-08-1229
68. Turitto VT (1982) Blood viscosity, mass transport, and thrombogenesis. Prog Hemost Thromb 6:139–177
69. Wendel M, Bazhenova L, Boshuizen R, Kolatkar A, Honnatti M, Cho EH, Marrinucci D, Sandhu A, Perricone A, Thistlethwaite P, Bethel K, Nieva J, Heuvel M, Kuhn P (2012) Fluid biopsy for circulating tumor cell identification in patients with early-and late-stage non-small cell lung cancer: a glimpse into lung cancer biology. Phys Biol 9(1):016005. doi:10.1088/1478-3967/9/1/016005
70. Weninger W, Biro M, Jain R (2014) Leukocyte migration in the interstitial space of non-lymphoid organs. Nat Rev Immunol 14(4):232–246. doi:10.1038/nri3641
71. Wirtz D, Konstantopoulos K, Searson PC (2011) The physics of cancer: the role of physical interactions and mechanical forces in metastasis. Nat Rev Cancer 11(7):512–522. doi:10.1038/nrc3080
72. Yamauchi K, Yang M, Jiang P, Xu M, Yamamoto N, Tsuchiya H, Tomita K, Moossa AR, Bouvet M, Hoffman RM (2006) Development of real-time subcellular dynamic multicolor imaging of cancer-cell trafficking in live mice with a variable-magnification whole-mouse imaging system. Cancer Res 66(8):4208–4214. doi:10.1158/0008-5472.CAN-05-3927
73. Yamauchi K, Yang M, Jiang P, Yamamoto N, Xu M, Amoh Y, Tsuji K, Bouvet M, Tsuchiya H, Tomita K, Moossa AR, Hoffman RM (2005) Real-time in vivo dual-color imaging of intracapillary cancer cell and nucleus deformation and migration. Cancer Res 65(10):4246–4252. doi:10.1158/0008-5472.CAN-05-0069
74. Yu M, Stott S, Toner M, Maheswaran S, Haber DA (2011) Circulating tumor cells: approaches to isolation and characterization. J Cell Biol 192(3):373–382. doi:10.1083/jcb.201010021
75. Zhang J, Johnson PC, Popel AS (2009) Effects of erythrocyte deformability and aggregation on the cell free layer and apparent viscosity of microscopic blood flows. Microvasc Res 77(3):265–272. doi:10.1016/j.mvr.2009.01.010

Modeling Proteolytically Driven Tumor Lymphangiogenesis

6

Georgios Lolas, Lasse Jensen, George C. Bourantas, Vasiliki Tsikourkitoudi, and Konstantinos Syrigos

Abstract

With the exception of a limited number of sites in the body, primary tumors infrequently lead to the demise of cancer patients. Instead, mortality and a significant degree of morbidity result from the growth of secondary tumors in distant organs. Tumor survival, growth and dissemination are associated with the formation of both new blood vessels (angiogenesis) and new lymph vessels (lymphagenesis or lymphangiogenesis). Although intensive research in tumor angiogenesis has been going on for the past four decades, experimental results in tumor lymphangiogenesis began to appear only in the last 10 years. In this chapter we expand the models proposed by Friedman, Lolas and Pepper on tumor lymphangiogenesis mediated by proteolytically and un-proteolytically processed growth factors (Friedman and Lolas G, Math Models Methods Appl Sci 15(01):95–107, 2005; Pepper and Lolas G, Selected topics in cancer modeling: genesis, evolution, immune competition, and therapy. In: The lymphatic vascular system in lymphangiogenesis invasion and metastasis a mathematical approach. Birkhäuser Boston, Boston, pp 1–22, 2008). The variables represent different cell densities and growth factors concentrations, and where possible the parameters are estimated from experimental and clinical data.

G. Lolas (✉) • V. Tsikourkitoudi • K. Syrigos
Oncology Unit, 3rd Department of Internal Medicine, Sotiria General Hospital, Athens School of Medicine, Athens, Greece
e-mail: glolas@med.uoa.gr; vasiatsik@med.uoa.gr; ksyrigos@med.uoa.gr

L. Jensen
Department of Microbiology, Tumor and Cell Biology (MTC), C1, Karolinska Institute, Stockholm, Sweden

Department of Medical and Health Sciences, Linköping University, 581 85 Linköping, Sweden
e-mail: lasse.jensen@liu.se

G.C. Bourantas
Faculty of Science, Technology and Communication, University of Luxembourg, Campus Kirchberg, 6,rue Richard Coudenhove-Kalergi L-1359, Luxembourg City, Luxembourg
e-mail: george.bourantas@gmail.com

K.A. Rejniak (eds.), *Systems Biology of Tumor Microenvironment*, Advances in Experimental Medicine and Biology 936, DOI 10.1007/978-3-319-42023-3_6

The results obtained from computational simulations carried out on the model equations produce dynamic heterogeneous ("anarchic") spatio-temporal solutions. More specifically, we observed coherent masses of tumor clusters migrating around and within the lymphatic network. Our findings are in line with recent experimental evidence that associate cluster formation with the minimization of cell loss favoring high local extracellular matrix proteolysis and thus protecting cancer invading cells from an immunological assault driven by the lymphatic network.

Keywords
Lymphangiogenesis • Proteolysis • Plasmin • Mature VEGF-C • Tumor clusters • Tumor heterogeneity

6.1 Introduction

It is well known that cancer growth and dissemination is an extremely complicated process that involves different inter-related biochemical and cellular interactions at many different spatial and temporal scales. Despite the existence of different types of tumours, solid tumours represent a large fraction of the clinical observed tumours and as such they are the main focus of the current chapter. The growth of solid tumors distinguishing them in two categories: the benign and malignant. Benign tumors remain confined to the tissue in which they arise [3]. Malignant tumors, invade and destroy the surrounding tissues and, by exploiting the blood or the lymphatic systems, establish new colonies, a process called *metastasis* [4–6].

Metastasis is the predominant cause of cancer death. There are four major routes of neoplastic dissemination: (1) local invasion; (2) direct seeding to body cavities; (3) hematogenous spread; and (4) lymphatic spread, preferentially to regional lymph nodes and later to distant sites [7].

For a primary tumor to grow, it needs a supply of nutrients, delivered by the blood. The tumor therefore secrets growth factors which induce the formation of new blood vessels, sprouting them from preexisting vessels and directing them toward the tumor. This is the process of *tumor angiogenesis* [8,9]. Targeting angiogenesis, namely, cutting of blood supply, is one off the strategies for blocking tumor growth and dissemination.

A similar, although far less well studied process, also occurs in the lymphatic system and is referred to as lymphangiogenesis or lymphagenesis [7, 10–15]. Traditionally the lymphatic system has not been considered to be actively involved in the process of metastasis, it has been suggested that cancer cells are passively carried into lymphatic vessels with interstitial fluid and proteins [16]. Surprisingly, almost all of the published literature focuses on the correlations between angiogenesis, microvessel density, metastatic spread, and tumor prognosis [7, 17]. However, recent studies have demonstrated the presence of enlarged lymphatic vessels in peritumoral regions of several human tumors, and the number of tumor-associated lymphatics has been correlated with lymph node metastases (reviewed by [18, 19]).

The lymphatic system comprises a vascular network of one-way, open-ended, thin-walled complex network of capillaries and larger vessels, collecting vessels, lymph nodes, trunks, and ducts that transport lymph and cells from body tissues back to the circulatory system [20–23]. The lymph is a protein-rich interstitial fluid drained from peripheral extracellular spaces, composed of interstitial fluid components, metabolites, and plasma proteins extravasated from blood capillaries, and cells from the immune system to the venous system [7, 17, 24].

Knowledge of lymphatics and lymphatic circulation progressed very slowly until the past 10 years [10,25,26]. One of the major limitations of research on lymphatic vessels and lymphangiogenesis was the lack of specific markers that could accurately differentiate Blood Endothelial Cells (BEC) from Lymphatic Endothelial Cells (LEC) [7]. A major impetus in understanding the physiopathology of the lymphatic system has come from the discovery of growth factors which induce the formation of new lymphatic capillaries, and the identification of molecular markers which allow one to differentiate between blood vascular and lymphatic endothelium. This in turn has led to the establishment of techniques for the isolation of pure populations of BEC and LEC [7, 11, 27].

Two growth factors have been identified that induce the growth of new lymphatic cappilaries, and both belong to the vascular endothelial growth factor (VEGF) family. These factors include VEGF-C and VEGF-D [28]. VEGF-C is the central regulator of lymphatic development [29] and is synthesized as a preprotein with a structural homology to the angiogenic growth factor VEGF [30]. VEGF-C was isolated from the human prostatic adenocarcinoma during a search for a specific lymphangiogenic ligand VEGFR-3 [30]. Although VEGFR-3 is expressed during early development by blood vascular endothelium, it becomes restricted mainly to lymphatic endothelial cells in normal adult tissues [31–33]. Other transmembrane proteins such as LYVE-1 (lymphatic vessel endothelial hyaluronan receptor-1) [34], podoplanin and desmoplakin have also been shown to be reliable lymphatic markers [35]. More recently, Prox-1 a homeobox gene has been identified as a major factor in the induction of the lymphatic development [36, 37].

Additionally, several forms of VEGF-C and VEGF-D are generated by proteolytic processing. VEGF-C (or VEGF-D) proteolytic processing increases its binding affinity to VEGFR-2 and VEGFR-3, cell surface-receptors that are predominantly expressed on blood- and lymphatic- vascular endothelia, respectively. Thus, proteolytically processed VEGF-C and VEGF-D have a dual role since they are able to induce both angiogenesis and lymphangiogenesis (see Fig. 6.1). Like the angiogenic specific cytokine VEGF, VEGF-C stimulates also the migration of blood endothelial cells and induces vascular permeability and endothelial-cell proliferation, but does it less effectively than VEGF in the sense that smaller concentrations of VEGF have the same effects as larger concentrations of VEGF-C.

Tumor angiogenesis and lymphangiogenesis provide new vessels that malignant cells can use to escape the confines of the primary tumor [38]. However, existing vessels already provide mechanism of escape, and therefore the relative importance of pre-existing versus new vessels is still to be determined [39]. Various studies have shown that angiogenesis is important for solid tumour growth and, presumably, also in hematogenous metastasis [40]. By contrast, the role of lymphatic vessels and the relevance of lymphangiogenesis to tumor pathology is less clear. Until recently only limited information concerning the molecular mechanisms and pathways involved in tumor lymphangiogenesis and tumor lymphatic invasion have been obtained [7, 41–46].

Unlike the blood vascular network, the lymphatic network does not provide nutrients to cells. It is, however, more conductive to tumor invasion and metastasis [18]. The lymphatic system is optimally suited for the entry and transport of cells (e.g., immune cells), and it also seems to have many advantages over the blood circulation as a pathway for tumor cell dissemination [21]. Even the smallest initial lymphatics are much larger than blood capillaries. In comparison to blood vessels, lymphatic vessels have thinner walls and poorly developed basement membrane, devoid of pericytes. Additionally, lymphatic vessel endothelium lacks tight junctions. Lymphatic endothelial cells are equipped with ovelapping junctions, which function as mechanical flap valves [20,47]. Fibers and collagen anchor the lymphatic endothelial cells into the surrounding connective tissue [48]. Upon increase in interstitial fluid pressure, the connecting tissue fibers become stretched, thereby widenning the lymphatic lumen. The lymphatic endothelial cells, which un-

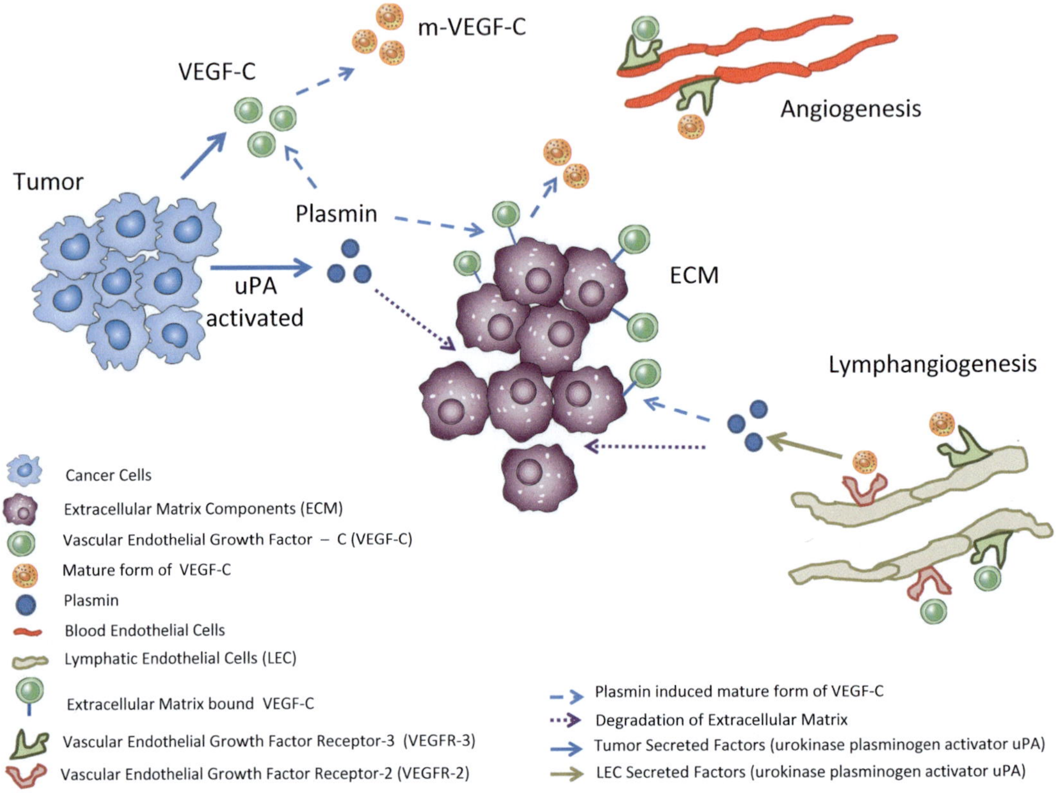

Fig. 6.1 The different effects that tumor secreted VEGF-C and VEGF-D have on blood angiogenesis versus lymphangiogenesis are dependent on their protelytic processing by extracellular matrix proteases

der normal conditions overlap, then move apart, effectively opening intercellular channels to allow fluid and macromolecules access to the lymphatic lumen [47]. It has been speculated that this may be the cause of the greater permeability of the lymphatic vessels. Flow velocities inside the lymphatic system are orders of magnitude lower than those observed in the blood system. Additionaly, lymph composition is nearly identical to the interstitial fluid and in this regard is able to promote the viability of invading tumor cells [7, 18].

By contrast, the bloodstream is a highly aggressive medium for neoplastic cells due to serum toxicity, high shear stresses, and mechanical deformation. Moreover, blood vessels have a more robust structure with a well-defined basement membrane and supporting cells [20]. As a consequence, hematogenous metastasis is a low-efficiency phenomenon, where a significant number of the neoplastic cells within bloodstream routes are quiescent or in apoptosis [49]. Thus, altogether, the lymphatic system provides a much better environment than the blood vasculature for both the survival and metastatic efficiency of disseminating tumor cells. In addition, metastatic cells may reach the bloodstream via lymphatic-venous shunts, high endothelial venules inside lymph nodes, or by lymphatic drainage into the thoracic duct [7, 15].

Cao and coworkers have found intratumor lymphatics in a large number of different tumors grown subcutaneously in mice [50, 51]. As shown in Fig. 6.2, in which lymphatic vessels are shown in blue, erythrocytes in red and tumor cells in green, such intratumoral lymphatics often exhibit highly irregular morphologies compared to the healthy lymphatics in for example the large intestine, and are in contrast to healthy lymphatics perfused with blood. This argues for an important

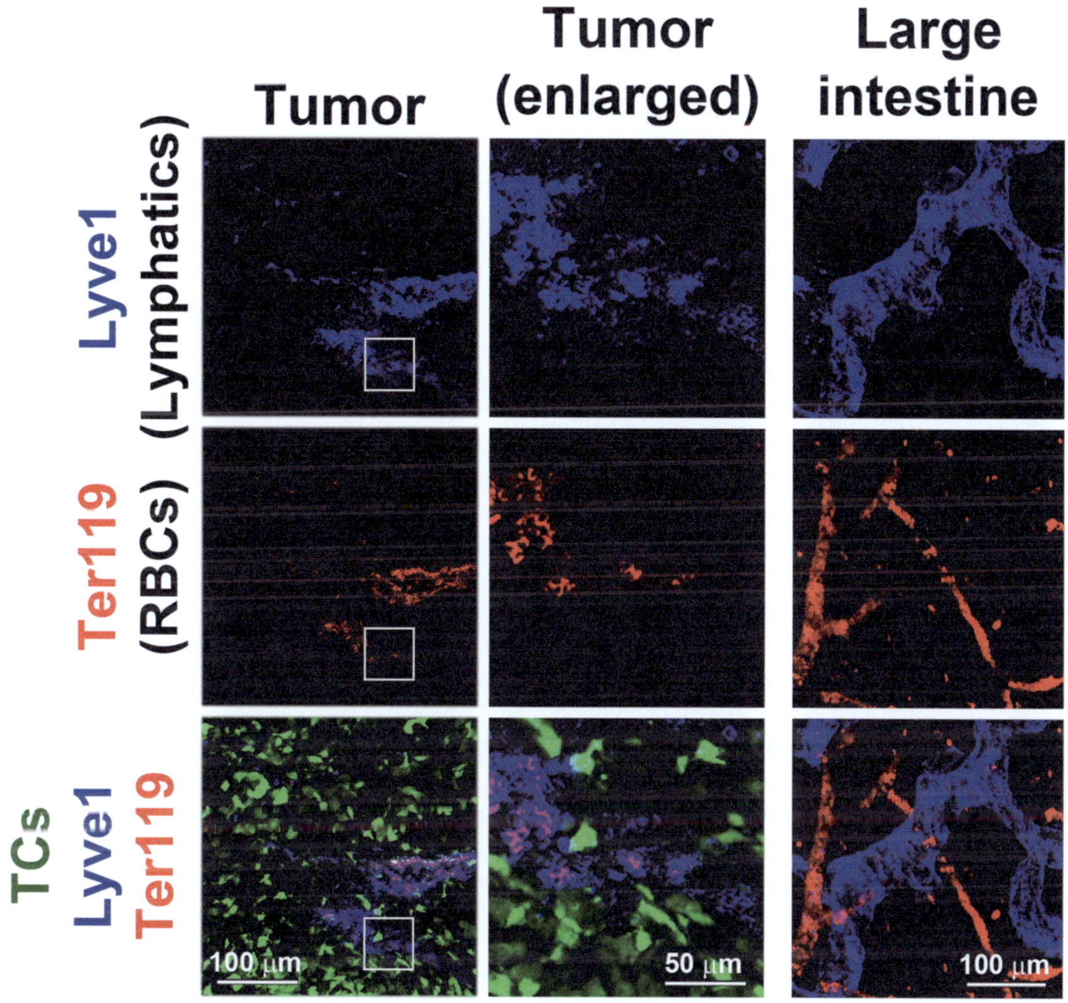

Fig. 6.2 Pathological nature of intra-tumoral lymphatic vessels. Whole-mount confocal imaging of cut samples from a GFP-labeled fibrosarcoma (*green*) grown subcutaneously or healthy large intestine from a mouse were stained with the lymphatic endothelial cells marker Lyve-1 (*blue*), erythrocyte marker Ter119 (*red*). Intratumoral lymphatics are shown to be highly disorganized and filled with Ter119-positive erythrocytes, whereas healthy lymphatics in the large intestine appear normal and without erythrocytes, which are instead restricted to the lumen of the intestinal blood vessels. TC: tumor cell, RBC: red blood cell. The *middle column* represent enlarged images of the areas marked by *white boxes* in the *left column*

role of the intratumoral lymphatics in draining liquid and cells that leak from the similarly highly disorganized tumor blood vasculature, as well as the tumor cells themselves. As platelets and other blood cells may attach to the surface of tumor cells and shield them from recognition by cytotoxic immune cells, the finding of blood in tumor lymphatics provide an additional level of pro-metastatic aid for the tumor cells.

Many lymphangiogenic factors are produced by tumor cells in high amounts [52], but as pro-factors that have very limited activity in their native form. These factors need to be proteolytically processed by proteases in the extracellular matrix before signaling is allowed. VEGF-C for example are cleaved by a ccbe1-ADAMTS3-complex in the extracellular matrix which enable signaling through VEGFR3 [53]. VEGF-A, VEGF-C and other lymphangiogenic factors are furthermore sequestered by heparin sulphate proteoglycans and other proteins in the extracellular matrix from which they can be mobilized by proteolytic degradation of these scaffold proteins [52]. In some cases such as signaling by the angiogenic and lymphangiogenic factor transforming growth factor-(TGF)beta, mobilization of already produced but extracellularly sequestered proteins are the signaling-governing event rather than upregulation of the amount of protein produced by the cells.

The most studied proteases involved in proteolytic activation and/or release of extracellular angiogenic factors belong to the matrix metalloproteinase (MMP) family, where especially MMP2 and MMP9 are known to be important for angiogenesis and lymphangiogenesis. Recently however, lymph vessel researchers have focused on a different mechanism in which ccbe1-ADAMTS3 proteolytic cleavage of pro-VEGF-C allow this factor to activate signaling through VEGFR3, which is necessary for lymphangiogenesis even when induced by other factors such as inflammatory factors [50, 51, 53]. As such, knockout of ccbe1 lead to the complete absence of lymphatic vessel development, massive edema and childhood lethality in a disease known as the hennekam syndrome. This is an example of the importance and rate-limiting actions of extracellular proteases for lymphangiogenesis and lymphatic metastasis. Tumor cells produce a large number of proteases including MMP-2 and MMP-9, especially if they have gained increased migratory/invasive capabilities through the so-called epithelial-to-mesenchymal-transition. Interestingly, these factors are also important for generation of endogenous angiogenesis and lymphangiogenesis inhibitors by proteolysis of extracellular matrix proteins such as fibronectin which releases their kringle domains known as angiostatin or endostatin that potently inhibit angiogenesis. As such the MMPs have dual functions on regulation of angiogenesis and lymphangiogenesis. This may explain why therapeutically targeting MMP-2 or -9 has been proven to be a non-viable approach for treatment of most cancers in spite of initial excitement about such treatments when they were first developed.

Another major protease system responsible for ECM degradation is the plasminogen activation system which generate the potent serine protease plasmin [54–56]. uPA is produced from several types of cells including macrophages, endothelial cells and tumor cells, as a single-chain proenzyme pro-uPA [57]. The binding of uPA to the cancer cell membrane-anchored receptor urokinase receptor (uPAR) accelerates plasminogen activation and, hence, generates plasmin, a serine protease capable of digesting basement membrane and extracellular matrix proteins [58, 59]. Plasmin itself is a broadly acting enzyme that not only catalyzes the breakdown of many of the known extracellular matrix (ECM) and basement membrane molecules, such as vitronectin, fibrin, laminin and collagens, but also may activate metalloproteinases [58].

On the modelling aspect, there is a significant gap between the amount of models developed for angiogenesis and those describing its sibling process lymphangiogenesis (for a comprehensive review see [60]). Therefore, modeling literature about lymphatic development although is a new and exciting area is still relative immature and sparse. Only a couple of specialized models have been published so far. Lolas and coworkers proposed mathematical models of tumor induced lymphangiogenesis enhance by proteolytically and unproteolytically processed lymphatic endothelial growth factors [1, 2]. The basis of the model discussed in this chapter is the model introduced in Lolas and Pepper in [2].

6.2 Mathematical Model Development

A system of nonlinear coupled partial differential equations (PDEs) describing the evolution in time and space of selected biological factors that are associated with tumor lymphangiogenesis, proteolysis and invasion was used to develop the proposed model. More specifically, the model is based on the following key physical variables: the tumour cell density (denoted by c); the lymphatic endothelial cell density (denoted by e); extracellular matrix protein density (ECM) (denoted by m); the urokinase plasminogen activator concentration (uPA) secreted by the lymphatic endothelial cells (denoted by u_e), and by the cancer cells (denoted by u_c); the plasmin concentration activated by LEC (denoted by p_e), and by the cancer cells (denoted by p_c); and the vascular endothelial growth factor-C concentration (VEGF-C) (denoted by f). Assumptions considered for developing the aforementioned in silico model are described below.

Cancer Cells: The controlled dissolution of the extracellular matrix by tumor cell-associated proteases allows tumor cells to pass through tissue barriers and gain access to the circulation [61,62]. The consequent digestion of ECM allows the cells to move into the spaces thereby created, and also set up tissue gradients, which the cells then exploit to move forwards [63–68]. Movement up concentration gradients of ECM has been reported as a mechanism enabling movement through tissues by a variety of cell types. Tumor cell motility toward high concentrations/densities of substratum-bound insolubilized components has been termed *haptotaxis* [69].

Another cause of tumor cell motility is *chemotaxis* whereby tumor cells undergo migration in the direction of a gradient of a soluble attractant [66, 70]. Tumor secretes a number of diffusible chemical substances such as VEGF-C and VEGF-D into the surrounding tissues and extracellular matrix to which cancer cells respond chemotactically [71,72] We assume that there is a change in cell number density due to dispersion: cells disperse from higher to lower densities in a random motion with diffusion coefficient D_c. The flux arising from this random motion is $\mathbf{J}_{\mathbf{random}} = -D_c \nabla c$ and we choose D_c to be constant.

The second important term that quantifies the change in cell number density is that of the directional flow of cells due to spatial gradients of environmental stimuli, such as those stimulating chemotactic or haptotactic responses mentioned above. To incorporate these responses into our mathematical model we take the cancer cell flux (due to gradients) to be

$$\mathbf{J}_{\mathbf{flux}} = \mathbf{J}_{\mathbf{chemo}} + \mathbf{J}_{\mathbf{hapto}}$$

where

$$\mathbf{J}_{\mathbf{chem}} = \zeta_c \, c \nabla u_c + \psi_c \, c \nabla p_c, \; \mathbf{J}_{\mathbf{hapt}} = \xi_c \, c \nabla m \,,$$

χ_c, ζ_c, and ψ_c are the chemotactic coefficients and ξ_c is the haptotactic coefficient; these coefficients are positive, and for simplicity we assume that they are constants.

Regarding the proliferation of cancer cells, we assume that satisfies a logistic growth law. Hence the resulting partial differential equation for the cancer cell motion is:

$$\frac{\partial c}{\partial t} = \underbrace{\nabla \cdot (D_c \nabla c)}_{\text{dispersion}} - \underbrace{\nabla \cdot (\zeta_c \, c \, \nabla u_c + \psi_c \, c \, \nabla p_c)}_{\textit{chemotaxis}} - \underbrace{\nabla \cdot (\xi_c \, c \, \nabla m)}_{\textit{haptotaxis}} + \underbrace{\mu_{11} \, c \, (1 - \frac{c}{c_o})}_{\textit{proliferation}} \,,$$

where μ_{11} is the proliferation rate of the cancer cells, and c_o is the maximum sustainable tumour cell density.

Lymphatic Endothelial Cells: We next derive an equation for the lymphatic endothelial cells (LEC). In response to the lymphangiogenic stimulus (e.g. VEGF-C), LEC in the neighboring lymphatic vessels are activated to secrete matrix degrading enzymes [73]. LEC then start to migrate into the extracellular space responding chemotactically to VEGF-C secreted by cancer cells [73, 74]. LEC directed migration is augmented through the release of matrix bound VEGF-C [75–77]. LEC through the secretion/activation of soluble liberating proteases creating a VEGF-C morphogen gradient. In this regard, LEC also respond chemotactically to the LEC secreted proteases and more specifically uPA and plasmin (which is activated by plasminogen), and they also respond haptotactically to extracellular matrix proteins (e.g., fibronectin, vitronectin) [55].

We assume that lymphatic endothelial cell mitosis is governed by logistic type growth $e\,(1-\dfrac{e}{e_o})$. The activation of both VEGFR-2 and VEGFR-3 that is a result of proteolytically and unproteolytically processed VEGF-C can lead to the proliferation and enlargement of peritumoral and intratumoral lymphatic vessels via the activation of two important signalling cascades [17, 21, 78]. Specifically Oh et al. [79] showed that the application of mature form of VEGF-C on the CAM of 13-day old chick embryos induced proliferation of lymphatic endothelial cells and formation of new lymphatics. We account for this sequence of events by assuming that the lymphatic endothelial cell proliferation is also increased by unproteolytically and proteolytically processed VEGF-C at a rate proportional to $e\ f$ and $e\ \ p_c\ f$ respectively. The above considerations lead to the following equation for the lymphatic endothelial cells:

$$\frac{\partial e}{\partial t} = \underbrace{\nabla\cdot(D_e\nabla e)}_{\text{dispersion}} - \underbrace{\nabla\cdot(\chi_e e\,\nabla f + \zeta_e e\,\nabla u_e + \psi_e e\,\nabla p_e)}_{\textit{chemotaxis}} - \underbrace{\nabla\cdot(\xi_e e\,\nabla m)}_{\textit{haptotaxis}} + \underbrace{\mu_{21}\,e\,(1-\frac{e}{e_o}) + \mu_{22}\,ef + \mu_{23}\,p_c f\,e}_{\textit{proliferation}}\,,$$

where D_e is a diffusion coefficient, χ_e, ζ_e, ψ_e and ξ_e are the chemotactic and haptotactic coefficients, respectively, while μ_{21}, μ_{22}, μ_{23} are proliferation coefficients. We assume that D_e, χ_e, ζ_e, ψ_e, ξ_e, μ_{21}, μ_{22} and μ_{23} are positive constants.

Extracellular Matrix: It is known that extracellular or interstitial matrix (ECM) does not diffuse and therefore we omit any diffusion term or "taxis" terms. However, ECM is not "static" in a sense that it is continually produced and remodelled. Therefore, based on experimental evidence that uPA activates plasminogen to its cancer cell-surface associated form plasmin, which in turn catalyses the breakdown of VN as well as other ECM constituents, we model the fact that plasmin (activated from LEC and CC cells respectively) degrades the VN upon contact at a degradation rate δ_{31}, δ_{32} respectively. Moreover, a logistic growth term accounts for the remodelling of the ECM by the cells present in the tissue. Using rate constant μ_{31} to describe the ECM renewal, we have the following equation for the ECM:

$$\frac{\partial m}{\partial t} = \underbrace{-\delta_{31}\,p_e\,m - \delta_{32}\,p_c\,m}_{\textit{proteolysis}} + \underbrace{\mu_{31}\,m\,(1-\frac{m}{m_o})}_{\textit{re-establishment}}\,. \tag{6.1}$$

Vascular Endothelial Growth Factor – C: VEGF-C with concentration $f(x,t)$ is released rapidly by the solid tumor and it diffuses into the surrounding tissue [30, 80]. As has already been mentioned in the Introduction, both VEGF-C and VEGF-D stimulate lymphangiogenesis in tissues as well as in tumors by activating the lymphatic endothelial cell surface tyrosine kinases receptors VEGFR-3 and VEGFR-2. For simplicity we shall focus our attention only on VEGF-C.

Proteolytic cleavage removes the propeptide forms of tumor secreted VEGF-C to generate mature forms that bind lymphatic endothelial cell surface receptors such as VEGFR-3 (or VEGFR-2) with much greater affinity than the full-length cancer-cell secreted VEGF-C (VEGF-D) forms (Joukov et al., 1997). In fact, several experimental evidence [81–83] highlighted that the proteolytic processing of VEGF-C and VEGF-D propeptides mainly by the serine protease plasmin generating a mature form exhibiting greatly enhanced binding and cross linking of VEGFR-2 and VEGFR-3 in comparison to full-length secreted VEGF-C and VEGF-D ($\alpha_{42}p_c cf$) [25]. Additionally, regarding the process of angiogenesis it has been demonstrated by [84] and [85] that the process of ECM degradation by plasmin leads to the release and cleavage of matrix bound latent cytokines, such as VEGF. Therefore, we assume that tumor-activated plasmin can also release and cleave matrix bound VEGF-C during the process of tumor ECM proteolysis ($\alpha_{43} p_c m f$).

We shall therefore assume that cancer cell regulated plasmin proteolytically cleaves tumor secreted and extracellular-matrix bound VEGF-C to generate its mature VEGF-C form.

We also assume that VEGF-C undergoes some form of decay either natural $\beta_{43} f$ or proportional to the neutralization of proteolytically and unproteolytically processed VEGF-C by its binding to the lymphatic endothelial cell-surface receptor VEGFR-3, $\beta_{41}\, f\, e$, $\beta_{42} p_c\, f\, e$. Summarizing, we have:

$$\frac{\partial f}{\partial t} = \underbrace{D_f \nabla^2 f}_{diffusion} + \underbrace{\underbrace{\alpha_{41}\, c}_{secreted} + \underbrace{\alpha_{42}\, p_c\, c f}_{mature\, form} + \underbrace{\alpha_{43}\, p_c\, m f}_{mature\, form}}_{production} - \underbrace{\beta_{41}\, e f - \beta_{42}\, p_c\, e f}_{neutrilization} - \underbrace{\beta_{43} f}_{natural\, decay},$$

uPA by LEC: Pepper and coworkers [85–87] demonstrated that angiogenic cytokines such as bFGF and VEGF increase uPA and uPAR expression in bovine lymphatic endothelial cells (BLE) in a manner very similar to that described for endothelial cells derived from the blood vascular system. In a more recent paper, Tille et al. [88] showed that this effect (i.e. regulation of the plasminogen activation system) is mediated by the binding of VEGF-C to its receptors.

Therefore, factors influencing the lymphatic endothelial cell regulated uPA concentration are assumed to be diffusion, uPA production and uPA neutralization and decay. During its secretion by the lymphatic endothelial cells ($\alpha_{51}\, e$) uPA diffuses throughout the extracellular matrix, with constant diffusion coefficient D_{u_e}, while uPA production is also enhanced by the binding of proteolytically and unproteolytically cleaved VEGF-C to the lymphatic endothelial cell-surface receptor VEGFR-3, and this is represented by terms $\alpha_{52}\, e f$ and $\alpha_{53}\, p_c\, e f$. Moreover, urokinase is decreased by natural decay ($\beta_{53}\, u_e$), by its neutralization via its binding to the lymphatic endothelial cell-surface receptor uPAR ($\beta_{51}\, u_e\, e$), and indirectly via its inhibition by PAI-1 that is activated by the process of plasmin degrading the ECM ($\beta_{52}\,(p_e\, m + p_c\, m)\, u_e$) . The equation governing the evolution of uPA concentration is therefore given by:

$$\frac{\partial u_e}{\partial t} = \underbrace{D_{u_e} \nabla^2 u_e}_{dispersion} + \underbrace{\alpha_{51}\, e + \alpha_{52}\, e f + \alpha_{53}\, p_c f\, e}_{production} - \underbrace{\beta_{51}\, e\, u_e}_{neutralization} - \underbrace{\beta_{52}\,(p_e\, m + p_c\, m)\, u_e}_{inhibition} - \underbrace{\beta_{53}\, u_e}_{natural\, decay},$$

uPA by CC: Cancers possess the ability to actively invade the peritumoral local tissue during the active process of lymphangiogenesis. Mimicking the proteolysis of the extracellular space by the lymphatic endothelial cells, cancer cells also secrete uPA ($\alpha_{61}\, c$) which diffuses throughout the peritumoral extracellular space, with constant diffusion coefficient D_{u_c} [71, 72]. Additionally, cancer cell secreted uPA undergoes some form of decay either natural ($\beta_{63}\, u_c$), or in proportion to its neutralization by its binding to the cancer cell-surface receptor(uPAR) ($\beta_{61}\, u_c\, c$). Cancer cells secreted uPA concentration is decreased even further by its inhibition of PAI-1 which in turn is activated by the cancer and lymphatic endothelial cells regulated plasmin degrading the peritumoral tissue ($\beta_{62}\,(p_c\,m + p_e\,m)\,u_c$). Under the aforementioned assumptions the equation governing the cancer cell secreted uPA concentration is given by:

$$\frac{\partial u_c}{\partial t} = \underbrace{D_{u_c}\,\nabla^2 u_c}_{dispersion} + \underbrace{\alpha_{61}\,c}_{production} - \underbrace{\beta_{61}\,c\,u_c}_{neutralization} - \underbrace{\beta_{62}\,(p_c\,m + p_e\,m)\,u_c}_{inhibition} - \underbrace{\beta_{63}\,u_c}_{natural\,decay},$$

Plasmin by LEC: As already mentioned above, VEGF-C increases the expression and activity of Plasminogen Activators (PAs). PAs convert plasminogen to plasmin, a protease responsible in lymphatic endothelial cell and cancer cell tissue remodelling and invasion [56, 71, 72, 89]. We assume that the binding of LEC secreted uPA to its LEC surface receptor (uPAR) provides the LEC with a potential proteolytic activity via activation of plasminogen and thus enhances the rate of plasmin formation ($\alpha_{71}\,u_e\,e$) [71, 72].

Moreover, the concentration of the lymphatic endothelial cells regulated plasmin is decreased as a result of natural decay ($\beta_{71}\,p_e$). Hence, in this model the equation for the LEC mediated plasmin dynamics is:

$$\frac{\partial p_e}{\partial t} = \underbrace{D_{p_e}\,\nabla^2 p_e}_{diffusion} + \underbrace{\alpha_{71}\,e\,u_e}_{production} - \underbrace{\beta_{71}\,p_e}_{natural\,decay}, \tag{6.2}$$

Plasmin by CC: In line with the lymphatic endothelial cell regulated plasmin formation that was just presented above, plasmin regulation by the cancer cells is assumed to diffuse with a constant diffusion coefficient D_{p_c}, and to be produced through the binding of cancer cell secreted uPA to its cancer cell-surface receptor uPAR ($\alpha_{81} c\,u_c$). Additionaly, cancer cell generated plasmin undergoes some form of natural decay ($\beta_{81}\,p_c$). Hence we have:

$$\frac{\partial p_c}{\partial t} = \underbrace{D_{p_c}\,\nabla^2 p_c}_{diffusion} + \underbrace{\alpha_{81}\,c\,u_c}_{production} - \underbrace{\beta_{81}\,p_c}_{natural\,decay}. \tag{6.3}$$

Nondimensionalization: Before we solve the system numerically, we first non-dimensionalize the equations as follows:

(1) Reference length scale, L, (e.g., the maximum distance of the lymphatic vessels to the cancer cells at a typical stage of lymphangiogenesis is 0.1–0.5 cm [40]),

(2) Reference time unit, $\tau = \frac{L^2}{D}$, where D is a reference chemical diffusion coefficient, e.g., $10^{-6}\,\mathrm{cm}^2\cdot\mathrm{s}^{-1}$ [90]. Hence we shall take τ to be 10^4s.

(3) Reference endothelial cell density e_o, tumour cell density c_o, extracellular matrix density m_o, VEGF-C concentration f_o, reference cancer cell generated uPA and plasmin concentration u_{c_o}, p_{c_o}, and reference lymphatic endothelial cell generated uPA and plasmin concentration u_{e_o}, p_{e_o} will be chosen later on.

The conversion from the dimensional to dimensionless units was done as follows:

$$[t^*, x^*] = [\frac{t}{\tau}, \frac{x}{L},]$$

$$[c^*, e^*, m^*, f^*, u^*_{e_o}, u^*_{c_o}, p^*_{e_o}, p^*_{c_o}] = [\frac{c}{c_o}, \frac{e}{e_o}, \frac{m}{m_o}, \frac{f}{f_o}, \frac{u_e}{u_{e_o}}, \frac{u_c}{u_{c_o}}, \frac{p_e}{p_{e_o}}, \frac{p_c}{p_{c_o}}],$$

and introduce new parameters via the following scaling:

$$
\begin{aligned}
D_e^* &= \frac{D_e}{D},\ D_c^* = \frac{D_c}{D},\ D_f^* = \frac{D_f}{D},\ D_{u_e}^* = \frac{D_{u_e}}{D},\ D_{u_c}^* = \frac{D_{u_c}}{D},\ D_{p_e}^* = \frac{D_{p_e}}{D},\\
D_{p_c}^* &= \frac{D_{p_c}}{D},\ \chi_e^* = \chi_e \frac{f_o}{D},\ \zeta_e^* = \zeta_e \frac{u_{e_o}}{D},\ \psi_e^* = \psi_e \frac{p_{e_o}}{D},\ \xi_e^* = \xi_e \frac{m_o}{D},\\
\zeta_c^* &= \zeta_c \frac{u_{c_o}}{D},\ \psi_c^* = \psi_c \frac{p_{c_o}}{D},\ \xi_c^* = \xi_c \frac{m_o}{D},\ \mu_{11}^* = \mu_{11}\tau,\ \mu_{21}^* = \mu_{21}\tau,\\
\mu_{22}^* &= \mu_{22} f_o\, \tau,\ \mu_{23}^* = \mu_{23} f_o\, p_{e_o} \tau,\ \delta_{31}^* = \delta_{31} p_{e_o} \tau,\ \delta_{32}^* = \delta_{32} p_{c_o} \tau,\\
\mu_{31}^* &= \mu_{31}\tau, p_{c_o}\, \alpha_{41}^* = \alpha_{41}\tau \frac{c_o}{f_o},\ \alpha_{42}^* = \alpha_{42}\tau p_{c_o},\ \alpha_{43}^* = \alpha_{43}\tau p_{c_o}\, m_o,\\
\beta_{41}^* &= \beta_{41}\tau\, e_o,\ \beta_{42}^* = \beta_{42}\tau\, p_{c_o}\, e_o,\ \beta_{43}^* = \beta_{43}\tau,\ \alpha_{51}^* = \alpha_{51}\tau \frac{e_o}{u_{e_o}},\\
\alpha_{52}^* &= \alpha_{52}\tau \frac{e_o f_o}{u_{e_o}},\ \alpha_{53}^* = \alpha_{53}\tau \frac{e_o f_o\, p_{e_o}}{u_{e_o}},\ \beta_{51}^* = \beta_{51}\tau\, e_o,\ \beta_{52}^* = \beta_{52}\tau \frac{p_{e_o}\, m_o}{u_{e_o}},\\
\beta_{53}^* &= \beta_{53}\tau,\ \alpha_{61}^* = \alpha_{61}\tau \frac{c_o}{u_{c_o}},\ \beta_{61}^* = \beta_{61}\tau\, c_o,\ \beta_{62}^* = \beta_{62}\tau \frac{p_{c_o}\, m_o}{u_{c_o}},\\
\beta_{63}^* &= \beta_{63}\tau,\ \alpha_{71}^* = \alpha_{71}\tau \frac{e_o u_{e_o}}{p_{e_o}},\ \beta_{71}^* = \beta_{71}\tau\, m_o,\ \beta_{72}^* = \beta_{72}\tau,\\
\alpha_{81}^* &= \alpha_{81}\tau \frac{c_o u_{c_o}}{p_{c_o}},\ \beta_{81}^* = \beta_{81}\tau\, m_o\, \beta_{82}^* = \beta_{82}\tau\,.
\end{aligned}
$$

For notational simplicity we shall henceforth omit the asterisks. The dimensionless equations can then be written in the following form:

$$
\begin{aligned}
\frac{\partial c}{\partial t} &= \underbrace{\nabla \cdot (D_c \nabla c)}_{\text{dispersion}} - \underbrace{\nabla \cdot (\zeta_c\, c \nabla u_c + \psi_c\, c \nabla p_c)}_{\textit{chemotaxis}} - \underbrace{\nabla \cdot (\xi_c\, c \nabla m)}_{\textit{haptotaxis}} + \underbrace{\mu_{11}\, c\,(1 - c)}_{\textit{proliferation}},\\
\frac{\partial e}{\partial t} &= \underbrace{\nabla \cdot (D_e \nabla e)}_{\text{dispersion}} - \underbrace{\nabla \cdot (\chi_e e\, \nabla f + \zeta_e e\, \nabla u_e + \psi_e e\, \nabla p_e)}_{\textit{chemotaxis}} - \underbrace{\nabla \cdot (\xi_e e\, \nabla m)}_{\textit{haptotaxis}} + \underbrace{\mu_{21}\, e\,(1 - e) + \mu_{22}\, e f + \mu_{23}\, p_e f\, e}_{\textit{proliferation}},\\
\frac{\partial m}{\partial t} &= \underbrace{-\delta_{31}\, p_e\, m - \delta_{32}\, p_c\, m}_{\textit{proteolysis}} + \underbrace{\mu_{31}\, m\,(1 - m)}_{\textit{re-establishment}},\\
\frac{\partial f}{\partial t} &= \underbrace{D_f\, \nabla^2 f}_{\textit{diffusion}} + \underbrace{\underbrace{\alpha_{41}\, c}_{\textit{secreted}} + \underbrace{\alpha_{42}\, p_c f\, c}_{\textit{mature form}} + \underbrace{\alpha_{43}\, p_c\, m f}_{\textit{mature form}}}_{\textit{production}} - \underbrace{\beta_{41}\, e f - \beta_{42}\, p_c\, e f}_{\textit{neutrilization}} - \underbrace{\beta_{43} f}_{\textit{natural decay}},\\
\frac{\partial u_e}{\partial t} &= \underbrace{D_{u_e}\, \nabla^2 u_e}_{\textit{dispersion}} + \underbrace{\alpha_{51}\, e + \alpha_{52}\, e f + \alpha_{53}\, p_c f\, e}_{\textit{production}} - \underbrace{\beta_{51}\, e\, u_e}_{\textit{neutrilization}} - \underbrace{\beta_{52}\,(p_e\, m + p_c\, m)\, u_e}_{\textit{inhibition}} - \underbrace{\beta_{53}\, u_e}_{\textit{natural decay}},
\end{aligned}
$$

$$\frac{\partial u_c}{\partial t} = \underbrace{D_{u_c}\nabla^2 u_c}_{dispersion} + \underbrace{\alpha_{61}\,c}_{production} - \underbrace{\beta_{61}\,c\,u_c}_{neutralization} - \underbrace{\beta_{62}\,(p_c\,m + p_e\,m)\,u_c}_{inhibition} - \underbrace{\beta_{63}\,u_c}_{natural\,decay},$$

$$\frac{\partial p_e}{\partial t} = \underbrace{D_{p_e}\nabla^2 p_e}_{diffusion} + \underbrace{\alpha_{71}\,e\,u_e}_{production} - \underbrace{\beta_{71}\,p_e}_{natural\,decay},$$

$$\frac{\partial p_c}{\partial t} = \underbrace{D_{p_c}\nabla^2 p_c}_{diffusion} + \underbrace{\alpha_{81}\,c\,u_c}_{production} - \underbrace{\beta_{81}\,p_c}_{natural\,decay}, \tag{6.4}$$

We complement the system (6.4) with boundary and initial conditions:

Boundary Conditions: Guided by the in vitro experimental protocol in which tumor invasion and lymphangiogenesis take place within an isolated system, we assume no-flux of all the diffusing concentrations. For simplicity we shall deal with the case of one-space dimension, taking the domain to be $0 \le x \le 1$. The boundary conditions are then given by the following equations,

$$-D_c\frac{\partial c}{\partial x} + c\,\zeta_c\frac{\partial u}{\partial x} + c\psi_c\frac{\partial p}{\partial x} + c\,\xi_c\frac{\partial m}{\partial x} = 0 \quad \text{at}\, x = 0,\ 1, \tag{6.5}$$

$$-D_c\frac{\partial e}{\partial x} + e\,\chi_e\frac{\partial f}{\partial x} + e\zeta_e\frac{\partial u_e}{\partial x} + e\,\psi_e\frac{\partial p_e}{\partial x} + e\,\xi_e\frac{\partial m}{\partial x} = 0 \ \text{at}\, x = 0,\ 1, \tag{6.6}$$

$$\frac{\partial f}{\partial x} = 0\ \text{at}\, x = 0,\ 1, \tag{6.7}$$

$$\frac{\partial u_e}{\partial x} = 0\ \text{at}\, x = 0,\ 1, \tag{6.8}$$

$$\frac{\partial u_c}{\partial x} = 0\ \text{at}\, x = 0,\ 1, \tag{6.9}$$

$$\frac{\partial p_e}{\partial x} = 0\ \text{at}\, x = 0,\ 1, \tag{6.10}$$

$$\frac{\partial p_c}{\partial x} = 0\ \text{at}\, x = 0,\ 1\,. \tag{6.11}$$

Initial Conditions: We assume that initially (i) there is a cluster of lymphatic endothelial cells already present and a cluster of cancer cells that have penetrated just a short distance into the extracellular matrix; (ii) the concentrations of the VEGF-C, the cancer cell secreted urokinase, and the cancer cell activated plasmin are proportional to the initial tumour density; and finally (iii) the LEC-secreted urokinase as well as LEC-activated plasmin have not yet been secreted and activated, respectivelly. More specifically,

$$\begin{aligned}
c(x,0) &= exp(\frac{-x^2}{\epsilon}),\ x \in [0,1] \text{ and } \epsilon > 0,\\
e(x,0) &= exp(\frac{-(x-1)^2}{\epsilon}),\ x \in [0,1] \text{ and } \epsilon > 0,\\
m(x,0) &= 1 - \frac{1}{2}\,exp(\frac{-x^2}{\epsilon}) - \frac{1}{2}\,exp(\frac{-(x-1)^2}{\epsilon}),\\
&\quad x \in [0,1] \text{ and } \epsilon > 0\,,\\
f(x,0) &= \frac{1}{10}\,exp(\frac{-x^2}{\epsilon}),\ x \in [0,1] \text{ and } \epsilon > 0\,,\\
u_e(x,0) &= 0,\ x \in [0,1]\,,\\
u_c(x,0) &= \frac{1}{20}\,exp(\frac{-x^2}{\epsilon}),\ x \in [0,1] \text{ and } \epsilon > 0\,,\\
p_e(x,0) &= 0,\ x \in [0,1]\,,\\
p_c(x,0) &= \frac{1}{20}\,exp(\frac{-x^2}{\epsilon}),\ x \in [0,1] \text{ and } \epsilon > 0\,,
\end{aligned} \tag{6.12}$$

where we take $\epsilon = 0.01$.

6.2.1 Estimation of Parameters

Whenever possible, parameter values are estimated from available experimental data. In cases where no experimental data could be found, parameter values were chosen to give the best qualitative numerical simulation results of experimentally observed events of proteolytically induced lymphangiogenesis. This is in line with previous papers successfully simulating proteolysis, tumor invasion and angiogenesis [56, 91–93].

6.2.2 Scaling Coefficients $c_0, e_0, m_0, f_0, u_{e_0}, u_{c_0}, p_{e_0}, p_{c_0}$

The cell diameter of most animal cells range from 10 to 30 μm [94]. Taking tumor cell diameter 10 μm $= 10 \times 10^{-3}$ mm $= 10^{-2}$ mm will give an approximate volume $V = \frac{4}{3}\pi r^3 = \frac{4}{3}\pi(\frac{1}{2}10^{-2})^3$ mm^3, so $V \approx 5.24 \times 10^{-7} \approx 5 \times 10^{-7}$ mm^3. In this regard if a cube of size 1 mm^3 was saturated with cancer cells with the aforementioned volume, there would be 0.2×10^7 cells. Moreover, if we assume the density of the cells to be 1g $\cdot$ ml^{-1} (the same as that of water), we have that a cancer cell weighs about 5×10^{-10}g. Hence, we choose $c_0 = 1 \times 10^{-3}$ g $\cdot$ cm^{-3}, and correspondingly $e_0 = 1 \times 10^{-4}$ g $\cdot$ cm^{-3}. Regarding ECM reference concentration Mochan and Keler [95] considered in their in vitro experiment the proteoglycan concentration to be 400 μg $\cdot$ ml^{-1} $= 4 \times 10^{-4}$g $\cdot$ cm^{-3} while in [96] they used a biological extracellular matrix of 1.5×10^{-3} g $\cdot$ cm^{-3} type I collagen. Therefore, we choose the reference ECM concentration to be in the c_0, e_0 range in other words we take $m_0 = 1 \times 10^{-3}$ g $\cdot$ cm^{-3}. Based on in vivo and in culture experimental evidence the reference concentration for several components of the VEGF family (e.g. VEGF, VEGF-C, VEGF-D) and more specifically for VEGF-C concentration is estimated to be between 22 and 100 ng $\cdot$ ml^{-1} $= 0.22 - 1 \times 10^{-7}g\cdot$cm^{-3}. [85, 87, 97–99]. Moreover, in [57, 100–102] they estimated the blood plasma concentration of uPA to be around 20 pM while [103] revealed the initial plasminogen activator to be 0.087 CTA units. The unit of urokinase activity, referred as the CTA unit and is based on the activity of working standard urokinase preparation which was independently assayed in several laboratories which releases 5×10^{-4} micromoles $\cdot$ (CTA)$^{-1}$ [104]. So we can approximately choose the reference uPA concentration to be $u_{c_o}, u_{e_o} \approx 1 \times 10^{-8}$g $\cdot$ cm^{-3}. Additionally, Ellis [105] in their experiments have used plasmin concentration of 0.5 nM and 100 pM respectively. Therefore, we assume that both the cancer and the LECs activated plasmin has a reference concentration of $p_{c_o}, p_{e_o} \sim 10^{-10}$ M. Taking into account plasmin molecular weight 76.5 kDa [106] we deduce that $p_{c_o}, p_{e_o} \approx 100 \times 10^{-12} \times 76{,}500$ gr $\cdot$ lt^{-1} so $p_{c_o}, p_{e_o} \approx 1 \times 10^{-8}$ gr $\cdot$ cm^{-3}.

6.2.2.1 Estimation of the Diffusion Coefficients D

We introduce a reference chemical diffusion coefficient $D \sim 10^{-6}$cm$^2 \cdot$ s^{-1} [90]. In a model of epidermal wound healing, Sherratt and Murray [107], used values of 5.9×10^{-11}cm$^2 \cdot$ s^{-1} – 3×10^{-9}cm$^2 \cdot$ s^{-1} for the random motility of epidermal cells. In a study of individual endothelial cells (ECs), Stokes et al. [108] calculated a random motility coefficient of $(7.1 \pm 2.7) \times 10^{-9}cm^2 \cdot$ s^{-1} for ECs migrating in a culture containing an angiogenic factor αFGF, heparin and fetal serum as well as a random motility coefficient of migrating endothelial cells, with agarose overlays of $(2.3 \pm 0.6) \times 10^{-9}cm^2 \cdot$ s^{-1} and without agarose overlays of $(6.9 \pm 2.6) \times 10^{-9}cm^2 \cdot$ s^{-1}. In agreement with the aforementioned measurements for cell dispersion, Bray [90] estimated the animal cell random motility coefficient to be $\sim 5 \times 10^{-10}$cm$^2 \cdot$s^{-1}. Correspondingly, our choice for both lymphatic endothelial and cancer cell dispersion will vary betweeen 10^{-11}cm$^2 \cdot$ s^{-1} and 10^{-9}cm$^2 \cdot$ s^{-1}, so that our nondimensional values will be between $D_c, D_e = 10^{-5} - 10^{-3}$.

Regarding the plasmin diffusion coefficient, Robbins et al. [109] and Fleury et al. [110], estimated the dimensional urokinase-activated plasmin diffusion coefficient to be 4.91×10^{-7} cm$^2 \cdot$ s^{-1} and 6.8×10^{-7} cm$^2 \cdot$ s^{-1} respectively. However, regarding the existence of several other diffusible factors such as uPA and VEGF-C, we make the assumption that both the lymphatic endothelial as well as the cancer cell activated plasmin diffusion coefficients are much smaller: we choose the dimensionless values for both D_{p_e} and D_{p_c} to be in the range of $1 \times 10^{-3} - 4.91 \times 10^{-1}$.

For the diffusion coefficient of a chemotattractant, Sherratt and Murray [107] choose values of $3.1 \times 10^{-7}\text{cm}^2 \cdot \text{s}^{-1} - 6.9 \times 10^{-6}\text{cm}^2 \cdot \text{s}^{-1}$ for an activator and inhibitor chemical respectively, while Chaplain et al. [111, 112] chose $3.3 \times 10^{-8}\text{cm}^2 \cdot \text{s}^{-1}$. Additionally, White et al. [113] estimated the urokinase diffusion coefficient to be $7.41 \times 10^{-7}\text{cm}^2 \cdot \text{s}^{-1}$, while Fleury et al. [110] considered the diffusion coefficient for a cell secreted protease and matrix-binding ligand to be between 80 μm² and 120 μm² respectively. More recently Shin et al. [114] estimated using a microfluidic assay the diffusion coefficient of VEGF to be $5.8 \times 10^{-7}\text{cm}^2 \cdot \text{s}^{-1}$. Assuming that the diffusion coefficient of a diffusible chemical is in the range $10^{-9} - 10^{-6}\text{cm}^2 \cdot \text{s}^{-1}$, we obtain a dimensionless estimate for D_{u_e}, D_{u_c}, and D_f in the range of 0.001–1.

6.2.2.2 Taxis Coefficients χ, ζ, ψ, ξ

Stokes et al. [108] estimated the chemotaxis coefficient of single individual endothelial cells migrating in a culture containing αFGF, to be $2600\,\text{cm}^2 \cdot \text{s}^{-1} \cdot \text{M}^{-1}$. However, since in this chapter we deal with densities of cells and also due to the presence of several chemo- and hapto-attractants we presume that the aforementioned coefficient could be much smaller (e.g. $26\ \text{cm}^2 \cdot \text{s}^{-1} \cdot \text{M}^{-1}$). Thus we choose the dimensional lymphatic endothelial and cancer cell chemotactic coefficients to be $\approx 26\ \text{cm}^2 \cdot \text{s}^{-1}\,\text{M}^{-1}$. The dimensional VEGF-C directed chemotactic coefficient (taking into account that VEGF-C molecular weight is 80 kDa) is estimated $\chi_e = 3.25 \times 10^{-2}$. In a similar manner (using uPA and plamsin molecular weight to be 54 kDa and 76.5 kDa) we deduce $\zeta_{c,e} = 4.81 \times 10^{-3}$ and $\psi_{c,e} = 3.4 \times 10^{-3}$.

In the absence of reliable experimental data, we consider the haptotactic coefficient of both the lymphatic endothelial and cancer cell to be in the chemotactic range; in other words to be between 26 and $2600\,\text{cm}^2 \cdot \text{s}^{-1}\text{M}^{-1}$. Taking the vitronectin molecular weight to be 75 kDa gives a non-dimensional value of ≈ 346. However, based on the presence of different exracellular matrix components with different molecular weight we chose the dimensionless haptotactic coefficients to be much smaller, thus we choose ξ_c and ξ_e to be between $3.46 \times 10^{-1} - 3.46 \times 10^{-5}$.

6.2.2.3 Proliferation Rate Constants, μ_{ii}

Yu et al. [115] estimated the doubling time of human epidermoid carcinoma cells (HEp3) from in vitro proliferation experiments to be 24 h. Furthermore, Shirasuna et al. [58] estimated the adenoid cystic carcinoma (AdCC) doubling time to be 86 h. Taking the proliferation rate as $\frac{ln2}{[24-86]}\ \text{h}^{-1}$ we get $\mu_{11} \approx [2.24{-}8.022] \times 10^{-6}\text{s}^{-1}$ and the non-dimensional $\mu_{11} \approx [2.24 - 8.022] \times 10^{-2}$.

Previously, Sherratt and Murray [107] estimated the epidermal cell growth rate constant to be $0.04\,\text{h}^{-1}$. In addition, Stokes and Lauffenburger [116] estimated the endothelial cell proliferation time to be 18 h. Nevertheless, Stokes and Lauffenburger [116], Chaplain et al. [117], as well as Orme and Chaplain [91], reduced the chosen value of the proliferation rate to be $0.02\,\text{h}^{-1}$ in order for them to compensate for the assumptions that fibronectin can inhibit endothelial cell proliferation, and that during the angiogenesis process the endothelial cell proliferation is mainly confined to a zone just proximal to the tips of the sprouting endothelial cell capillaries. In a similar manner in [118, 119] estimated the LEC doubling time between 24 and 30 h, deducing that these growth rates could been prolonged for even longer time periods. Regarding the LEC proliferation rate μ_{21} we choose to be between $5.55 \times 10^{-6}\ \text{s}^{-1} - 1.07 \times 10^{-5}\text{s}^{-1}$ and the non-dimensional one to be $5.55 \times 10^{-2} - 1.07 \times 10^{-1}\text{s}^{-1}$

ECM-bound VEGF-C can be released into a soluble form by plasmin resulting in an increased proliferation rate of LEC as it has been investigated for VEGF-bound ECM and endothelial cells from [120]. To obtain estimates of the aforementioned rate μ_{22}, we consider just the

Fig. 6.3 A schematic representation of the VEGF-C/VEGFR-3 dynamics

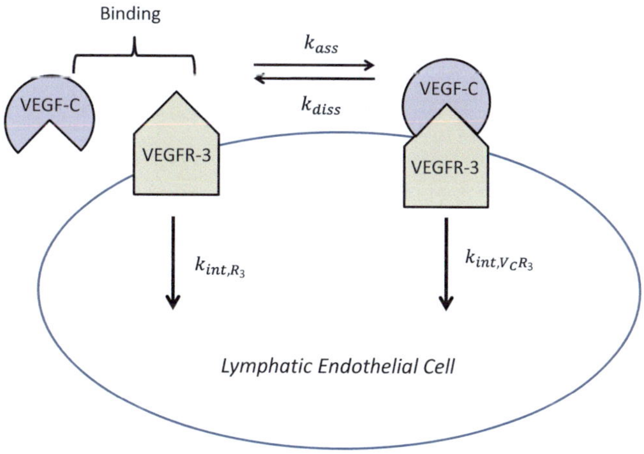

part of the LEC equation $de/dt = \mu_{22}ef$, whose solution is given by

$$e(t) = e(0) \cdot exp(\mu_{22}ft) \tag{6.13}$$

In [120, Figure 6,7] endothelial cell proliferation is measured under several ECM conditions from cells expressing VEGF isoforms. Moreover, [120] specifies that 7000 cells/well were plated and the final cell numbers were determined 5 days later. Then by taking $e(0) =$ 7000cells/well, we get the dimensional estimate for $\mu_{22} = 1.06 \times 10^3 \text{cm}^3 \cdot \text{gr}^{-1} \cdot \text{s}^{-1}$ and the non-dimensional $\mu_{22} = 1.06$.

6.2.2.4 VEGF-C Parameters

In [121, Figure 5] the authors reported that 293 cells tranfected with the rat VEGF-C gene plated in a 6-well ($\approx 10^6$ cells/well $\approx 10^{-4}$gr ml^{-1}) produced 50 ng $\cdot$ ml^{-1} or 50×10^3pg $\cdot$ mm^{-3} of VEGF-C after 6 days of culture. Then $\alpha_{41} \approx 9,645 \times 10^{-10} \text{s}^{-1}$. Thus the dimensional value for α_{41} is $\alpha_{41} \approx 9.645 \times 10^{-17} \text{gr} \cdot \text{cm}^{-3} \cdot \text{s}^{-1} \cdot \text{cell}^{-1}$ and the non-dimensional value is $\alpha_{41} = 9.645 \times 10^{-2}$.

In a similar manner Bocci et al. [122] investigated the production and autocrine action of VEGF by placental endothelial cells. They showed that VEGF is secreted in placental endothelial cells reaching a plateau from day 24 (68.74 ± 7.52 pg $\cdot$ ml^{-1}) to day 27 (67.20 ± 6.28 pg$\cdot$ml^{-1}). Taking an average of the aforementioned maximum values (considering the days 24–27) we deduce from [121, Figure 5] that at day 25 endothelial cells occupy an area of 23 mm^2 and secreted 74.77 pg $\cdot$ ml^{-1} of VEGF. Assuming an endothelial cell area to be $10 \times 10\,\mu$m we deduce that an area of 23 mm^2 is occupied from 23×10^4 endothelial cells which is $\approx 1.15 \text{gr} \cdot \text{cm}^{-3}$. Then $\alpha_{41} \approx 1.88 \times 10^{-12}\, \text{s}^{-1}$ and the non-dimensional value to be $\alpha_{41} \approx 1.88 \times 10^{-4}$ (Fig. 6.3).

To estimate a_{42} we use [120, Figure 7,8]. Taking into account only a specific part of the VEGF-C equation, namely $\frac{df}{dt} = a_{43}p_cfm$ we deduce that the dimensionless value of a_{43} is ≈ 0.01732. In a similar way from [120, Figure 6] we deduce that $a_{42} = ln(\frac{7}{0.3})\frac{1}{0.0784}\text{L}\cdot\text{h}^{-1}\cdot\mu\text{mol}^{-1}\cdot \text{h}^{-1} \approx 1.34 \times 10^2 \text{cm}^3 \cdot \text{gr}^{-1} \cdot \text{s}^{-1}$ Regarding the consumption/internalization of unproteolytically and proteolytically processed VEGF-C to lymphatic endothelial cell surface receptor VEGFR-3 we will follow the methodology proposed by Bianchi et al. [123] and Popel and coworkes [124]. We recall the schematic representation employed in [123] regarding the ligand-receptor dynamics together with the associated kinetic parameters estimated in [41, 74, 125] and presented in Table 6.1.

Based on the aforementioned reaction scheme, the corresponding equations for VEGF-C and the complex VEGF-C/VEGFR-3 (for a single cell) are:

$$\frac{dV_C}{dt} = -k_{ass} \cdot V_C \cdot R_3 + k_{diss} \cdot [V_CR_3] \tag{6.14}$$

Table 6.1 Kinetic constants for receptor VEGFR-3 bound VEGF-C/D

Parameters	Symbol	Estimated range	Sources
Association constant	k_{ass}	$13.6 \times 10^4 \mathrm{s}^{-1} \cdot \mathrm{M}^{-1}$	[74]
		$0.35 \times 10^4 \mathrm{M}^{-1} \mathrm{s}^{-1}$	[74]
		$[0.8-1.8] \times 10^4 \mathrm{M}^{-1} \mathrm{s}^{-1}$	[125]
Dissociation constant	k_{diss}	$6.05 \times 10^{-4}\ \mathrm{s}^{-1}$	[74]
		$4.0 \times 10^{-4}\ \mathrm{s}^{-1}$	[74]
		$[7.0 - 12.0] \times 10^{-4}\ \mathrm{s}^{-1}$	[125]
Internalization constant	k_{int}	[127, Figure 3e,3f]	[127]
		$\approx 5.77 \times 10^{-4}\ \mathrm{s}^{-1}$	

$$\frac{d[V_C R_3]}{dt} = +k_{ass} \cdot V_C \cdot R_3 - k_{diss} \cdot [V_C R_3] - k_{int,V_C R_3} \cdot [V_C R_3] \qquad (6.15)$$

Solving the equation for the complex VEGF-C/VEGFR3 at equilibrium we deduce that:

$$+k_{ass} \cdot V_C \cdot R_3 - k_{diss} \cdot [V_C R_3] - k_{int,V_C R_3} \cdot [V_C R_3] = 0$$

$$\Leftrightarrow [V_C R_3]_{eq} = \frac{k_{ass} R_3}{k_{diss} + k_{int,V_C R_3}} \cdot V_C \qquad (6.16)$$

To determine the molar concentration of VEGFR-3, we note from [126] that the VEGFR-3 concentration was estimated to be between 0.2 and 40 ng/ml, using primary human dermal microvascular endothelial cells (HDMEC) and primary human umbilical vein endothelial cells (HUVEC). The cells were grown to 80 % confluence in 75 cm^2 tissue culture flasks $\approx$ 8.4 $\times$ 10^{6}cells. So by considering the 80 % of the aforementioned cell population ($\approx 6.72 \times 10^6$cells $\approx 2.24 \times 10^{-4}$ gr $\cdot$ cm^{-3}) and VEGFR-3 molecular weight to be 152,757 Da we get $R \approx 1.17 \times 10^{-9} \mathrm{mol} \cdot \mathrm{l}^{-1} \cdot \mathrm{cm}^3 \cdot \mathrm{gr}^{-1}$ and subsequently

$$[V_C R_3]_{eq} = \frac{k_{ass} R_3}{k_{diss} + k_{int,V_C R_3}} \cdot V_C \approx 1.346 \times 10^{-1} \cdot V_C\ \mathrm{cm}^3 \cdot \mathrm{gr}^{-1} \cdot \mathrm{M}^{-1}$$

By substituting the aforementioned value of $[V_C R_3]_{eq}$ in Eq. 6.14 we deduce:

$$\frac{d[V_C R_3]}{dt} = -\underbrace{\left(k_{ass} \cdot R_3 - k_{diss} \cdot 1.346 \times 10^{-1}\right)}_{7.7687 \times 10^{-5} \cdot \mathrm{s}^{-1} \cdot \mathrm{cm}^3 \cdot \mathrm{gr}^{-1}} \cdot V_C \qquad (6.17)$$

Then the dimensional and dimesnionless values of β_{41} are $\approx 7.7687 \times 10^{-5} \mathrm{s}^{-1} \cdot \mathrm{cm}^3 \cdot \mathrm{gr}^{-1}$ and $\approx 7.7678 \times 10^{-5}$ respectively.

In [74, 80, 97, 128] estimated that the proteolytic processing of VEGF-D and VEGF-C respectively increases their binding affinity to VEGFR-3 up to 400-fold. Therefore, we choose the non-dimensional β_{42} value to be $\beta_{42} = 400 \times \beta_{41} = 3.10748 \times 10^{-2}$.

Enholm et al. [129] have shown that the VEGF-C half life is 5.5–6 h in the presence of serum, while Ristimaki et al. [130] estimated the VEGF-C half-life to be 3.5 h or much greater than 4 h depending on the absence or the presence of the cytokine interleukin(IL)-1β. Therefore, by using the fact that half-life $t_{1/2}$ is given by $\frac{ln2}{\beta_{43}}$ [131] and estimating the VEGF-C half-life to be between 3.5 and 8 h, the VEGF-C decay rate is taken to be $\beta_{43} \approx \frac{ln2}{3.5 - 5\mathrm{h}} \approx 3.8 - 5.5 \times 10^{-5} \mathrm{s}^{-1}$ and the associate dimensionless value $\approx 3.8 \times 10^{-1} - 5.5 \times 10^{-1}$.

6.2.2.5 uPA Parameters

We estimate β_{53}, β_{63} by using experimental data on uPA-bound to the cell-surface receptor uPAR. In [132] they have shown that receptor-bound uPA remains exposed at the cell-surface with a half-life of $t_{1/2} = 4 - 5$ h, while [57] calculated that bound uPA does not dissociate rapidly from the cell surface ($t_{1/2} > 5$ h), nor is it appreciably endocytosed and/or degraded during that time. These results extend the observations of [133] and [134] who had reported that a large fraction of uPA is accessible to trypsin or acid elution 2–4 h after binding to the cells. Assuming the uPA half-life time to be between 2 and 6 h, we deduce the dimensionless β_{53}, $\beta_{63} \approx 0.48 - 0.96$.

Table 6.2 Values of the model variables at $t = 0$

Parameters	Symbol	Estimated range	Sources
Association constant	k_{ass}	$4 \pm 1 \times 10^6 \text{min}^{-1} \cdot \text{M}^{-1}$	[135]
		$5.92 \times 10^5 \text{M}^{-1}\text{s}^{-1}$	[136]
Dissociation constant	k_{diss}	$6.2 \pm 1.4 \times 10^{-3}\ \text{min}^{-1}$	[135]
		$6.2 \pm 1.8 \times 10^{-3}\ \text{s}^{-1}$	[136]
Internalization constant	k_{int}		[137]
Binding sites/cell	B_{max}	$2.2 \pm 0.1 \times 10^5$ sites/cell	[135]
		$2.7 \pm 0.55 \times 10^5$ sites/cell	[136]
		$1.9 \times 10^4 - 1.87 \times 10^5$	[138]

The rates β_{51}, β_{61}, are associated with uPA binding to the LEC and cancer cell surface receptors, such as uPAR.

We will base our estimations on the methodology employed for VEGF-C/VEGFR3 internalization procedure. Therefore, by setting the uPA as U, uPAR as R and the complex uPA/uPAR as [UR], we deduce that:

$$\Leftrightarrow [\text{UR}]_{eq} = \frac{k_{ass} \cdot \text{R}}{k_{diss} + k_{int,\text{UR}}} \cdot \text{U} \qquad (6.18)$$

We will estimate the uPA binding constants to uPAR LEC and cancer cells receptors by using the parameters given in Table 6.2. Regarding the internalization of the uPA/uPAR complex and since this is mediated by the binding of PAI-1 which we do not consider in the current model we will use the internalization constants given for the triplex uPA/uPAR/PAI-1 from [137, Figure1B]. Assuming the the uPAR (R) receptors to be between 10^3 and 10^6 receptors/cell (choosing $\text{R} = 10^3 - 10^5$receptors/cell), and the cell volume to be 500 μm^3 we deduce that $\text{R} = 3.32[\times 10^{-7} - \times 10^{-5}]$ M/cell and $[\text{UR}_{\text{LEC}}]_{eq} \approx 55.12 - 5512$. Substituting the aforementioned value for the uPA equation (in a similar manner as the VEGF-C equation) we have:

$$\frac{d\text{U}}{dt} = -\underbrace{\left(k_{ass} \cdot \text{R} - k_{diss}[\text{UR}_{\text{LEC}}]_{eq}\right)}_{2.067 \times [10^{-2} - 1]\text{s}^{-1} \cdot \text{cell}^{-1}} \cdot \text{U}$$

Further assuming that 1cell $\approx 1\ \text{gr} \cdot \text{cm}^{-3}$ [139] we deduce that the internalization per cell constant is $\beta_{51} = 2.067 \cdot \times [10^{-2} - 1]\ \text{s}^{-1} \cdot \text{gr} \cdot \text{cm}^{-3}$ while the non-dimensional value is $\beta_{51} = 2.067 \cdot \times [10^{-2} - 1]$.

In a similar manner we deduce the internalization constant for cancer cells by employing the parameters estimated by [136] presented also in Table 6.2 estimating the internalization constant from [137, Figure1B] ($k_{int} = \frac{ln2}{30 \cdot 60} \text{s}^{-1} = 0.38 \times 10^{-3} \text{s}^{-1}$). Thus, we get $[\text{UR}_{CC}]_{eq} = 2.345 \times [10^1 - 10^3] \cdot \text{U}$ and

$$\frac{d\text{U}_{CC}}{dt} = -\underbrace{\left(k_{ass} \cdot \text{R}_{CC} - k_{diss}[\text{UR}_{CC}]_{eq}\right)}_{0.885 \times [10^{-1} - 1]\text{s}^{-1} \cdot \text{cell}^{-1}} \cdot \text{U}_{CC}$$

so regarding the dimensional and non-dimensional internalization constant per cell ($cell \approx 1\text{gr} \cdot \text{cm}^{-3}$) we have $\beta_{61} = 0.885 \times [10^{-1} - 1]\text{cm}^3 \cdot \text{s}^{-1} \cdot \text{gr}^{-1}$ and $\beta_{61} = 0.885 \times [1 - 10]$ respectively.

The secretion of uPA by several explants of human malignant tumors was studied in short term organ culture by Markus and coworkers in [103, 140, 141]. We choose the maximum urokinase secretion rate estimated from [103, Table 1]for primary colon tumor to be 13.37 CTA units$\cdot \text{g}^{-1} \cdot \text{h}^{-1}$. The unit of urokinase activity, referred as the CTA unit and is based on the activity of working standard urokinase preparation which was independently assayed in several laboratories which releases 5×10^{-4}micromoles CTA^{-1} [104]. Thus, $a_{51} \sim 13.37$ CTA units $\cdot\ \text{g}^{-1} \cdot \text{h}^{-1} \approx 13.37 \times 5 \times 10^{-4}$ micromoles$\cdot\ \text{g}^{-1} \cdot \text{h}^{-1}$. Taking into account that the uPA molecular weight is ($\approx$54,000 Da) [104] then we get $a_{61} \approx 0.1 \times 10^{-3}\ \text{gr} \cdot \text{cm}^{-3} \cdot \text{s}^{-1}$. Considering that the cultured tissue was 18mg $\approx 10^6$cells. Therefore, we conclude that $\alpha_{61} \approx 0.1 \times 10^{-12} \cdot \text{s}^{-1} \cdot \text{cells}^{-1}$ and the non-dimensional value to be $\alpha_{61} \approx 0.1 \times 10^{-3}$. In a similar manner and taking into account that the secretion rate for normal colon was estimated in [103, Table 1] to be 1.21 CTA units$\cdot\ \text{g}^{-1} \cdot \text{h}^{-1}$ we deduce that the LEC urokinase secretion rate is $a_{51} \approx 0.9075 \times 10^{-14}\ \text{s}^{-1} \cdot \text{cells}^{-1}$ and dimensionless value of it $a_{51} \approx 0.9075 \times 10^{-6}$. It needs to be considered that the most important observation made in the studies of Markus and coworkers was that the secretion rate of metastatic tumors were significantly lower than those of the primary ones. Furthermore, during the 150 h culturing the tumor produced 172 times more enzyme than it contained in zero time.

Regarding the parameters α_{52}, α_{53} we will estimate them through the binding constant of VEGF-C to LEC receptor VEGFR-3. Thus, from Table 6.1 without loss of generality we will use the binding constant $0.35 \times 10^4 \mathrm{M}^{-1} \cdot \mathrm{s}^{-1}$ and we deduce that $\alpha_{52} = 0.04375 \times 10^3 \mathrm{cm}^3 \cdot \mathrm{gr}^{-1} \cdot \mathrm{s}^{-1}$ and the non-dimensional value of it 437.5. As it has already been stated the proteolyticaly processed mature VEGF-C binds with higher affinity (50 − 400-fold) to associated LEC VEGFR-3 receptor therefore we take $\alpha_{52} = 0.021875 \times 10^4 \mathrm{cm}^3 \cdot \mathrm{gr}^{-1} \cdot \mathrm{s}^{-1}$ and thus $\alpha_{52} = 21{,}875$. However, taking into account the abundance presence of PAI-1 (although we did not directly consider it here) in the peritumoral space we expect the aforementioned values to be substantially reduced in other words we choose $\alpha_{51} =\approx 437.5 \times [10^{-5} - 10^{-3}]$ and $\alpha_{52} =\approx 21{,}875 \times [10^{-6} - 10^{-3}]$ respectively.

6.2.2.6 Degradation of the ECM

An extensive overview of the role of matrix degrading enzymes such as plasmin in the degradation of pericellular matrix by neoplastic as well normal cell can be found in the work of [142,143]

In [144, Table 1] the effects of Plasminogen treatment on the degradation of several components of the ECM by macrophages is presented; they are summarized here in Table 6.3.

Recalling part of the degradation components of the ECM equation we have:

$$\frac{dm}{dt} = \delta_{31} p_e m \tag{6.19}$$

Table 6.3 Rates of degradation of glycoproteins by macrophages (5×10^5 cells) for 48 h.s

Matrix treatment	Plasminogen added $\mu g \cdot ml$	Matrix component digested %
Control	0	7.6 ± 5.0
Plg	10	60.6 ± 11.8

then by integrating by parts we deduce $m(t) = c\, e^{\delta_{31} p t}$, where c is a constant that can be estimated by putting $t = 0$ and $p(0) = 0$ in the last equation. For $t = 48$ h, we have $m(48) = c e^{\delta_{32} p_c 48} \Leftrightarrow \delta_{31} \approx 0.2465\ \mathrm{cm}^3 \cdot \mathrm{gr}^{-1} \cdot \mathrm{s}^{-1}$ and the nondimensional value to be $\delta_{31} = 0.2465 \times 10^{-4}$. In [143, Figure 1, Figure3] they estimated the degradation of extracellular matrix by neoplastic urothelial cells as well as normal urothelial cells when $4\,\mu\mathrm{gr} \cdot \mathrm{ml}^{-1}$ of plasminogen was added. Their results showed that normal urothelial cells degraded all of the matrix constituents including elastin to a greater extent than the neoplastic cells. Similarly with the approach presented above we will estimate the ECM degradation rate induced by LEC from [143, Figure 6]. Thus, from [143, Figure 6] we get:

$$\delta_{32} \approx 0.0395\ \mathrm{cm}^3 \cdot \mathrm{gr}^{-1} \cdot \mathrm{s}^{-1}$$

and the dimensionless value to be $\delta_{32} \approx 0.0395 \times 10^{-4}$

Results associated with the degradation of the extracellular matrix could be also be obtained from experimental results presented by Pins et al. [145] where they calculated that within 16h of fibroblast populated collagen lattices treated with plasmin (100 nM) contracted from approximately 20 mm to less than 2 mm resulting in other words on a loss of 90 % of the underlying matrix. This gives a degradation range of $\delta \approx 79.95\ \mathrm{cm}^3 \cdot \mathrm{gr}^{-1} \cdot \mathrm{s}^{-1}$ and the dimensionless one to be $\approx 79.95 \times 10^{-4} \mathrm{cm}^3 \cdot \mathrm{gr}^{-1} \cdot \mathrm{s}^{-1} \cdot \mathrm{cell}^{-1}$ and the dimensionless one to be $\approx 79.95 \times 10^{-4}$.

However due to the fact that we deal we aggressive processes such as tumor invasion and tumor lymphangiogenesis we anticipate that degradation rates for both LEC and cancer cell plsmin proteolyis to be much bigger then the ones obtained before.

Regarding the production/remodeling of ECM components we used the experimental data provided in [58, 146] and more specifically in [146, Figure 3]. The work by [146] revealed that extracellular material began to appear 6–7 days after cell seeding and rapidly increased in quantity. In their experiments 3 mg of matrix proteins formed per 35 mm culture dish and maintained for up to 10 weeks. From [146, Figure 3] we deduce that ECM proteins doubling time (from 200 to 400 μm) is 4 days so $\mu_{31} = 2 \times 10^{-6}\text{s}^{-1}$ and the non-dimensional value to be $\mu_{31} = 2 \times 10^{-2}$.

6.2.2.7 Plasmin Parameters

For α_{71} and α_{81} respectively, it seems reasonable to take the rate at which uPA binds to LEC and cancer cell surface receptors. Now, from Table 6.2 we recall that the uPA/uPAR association constant is $0.083 \times 10^6 \text{M}^{-1} \cdot \text{s}^{-1}$ which gives the associated plasmin secretion/activation rate to be $\alpha_{71} = 0.1085 \times 10^3\, \text{cm}^3 \cdot \text{gr}^{-1} \cdot \text{s}^{-1}$ and its non-dimensional value is $\alpha_{71} = 0.1085 \times 10^4$. Similarly using from Table 6.2 the binding affinity constant of $5.92 \times 10^5 \text{M}^{-1} \cdot \text{s}^{-1}$ we deduce $\alpha_{81} = 10.96 \times 10^3\, \text{cm}^3 \cdot \text{gr}^{-1} \cdot \text{s}^{-1}$ and the dimensionless value of it to be $\alpha_{81} = 10.96 \times 10^4$. Regarding the plasmin decay rate in [147] they have estimated plasminogen half-life 2.24 ± 0.29 days, while digested forms of plasminogen have a half-life of 0.8 days. Taking into consideration the aforementioned values we have $\beta_{71}, \beta_{81} \approx \frac{ln2}{2.53-0.8}\, \text{days}^{-1}$ and we get $\beta_{71}, \beta_{81} = [3.17 - 10] \times 10^{-6}\text{s}^{-1}$ and the non-dimensional values to be $\beta_{71}, \beta_{81} = [3.17 - 10] \times 10^{-2}$. However, due to the presence of anti-plasmin and its extremely fast regulation we anticipate that the values of β_{71} and β_{81} could increase and take values between 3.17 and 10.

The range of all parameters in both dimensional and dimensionless form, is summarized in Table 6.4. It is the later values that are used in the simulations.

6.3 Numerical Results for the PDE Model

To compute the numerical solutions in one space dimension we use the Matlab command pdepe, which uses the method in [148] for the spatial discretization of the model equations, thereby reducing them to a system of (time-dependent) ordinary differential equations which are readily integrated (using the Matlab routine of ode15s). Typically we take 2001 spatial points in $[0, 1]$ while the accuracy in the time integration is 1.0×10^{-5}.

In the simulations of Figs. 6.4, 6.5, and 6.6, the parameter values used were as follows: $D_e = 9.75 \times 10^{-5}$, $D_c = 9.75 \times 10^{-5}$, $D_f = 5 \times 10^{-3}$, $D_{u_e} = 2.75 \times 10^{-3}$, $D_{u_c} = 2.75 \times 10^{-3}$, $D_{p_e} = 1.5 \times 10^{-3}$, $D_{p_c} = 1.5 \times 10^{-3}$, $\chi_e = 3.25 \times 10^{-2}$, $\xi_e = 340.0 \times 10^{-5}$, $\zeta_e = 4.815 \times 10^{-2}$, $\psi_e = 3.410^{-2}$, $\xi_c = 340.010^{-5}$, $\zeta_c = 4.815 \times 10^{-2}$, $\psi_c = 3.410^{-2}$, $\delta_{31} = 246.5 \times 10^{-1}$, $\delta_{32} = 39.5 \times 10^{-1}$, $\alpha_{41} = 9.645 \times 10^{-1}$, $\alpha_{42} = 1.34 \times 10^{-2}$, $\alpha_{43} = 1.732 \times 10^{-2}$, $\beta_{41} = 7.7678 \times 10^{-5}$, $\beta_{42} = 3.10748 \times 10^{-2}$, $\beta_{43} = 0.55$, $\alpha_{51} = 0.9075 \times 10^{-1}$, $\alpha_{52} = 43.75 \times 10^{-2}$, $\alpha_{53} = 21.875 \times 10^{-2}$, $\beta_{51} = 2.067 \times 10^{-1}$, $\beta_{52} = 1.05 \times 10^{-1}$, $\beta_{53} = 9.62 \times 10^{-1}$, $\alpha_{61} = 0.1 \times 10^{-1}$, $\beta_{61} = 8.85 \times 10^{-1}$, $\beta_{62} = 1.5 \times 10^{-2}$, $\beta_{63} = 4.8 \times 10^{-1}$, $\alpha_{71} = 10.85 \times 10^{-1}$, $\beta_{71} = 69.0 \times 10^{-3}$, $\alpha_{81} = 10.96 \times 10^{-1}$, $\beta_{81} = 69.0 \times 10^{-3}$, $\mu_{11} = 8.022 \times 10^{-2}$, $\mu_{21} = 1.07 \times 10^{-1}$, $\mu_{22} = 10.6 \times 10^{-1}$, $\mu_{23} = 10.6 \times 10^{-2}$, $\mu_{31} = 20.00 \times 10^{-3}$

In Fig. 6.4, at $t = 5$ (∼14 h), we note that large clusters of both the lymphatic endothelial and the cancer cells have migrated a short distance into the extracellular space; cancer cells secrete uPA and VEGF-C while they activate plasmin in order to degrade the peritumoral space. By $t = 15$ (∼1.7 days), several clusters of lymphatic endothelial cells have already been formed at the leading edge and rear edge of the lymphatic network migrating chemotactically mainly to cancer cells secreted VEGF-C. Between $t = 10$ (≈ 1day) and $t = 15$ (≈ 1.7 days) tumor cluster cells have been formed merging together in order to increase their metastatic probabilities under the immuno-lymphatic attack.

Our results are in line with experimental evidence of tumor invasion where cancer cells migrated large distance ranging between 0.1 and 0.5 mm. More specifically in [149] they investigated tumor migration where tumor cells formed tracks-depressions or channels-in the matrix substratum. Depending on the initial cell density, the tracks eventually merged into complex interconnected meshworks in the gel. The channels were linear but random in direction and approximately as wide as the cell diameter. It

Table 6.4 Table of parameters for differential equations. Parameters estimated from this work are presented as "tw" (this work). In addition, when a reference is given in parentheses with tw, these parameters have been estimated from information given in the corresponding citation

Dimensional parameter	Dimensional range	Dimensionless value	Source
L	0.1cm	1	tw
D_0	$10^{-6}\ cm^2 \cdot s^{-1}$	1	[90]
c_0	$1 \times 10^{-3} g \cdot cm^{-3}$	1	[94]
e_0	$1 \times 10^{-4} g \cdot cm^{-3}$	1	[94]
m_0	$1 \times 10^{-3} g \cdot cm^{-3}$	1	[95, 96]
f_0	$1 \times 10^{-7} g \cdot cm^{-3}$	1	[85, 87, 97–99]
u_{c0}, u_{e0}	$1 \times 10^{-8} g \cdot cm^{-3}$	1	[57, 100–103]
p_{c0}, p_{e0}	$1 \times 10^{-8} g \cdot cm^{-3}$	1	[95],tw
D_c, D_e	5.9×10^{-11} - $9.8 \times 10^{-9} cm^{-2} \cdot s^{-1}$	$10^{-5} - 10^{-3}$	[90, 107, 108]
D_f	$10^{-9} - 10^{-6} cm^{-2} \cdot s^{-1}$		[110, 114]
D_{u_c}, D_{u_e}	$10^{-9} - 10^{-6} cm^{-2} \cdot s^{-1}$	$10^{-3} - 1$	[54, 56, 89, 107, 112, 113]
D_{p_c}, D_{p_e}	$10^{-9} - 4.91 \times 10^{-7} cm^{-2} \cdot s^{-1}$	$1 \times 10^{-3} - 4.91 \times 10^{-1}$	[54, 56, 89, 109, 110]
χ_e	$26 cm^{-2} \cdot s^{-1} \cdot M^{-1}$	3.25×10^{-2}	[108],tw
ζ_c, ζ_e	$26 cm^{-2} \cdot s^{-1} \cdot M^{-1}$	4.815×10^{-3}	[108],tw
ψ_c, ψ_e	$26 cm^{-2} \cdot s^{-1} \cdot M^{-1}$	3.4×10^{-3}	[108],tw
ξ_c, ξ_e	$26 cm^{-2} \cdot s^{-1} \cdot M^{-1}$	3.46×10^{-1}	[108],tw
μ_{11}	$2.24 \times 10^{-6} - 8.022 \times 10^{-6}\ s^{-1}$	$2.24 \times 10^{-2} - 8.022 \times 10^{-2}$	[58, 91, 115]
μ_{21}	$5.55 \times 10^{-6} - 1.07 \times 10^{-5}\ s^{-1}$	$5.55 \times 10^{-2} - 1.07 \times 10^{-1}\ s^{-1}$	[91, 107, 108]
μ_{22}	$1.06 \times 10^3 cm^3 \cdot gr^{-1} \cdot s^{-1}$	$\mu_{22} = 1.06$	[120]
μ_{31}	$2.0 \times 10^{-6}\ s^{-1}$	2.0×10^{-2}	[58, 146]
δ_{31}	$0.2465 \times 10^{-4} cm^3 \cdot gr^{-1} \cdot s^{-1}$	0.2465×10^{-4}	[144]
δ_{32}	$0.2465 \times 10^{-4} cm^3 \cdot gr^{-1} \cdot s^{-1}$	0.2465×10^{-4}	[144]
α_{41}	$9.645 \times 10^{-10} s^{-1}$	$a_{41} = 9.645 \times 10^{-2}$	[121]
α_{42}	$a_{42} = 1.732 \times 10^2 cm^3 \cdot gr^{-1} \cdot s^{-1}$	$a_{42} = 1.732 \times 10^{-2}$	[120]
α_{43}	$1.34 \times 10^2 cm^3 \cdot gr^{-1} \cdot s^{-1}$	$a_{43} = 1.34 \times 10^{-2}$	[120]
β_{41}	$7.7687 \times 10^{-5} cm^3 \cdot gr^{-1} \cdot s^{-1}$	7.7687×10^{-5}	[74, 125, 127]
β_{42}	$[50 - 400] \times \beta_{41}$	$3.88435 \times 10^{-3} - 3.10748 \times 10^{-2}$	[30, 74, 97, 128]
β_{43}	$3.8 - 5.5 \times 10^{-5}\ s^{-1}$	$3.8 - 5.5 \times 10^{-1}$	[129, 130]
α_{51}	$0.9075 \times 10^{-14} s^{-1} \cdot cells^{-1}$	0.9075×10^{-6}	[103]
α_{52}	$0.04375 \times 10^3\ cm^3 \cdot gr^{-1} \cdot s^{-1}$	$437.5 \times [10^{-5} - 10^{-3}$	[103]
α_{53}	$0.021875 \times 10^4\ cm^3 \cdot gr^{-1} \cdot s^{-1}$	$21{,}875 \times [10^{-6} - 10^{-3}$	[103]
β_{51}	$2.067 \times [10^{-2} - 1]\ cm^3 \cdot gr^{-1} \cdot s^{-1}$	$2.067 \times [10^{-2} - 1]$	tw
β_{52}, β_{62}		1.05×10^{-1}	tw
β_{53}, β_{63}	$[4.8 - 9.62] \times 10^{-5}\ s^{-1}$	$[4.8 - 9.62] \times 10^{-1}$	[57, 132–134]
α_{61}	$0.1 \times 10^{-12} s^{-1} \cdot cells^{-1}$	0.1×10^{-3}	[103]
β_{61}	$0.885 \times [10^{-1} - 1]\ cm^3 \cdot gr^{-1} \cdot s^{-1}$	$0.885 \times [1 - 10]$	[136–138],tw
α_{71}	$0.1085 \times 10^3\ cm^3 \cdot gr^{-1} \cdot s^{-1}$	0.1085×10^4	[74, 125, 127]
α_{81}	$10.96 \times 10^3\ cm^3 \cdot gr^{-1} \cdot s^{-1}$	10.96×10^4	[74, 125, 127]
β_{71}, β_{81}	$[3.17 - 10] \times 10^{-6} s^{-1} \cdot cells^{-1}$	$[3.17 - 10] \times 10^{-2}$	[147]

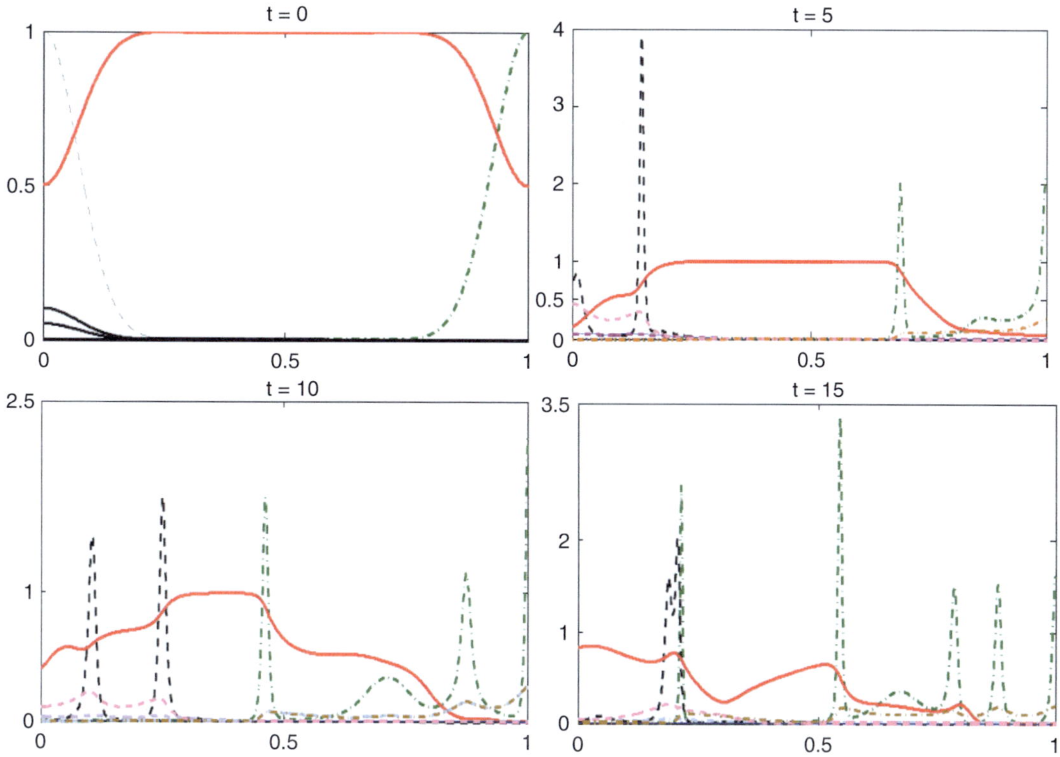

Fig. 6.4 Sequence of profiles showing the evolution of the tumour cell density $c(x,t)$ (*solid dashed black line*), the lymphatic endothelial cell density $e(x,t)$ (*solid dotted-dashed green line*), the ECM density $m(x,t)$ (*solid red line*), the VEGF-C concentration $f(x,t)$ (*dashed black line*), the cancer cells secreted uPA concentration $u_c(x,t)$ (*dot-dashed black line*),the cancer cells activated plasmin $p_c(x,t)$ (*dotted black line*), the LEC secreted uPA concentration $u_e(x,t)$ (*dot-dashed red line*), and LEC activated plasmin concentration $p_e(x,t)$ (*dotted red line*)

was common to observe a single tumor cell migrate in one direction, creating a channel, then suddenly reverse direction and move back along the original migration track. Frequently, solitary cells were seen to diverge from the established paths and form new channels. After 2–4 days the network of channels covered the entire surface of the matrix substratum. During this period the tumor cells proliferated, and the daughter cells apparently remained in the migratory tracks of their progenitors [149]. Within several hours after seeding (HT1080 fibrosarcoma cells) the tumor cells initiated a random migration, leaving behind channels etched in the surface of the matrix. Eventually the channels became interconnected into a complex network.

Similar interconnected cell migrated channels we also observe in Fig. 6.5 where at $t = 45$ ($\approx$ 5.2 days) clusters of LE cells of the lymphatic network continue to form while three different clusters of cancer cells start migrating in and around the lymphatic network. By $t = 85$ ($\approx$ 10 days), large clumps of cancer cells have been formed moving through the evolved lymphatic network increasing their metastatic potential under the immunological threat .

As time evolves, at $t = 160$ ($\approx$18.5 days) in Fig. 6.6, several clusters of lymphatic endothelial and cancer cells have formed. The spatio-temporal heterogeneity of cancer cells and LECs is still evident by $t = 250$ ($\approx$ 29 days) as a result of the dynamic activity of the interactions between the lymphatic growth factor, the proteases, the extracellular matrix, the cancer, and the lymphatic endothelial cells [150]. The simulations plotted in Fig. 6.6 reveal how the proteolytically

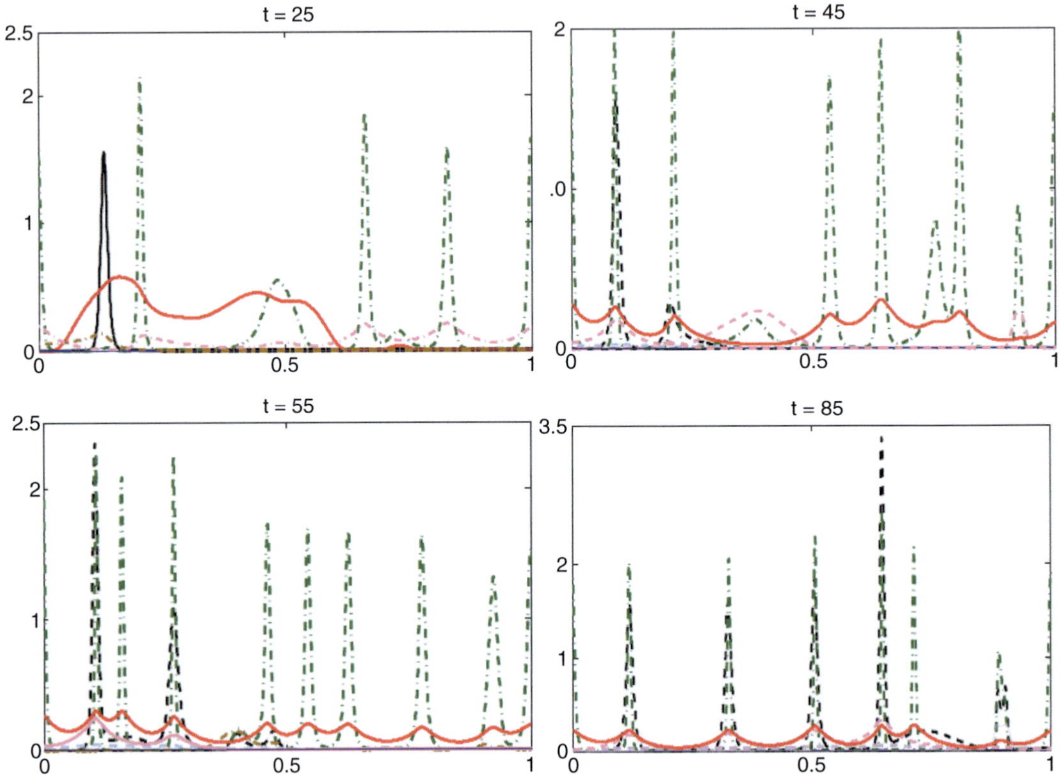

Fig. 6.5 Sequence of profiles showing the evolution of the tumour cell density $c(x, t)$ (*dashed black line*), the lymphatic endothelial cell density $e(x, t)$ (*dashed-dotted green line*), the ECM density $m(x, t)$ (*solid red line*), the VEGF-C concentration $f(x, t)$ (*dashed black line*), the cancer cells secreted uPA concentration $u_c(x, t)$ (*dot-dashed black line*),the cancer cells activated plasmin $p_c(x, t)$ (*dotted black line*), the LEC secreted uPA concentration $u_e(x, t)$ (*dot-dashed red line*), and LEC activated plasmin concentration $p_e(x, t)$ (*dotted red line*)

driven tumor-lymphangiogenesis resulting in spatio-temporal ("anarchic") patterning.

Similar observations have been addressed experimentally where the dissemination of coherent masses of tumor satellites is histologically observed in tumors of epithelial origin (e.g. invasive oral carcinoma and adenocarcinoma) or melanoma [151–153]. The leading edge of such clusters is composed of highly motile cells (pathfinder cells) creating driving force and guide the migration. A similar coordinated cluster migration has also been observed in an in vitro model of Xiphophorus (*platyfish*) melanoma [154]. Liotta et al. [155] have indicated that tumor cell clumps produce a significantly greater number of metastatic foci than does a similar number of single tumor cells. In this regard, it is speculated that clustered tumorigenic migration could also explain the spreading of heterogeneous sets of tumor cells, in order to minimize cell loss, favoring high local proteolysis [156] and thus protecting cells from immunological assault [152]. Thus, locomoting cell clusters could be a novel and potentially important mechanism of cancer cell invasion and metastasis. It is reasonable to speculate that the dissemination of cell clusters may represent an efficient mechanism of cancer spread in the tissue, relative resistance to immunological assault, and cancer cell survival in the lymphatic circulation [151].

Furthermore, our model can be used to simulate the effects of anti-lymphangiogenic drugs on the progress of tumor lymphangiogenesis and tumor invasion into the lymphatic system.

Several molecules which have been shown to be effective in inhibiting tumor lymphangio-

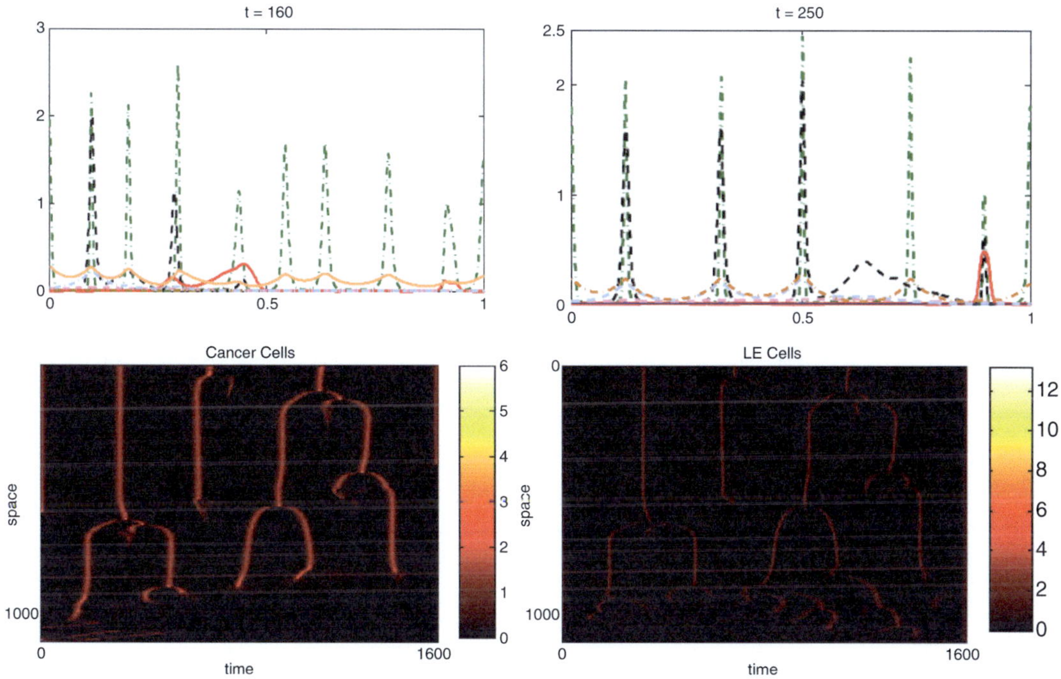

Fig. 6.6 Sequence of profiles showing the evolution of the tumour cell density $c(x,t)$ (*dashed black line*), the lymphatic endothelial cell density $e(x,t)$ (*dashed-dotted green line*), the ECM density $m(x,t)$ (*solid red line*), the VEGF C concentration $f(x,t)$ (*dashed black line*), the cancer cells secreted uPA concentration $u_c(x,t)$ (*dot-dashed black line*),the cancer cells activated plasmin $p_c(x,t)$ (*dotted black line*), the LEC secreted uPA concentration $u_e(x,t)$ (*dot-dashed red line*), and LEC activated plasmin concentration $p_e(x,t)$ (*dotted red line*). An example of spatio-temporal proteolytically driven patterning. A number of cellular aggregations evolving in an arrhythmic sequence of emerging and merging events. CC and LEC plotted as a function of time (*horizontal axis*) and space (*vertical axis*). The colorscale bar indicates increasing CC and LEC cell density from 0 (*black*) to 6 (*white*)

genesis and lymph node metastasis include a soluble VEGFR-3-IgG fusion protein and neutralizing anti-VEGF-C/D antibodies. These tools provide a glimpse at what could potentially be a novel therapeutic opportunity for the prevention of tumor cell dissemination and the formation of metastases. However, it will be important to clearly define an appropriate therapeutic window in cancer patients, since it is likely that tumor cell dissemination will already have occurred in many patients at the time of initial presentation. Therapy could then be aimed at preventing further dissemination from existing lymph node metastases (the bridgehead theory [35]). In this regard, maintaining metastases in a state of dormancy would appear to be an additional therapeutic option.However in the current model we will focus on the proteolytically processed VEGF-C as our major aim of anti-lympangiogenic strategies. We shall demonstrate the following example.

It has been suggested that the secretion of VEGF-C by the cancer cells, as well as its proteolytic cleavage by cancer cells activated plamsin is a special candidate for anti-lymphangiogenic strategies. In order to reduce the chemotactic effect of VEGF-C at the parental lymphatic vessel, we may inject a soluble form of the lymphatic endothelial cell receptor-3 (VEGFR-3) and anti-plasmin into the peritumoral space. We model this by setting $\alpha_{42} = 1.34 \times 10^{-5}$ and $\alpha_{43} = 1.732 \times 10^{-5}$.

The forced decay of proteolytically processed VEGF-C would also result in a decrease in LEC driven uPA activation. We model this by taking $\alpha_{52} = 43.75 \times 10^{-4}$, and $\alpha_{53} = 21.875 \times 10^{-4}$. The, proreolytically processed VEGF-C forced

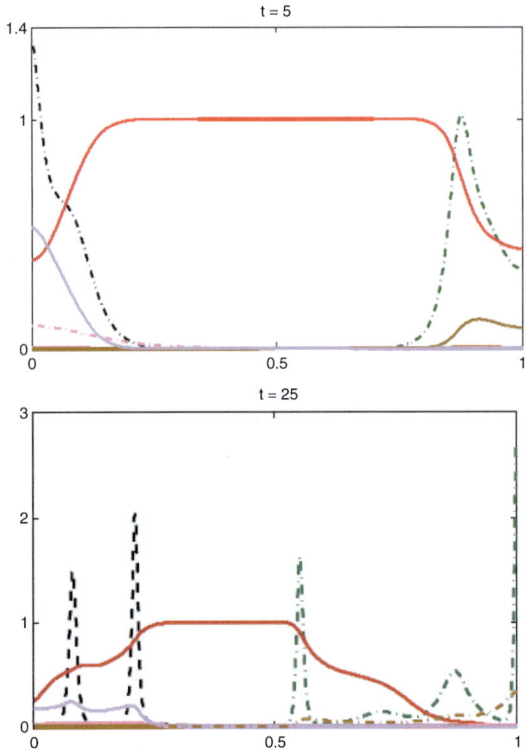

Fig. 6.7 Sequence of profiles showing the evolution of the tumour cell density $c(x,t)$ (*dashed black line*), the lymphatic endothelial cell density $e(x,t)$ (*dashed-dotted green line*), the ECM density $m(x,t)$ (*solid red line*), the VEGF-C concentration $f(x,t)$ (*dashed black line*), the cancer cells secreted uPA concentration $u_c(x,t)$ (*dot-dashed black line*),the cancer cells activated plasmin $p_c(x,t)$ (*dotted black line*), the LEC secreted uPA concentration $u_e(x,t)$ (*dot-dashed red line*), and LEC activated plasmin concentration $p_e(x,t)$ (*dotted red line*)

reduction through the anti-plasmin use will also affect the effect that proteolytically and unproteolytically processed VEGF-C have in lymphatic endothelial cells proliferation, and we therefore chose $\mu_{22} = 10.6 \times 10^{-2}$, $\mu_{23} = 20 \times 10^{-3}$.

Figure 6.7, shows the result of the above changes in the model parameters caused by injecting soluble forms of VEGFR-3 and anti-plasmin. Comparing Fig. 6.7 with Figs. 6.4 and 6.5 we see, for example, that in Fig. 6.4, at $t = 15(\sim 4$ days) the lymphatic endothelial cells have already migrated through the whole domain and reached the initial cluster of cancer cells, while in Fig. 6.7 they have migrated appreciably less. As time evolves at $t = 45(\sim 10$ days) in Fig. 6.5 the profiles of LEC is highly heterogeneous, whereas in Fig. 6.7 the LEC have not reached the initial tumor yet while also part of the ECM remained intact.

6.4 Conclusions

The lymphatic system acts to drain and return interstitial fluids to the main circulatory system through a complex network of vessels. Recent experimental evidence established connection between tumor and lymphangiogenesis in its vicinity. A very importnant role in our proposed model has been played by the tumor-activated plasmin. Tumor-activated plasmin not only proteolytically cleaves tumor-secreted VEGF-C and therefore produces its mature form which binds with high affinity to the lymphatic specific endothelial cell surface receptor VEGFR-3, but it can also, via its role as an extracellular matrix degrading protease release matrix-bound inactive VEGF-C.

In the proposed continuum model, we have worked on the basis that lymphangiogenesis is occuring in and around tumors. We do however recognize that many tumors utilize preexisting lymphatics as a means to disseminate tumor cells. In this case, VEGF-C may also be important in that it may alter the function of preexisting endothelial cells (for example by altering endothelial to tumor cell adhesion or by producing chemokines), as may the role of uPA and plasmin in activating VEGF-C.

Tumors are heterogeneous from the fact that they consist of subpopulations of neoplastic cells with different metastatic, migrating, and invading potential to the newly formed lymphatic vessels. Our simulation captures this tumor heterogeneity: Mimicking the lymphatic endothelial heterogeneity, we observe proliferative cancer cells near the lymphatic endothelial cells already within 20 days. Thus although lymphangiogenesis is essentially a two- (in vitro) or three-dimensional (in vivo) process, our one-dimensional results already successfully uncover the complex dynamics following tumor lymphangiogenesis.

Our observations support the concept that malignant tissues can develop patterns similar to those that have been observed in embryological morphogenesis. It is reasonable to speculate that tumor dissemination cell migration, lymphangiogenesis, invasion and pericellular proteolysis although can be dissociated and may continue independently, these processes provide concurrent, spatially controlled and synchronized contribution to the overall invasion and cancer dissemination process. The dynamic coupling of the aforementioned processes is indispensable for successful lymphatic invasion and metastasis.

References

1. Friedman A, Lolas G (2005) Analysis of a mathematical model of tumor lymphangiogenesis. Math Models Methods Appl Sci 15(01):95–107
2. Pepper MS, Lolas G (2008) Selected topics in cancer modeling: genesis, evolution, immune competition, and therapy. In: The lymphatic vascular system in lymphangiogenesis invasion and metastasis a mathematical approach. Birkhäuser Boston, Boston, pp 1–22
3. Alberts B, Johnson A, Lewis J, Morgan D, Raff M, Roberts K, Walter P (2014) Molecular biology of the cell. Garland, New York
4. Hanahan D, Weinberg RA (2000) The hallmarks of cancer. Cell 100(1):57–70
5. Fidler IJ (2003) The pathogenesis of cancer metastasis: the 'seed and soil' hypothesis revisited. Nat Rev Cancer 3(6):453–458
6. Hanahan D, Weinberg RA (2011) Hallmarks of cancer: the next generation. Cell 144(5):646–674
7. Pepper MS (2001) Lymphangiogenesis and tumor metastasis: myth or reality? Clin Cancer Res 7(3):462–468
8. Folkman J (1971) Tumor angiogenesis: therapeutic implications. N Engl J Med 285(21): 1182–1186
9. Carmeliet P, Jain RK (2000) Angiogenesis in cancer and other diseases. Nature 407(6801):249–257
10. Plate K (2001) From angiogenesis to lymphangiogenesis. Nat Med 7(2):151–152
11. Alitalo K, Carmeliet P (2002) Molecular mechanisms of lymphangiogenesis in health and disease. Cancer Cell 1(3):219–227
12. Brown P (2005) Lymphatic system: unlocking the drains. Nature 436(7050):456–458
13. Holopainen T, Bry M, Alitalo K, Saaristo A (2011) Perspectives on lymphangiogenesis and angiogenesis in cancer. J Surg Oncol 103(6):484–488
14. Duong T, Koopman P, Francois M (2012) Tumor lymphangiogenesis as a potential therapeutic target. J Oncol 2012:204946
15. Stacker SA, Williams SP, Karnezis T, Shayan R, Fox SB, Achen MG (2014) Lymphangiogenesis and lymphatic vessel remodelling in cancer. Nat Rev Cancer 14(3):159–172
16. Hartveit E (1990) Attenuated cells in breast stroma: the missing lymphatic system of the breast. Histopathology 16(6):533–543
17. Karkkainen MJ, Makinen T, Alitalo K (2002) Lymphatic endothelium: a new frontier of metastasis research. Nat Cell Biol 4(1):2–5
18. Pepper MS, Tille JC, Nisato R, Skobe M (2003) Lymphangiogenesis and tumor metastasis. Cell Tissue Res 314(1):167–177
19. Nisato RE, Tille JC, Pepper MS (2003) Lymphangiogenesis and tumor metastasis. Thromb Haemost 90(4):591–597
20. Jussila L, Alitalo K (2002) Vascular growth factors and lymphangiogenesis. Physiol Rev 82(3):673–700
21. Oliver G, Detmar M (2002) The rediscovery of the lymphatic system: old and new insights into the development and biological function of the lymphatic vasculature. Genes Dev 16(7):773–783
22. Oliver G (2004) Lymphatic vasculature development. Nat Rev Immunol 4(1):35–45
23. Schulte-Merker S, Sabine A, Petrova TV (2011) Lymphatic vascular morphogenesis in development, physiology, and disease. J Cell Biol 193(4):607–618
24. Reis-Filho JS, Schmitt FC (2003) Lymphangiogenesis in tumors: what do we know? Microsc Res Tech 60(2):171–180
25. Shayan R, Achen MG, Stacker SA (2006) Lymphatic vessels in cancer metastasis: bridging the gaps. Carcinogenesis 27(9):1729–1738

26. Cao Y (2008) Why and how do tumors stimulate lymphangiogenesis? Lymphat Res Biol 6(3–4):145–148
27. Tammela T, Alitalo K (2010) Lymphangiogenesis: molecular mechanisms and future promise. Cell 140(4):460–476
28. Kaipainen A, Korhonen J, Mustonen T, van Hinsbergh VW, Fang GH, Dumont D, Breitman M, Alitalo K (1995) Expression of the fms-like tyrosine kinase 4 gene becomes restricted to lymphatic endothelium during development. Proc Natl Acad Sci USA 92(8):3566–3570
29. Karkkainen MJ, Haiko P, Sainio K, Partanen J, Taipale J, Petrova TV, Jeltsch M, Jackson DG, Talikka M, Rauvala H, Betsholtz C, Alitalo K (2004) Vascular endothelial growth factor C is required for sprouting of the first lymphatic vessels from embryonic veins. Nat Immunol 5(1):74–80
30. Joukov V, Pajusola K, Kaipainen A, Chilov D, Lahtinen I, Kukk E, Saksela O, Kalkkinen N, Alitalo K (1996) A novel vascular endothelial growth factor, VEGF-C, is a ligand for the Flt4 (VEGFR-3) and KDR (VEGFR-2) receptor tyrosine kinases. EMBO J, 15(2):290–298
31. Lymboussaki A, Partanen TA, Olofsson B, Thomas-Crusells J, Fletcher CD, de Waal RM, Kaipainen A, Alitalo K (1998) Expression of the vascular endothelial growth factor C receptor VEGFR-3 in lymphatic endothelium of the skin and in vascular tumors. Am J Pathol 153(2):395–403
32. Partanen TA, Paavonen K (2001) Lymphatic versus blood vascular endothelial growth factors and receptors in humans. Microsc Res Tech 55(2):108–121
33. Su JL, Yen CJ, Chen PS, Chuang SE, Hong CC, Kuo IH, Chen HY, Hung MC, Kuo ML (2007) The role of the VEGF-C/VEGFR-3 axis in cancer progression. Br J Cancer 96(4):541–545
34. Jackson DG, Prevo R, Clasper S, Banerji S (2001) LYVE-1, the lymphatic system and tumor lymphangiogenesis. Trends Immunol 22(6):317–321
35. Sleeman JP, Krishnan J, Kirkin V, and Baumann P (2001) Markers for the lymphatic endothelium: in search of the holy grail? Microsc Res Tech 55(2):61–69
36. Wigle JT, Harvey N, Detmar M, Lagutina I, Grosveld G, Gunn MD, Jackson DG, Oliver G (2002) An essential role for Prox1 in the induction of the lymphatic endothelial cell phenotype. EMBO J 21(7):1505–1513
37. Hong YK, Detmar M (2003) Prox1, master regulator of the lymphatic vasculature phenotype. Cell Tissue Res 314(1):85–92
38. Adams RH, Alitalo K (2007) Molecular regulation of angiogenesis and lymphangiogenesis. Nat Rev Mol Cell Biol 8(6):464–478
39. Ruoslahti E (1996) How cancer spreads. Sci Am 275(3):72–77
40. Chang L, Kaipainen A, Folkman J (2002) Lymphangiogenesis new mechanisms. Ann N Y Acad Sci 979:111–119
41. Stacker SA, Caesar C, Baldwin ME, Thornton GE, Williams RA, Prevo R, Jackson DG, Nishikawa S, Kubo H, Achen MG (2001) VEGF-D promotes the metastatic spread of tumor cells via the lymphatics. Nat Med 7(2):186–191
42. Stacker SA, Hughes RA, Achen MG (2004) Molecular targeting of lymphatics for therapy. Curr Pharm Des 10(1):65–74
43. Achen MG, McColl BK, Stacker SA (2005) Focus on lymphangiogenesis in tumor metastasis. Cancer Cell 7(2):121–127
44. Achen MG, Stacker SA (2006) Tumor lymphangiogenesis and metastatic spread-new players begin to emerge. Int J Cancer 119(8):1755–1760
45. Ji RC (2006) Lymphatic endothelial cells, tumor lymphangiogenesis and metastasis: new insights into intratumoral and peritumoral lymphatics. Cancer Metastasis Rev 25(4):677–694
46. Achen MG, Stacker SA (2008) Molecular control of lymphatic metastasis. Ann N Y Acad Sci 1131:225–234
47. Swartz MA, Skobe M (2001) Lymphatic function, lymphangiogenesis, and cancer metastasis. Microsc Res Tech 55(2):92–99
48. Podgrabinska S, Braun P, Velasco P, Kloos B, Pepper MS, Skobe M (2002) Molecular characterization of lymphatic endothelial cells. Proc Natl Acad Sci USA 99(25):16069–16074
49. Mehes G, Witt A, Kubista E, Ambros PF (2001) Circulating breast cancer cells are frequently apoptotic. Am J Pathol 159(1):17–20
50. Cao R, Ji H, Feng N, Zhang Y, Yang X, Andersson P, Sun Y, Tritsaris K, Hansen AJ, Dissing S, Cao Y (2012) Collaborative interplay between FGF-2 and VEGF-C promotes lymphangiogenesis and metastasis. Proc Natl Acad Sci USA 109(39): 15894–15899
51. Ji H, Cao R, Yang Y, Zhang Y, Iwamoto H, Lim S, Nakamura M, Andersson P, Wang J, Sun Y, Dissing S, He X, Yang X, Cao Y (2014) TNFR1 mediates TNF-α-induced tumour lymphangiogenesis and metastasis by modulating VEGF-C-VEGFR3 signalling. Nat Commun 5:4944
52. Cao Y (2005) Opinion: emerging mechanisms of tumour lymphangiogenesis and lymphatic metastasis. Nat Rev Cancer 5(9):735–743
53. Le Guen L, Karpanen T, Schulte D, Harris NC, Koltowska K, Roukens G, Bower NI, van Impel A, Stacker SA, Achen MG, Schulte-Merker S, Hogan BM (2014) Ccbe1 regulates Vegfc-mediated induction of Vegfr3 signaling during embryonic lymphangiogenesis. Development 141(6):1239–1249
54. Lolas G (2003) Mathematical modelling of the urokinase plasminogen activation system and its role in cancer invasion of tissue. Ph.D. thesis, Department of Mathematics, University of Dundee
55. Sidenius N, Blasi F (2003) The urokinase plasminogen activator system in cancer: recent advances and implication for prognosis and therapy. Cancer Metastasis Rev 22(2–3):205–222

56. Chaplain MAJ, Lolas (2006) Mathematical modelling of cancer invasion of tissue: dynamic heterogeneity. Netw Heterog Media 1(3):399–439
57. Blasi F, Vassalli JD, DanøK (1987) Urokinase-type plasminogen activator: proenzyme, receptor, and inhibitors. J Cell Biol 104(4):801–804
58. Shirasuna K, Saka M, Hayashido Y, Yoshioka H, Sugiura T, Matsuya T (1993) Extracellular matrix production and degradation by adenoid cystic carcinoma cells: participation of plasminogen activator and its inhibitor in matrix degradation. Cancer Res 53(1):147–152
59. Wolf K, Friedl P (2011) Extracellular matrix determinants of proteolytic and non-proteolytic cell migration. Trends Cell Biol 21(12):736–744
60. Scianna M, Bell CG, Preziosi L (2013) A review of mathematical models for the formation of vascular networks. J Theor Biol 333:174–209
61. Aznavoorian S, Murphy AN, Stetler-Stevenson WG, Liotta LA (1993) Molecular aspects of tumor cell invasion and metastasis. Cancer 71(4):1368–1383
62. Wolf K, Friedl P (2009) Mapping proteolytic cancer cell-extracellular matrix interfaces. Clin Exp Metastasis 26(4):289–298
63. McCarthy JB, Palm SL, Furcht LT (1983) Migration by haptotaxis of a Schwann cell tumor line to the basement membrane glycoprotein laminin. J Cell Biol 97(3):772–777
64. McCarthy JB, Furcht LT (1984) Laminin and fibronectin promote the haptotactic migration of B16 mouse melanoma cells in vitro. J Cell Biol 98(4):1474–1480
65. McCarthy JB, Hagen ST, Furcht LT (1986) Human fibronectin contains distinct adhesion- and motility-promoting domains for metastatic melanoma cells. J Cell Biol 102(1):179–188
66. Taraboletti G, Roberts DD, Liotta LA (1987) Thrombospondin-induced tumor cell migration: haptotaxis and chemotaxis are mediated by different molecular domains. J Cell Biol 105(5):2409–2415
67. Aznavoorian S, Stracke ML, Krutzsch H, Schiffmann E, Liotta LA (1990) Signal transduction for chemotaxis and haptotaxis by matrix molecules in tumor cells. J. Cell Biol 110(4):1427–1438
68. Aznavoorian S, Stracke ML, Parsons J, McClanahan J, Liotta LA (1996) Integrin alphavbeta3 mediates chemotactic and haptotactic motility in human melanoma cells through different signaling pathways. J Biol Chem 271(6):3247–3254
69. Carter SB (1967) Haptotaxis and the mechanism of cell motility. Nature 213(5073):256–260
70. Carter SB (1965) Principles of cell motility: the direction of cell movement and cancer invasion. Nature 208(5016):1183–1187
71. Andreasen PA, Kjøller L, Christensen L, Duffy MJ (1997) The urokinase-type plasminogen activator system in cancer metastasis: a review. Int J Cancer 72(1):1–22
72. Andreasen PA, Egelund R, and Petersen HH (2000) The plasminogen activation system in tumor growth, invasion, and metastasis. Cell Mol Life Sci 57(1):25–40
73. Wiig H, Keskin D, Kalluri R (2010) Interaction between the extracellular matrix and lymphatics: consequences for lymphangiogenesis and lymphatic function. Matrix Biol 29(8):645–656
74. Makinen T, Veikkola T, Mustjoki S, Karpanen T, Catimel B, Nice EC, Wise L, Mercer A, Kowalski H, Kerjaschki D, Stacker SA, Achen MG, Alitalo K (2001) Isolated lymphatic endothelial cells transduce growth, survival and migratory signals via the VEGF-C/D receptor VEGFR-3. EMBO J 20(17):4762–4773
75. Helm CL, Fleury ME, Zisch AH, Boschetti F, Swartz MA (2005) Synergy between interstitial flow and VEGF directs capillary morphogenesis in vitro through a gradient amplification mechanism. Proc Natl Acad Sci USA 102(44): 15779–15784
76. Rutkowski JM, Swartz MA (2007) A driving force for change: interstitial flow as a morphoregulator. Trends Cell Biol 17(1):44–50
77. Shields JD, Fleury ME, Yong C, Tomei AA, Randolph GJ, Swartz MA (2007) Autologous chemotaxis as a mechanism of tumor cell homing to lymphatics via interstitial flow and autocrine CCR7 signaling. Cancer Cell 11(6):526–538
78. Kodama M, Kitadai Y, Tanaka M, Kuwai T, Tanaka S, Oue N, Yasui W, Chayama K (2008) Vascular endothelial growth factor C stimulates progression of human gastric cancer via both autocrine and paracrine mechanisms. Clin Cancer Res 14(22):7205–7214
79. Oh SJ, Jeltsch MM, Birkenhager R, McCarthy JE, Weich HA, Christ B, Alitalo K, Wilting J (1997) VEGF and VEGF-C: specific induction of angiogenesis and lymphangiogenesis in the differentiated avian chorioallantoic membrane. Dev Biol 188(1):96–109
80. Joukov V, Sorsa T, Kumar V, Jeltsch M, Claesson-Welsh L, Cao Y, Saksela O, Kalkkinen N, Alitalo K (1997) Proteolytic processing regulates receptor specificity and activity of VEGF-C. EMBO J 16(13):3898–3911
81. McColl BK, Baldwin ME, Roufail S, Freeman C, Moritz RL, Simpson RJ, Alitalo K, Stacker SA, Achen MG (2003) Plasmin activates the lymphangiogenic growth factors VEGF-C and VEGF-D. J Exp Med 198(6):863–868
82. Harris NC, Paavonen K, Davydova N, Roufail S, Sato T, Zhang YF, Karnezis T, Stacker SA, Achen MG (2011) Proteolytic processing of vascular endothelial growth factor-D is essential for its capacity to promote the growth and spread of cancer. FASEB J 25(8):2615–2625
83. Harris NC, Achen MG (2014) The proteolytic activation of angiogenic and lymphangiogenic growth factors in cancer–its potential relevance for therapeutics and diagnostics. Curr Med Chem 21(16):1821–1842

84. Oh CW, Hoover-Plow J, Plow EF (2003) The role of plasminogen in angiogenesis in vivo. J Thromb Haemost 1(8):1683–1687
85. Pepper MS, Mandriota SJ, Jeltsch M, Kumar V, Alitalo K (1998) Vascular endothelial growth factor (VEGF)-C synergizes with basic fibroblast growth factor and VEGF in the induction of angiogenesis in vitro and alters endothelial cell extracellular proteolytic activity. J Cell Physiol 177(3):439–452
86. Pepper MS, Ferrara N, Orci L, Montesano R (1991) Vascular endothelial growth factor (VEGF) induces plasminogen activators and plasminogen activator inhibitor-1 in microvascular endothelial cells. Biochem Biophys Res Commun 181(2):902–906
87. Pepper MS, Wasi S, Ferrara N, Orci L, Montesano R (1994) In vitro angiogenic and proteolytic properties of bovine lymphatic endothelial cells. Exp Cell Res 210(2):298–305
88. Tille JC, Wang X, Lipson KE, McMahon G, Ferrara N, Zhu Z, Hicklin DJ, Sleeman JP, Eriksson U, Alitalo K, Pepper MS (2003) Vascular endothelial growth factor (VEGF) receptor-2 signaling mediates VEGF-C(deltaNdeltaC)- and VEGF-A-induced angiogenesis in vitro. Exp Cell Res 285(2):286–298
89. Chaplain MAJ, Lolas G (2005) Mathematical modelling of cancer cell invasion of tissue: the role of the urokinase plasminogen activation system. Math Models Methods Appl Sci 15(11):1685–1734. dc.publisher: World Scientific
90. Bray D (2000) Cell movements: from molecules to motility. Garland Science, New York
91. Orme ME, Chaplain MA (1997) Two-dimensional models of tumour angiogenesis and anti-angiogenesis strategies. IMA J Math Appl Med Biol 14(3):189–205
92. Anderson AR, Chaplain MA (1998) Continuous and discrete mathematical models of tumor-induced angiogenesis. Bull. Math. Biol. 60(5):857–899
93. Chaplain MAJ, Lolas G (2005) Mathematical modelling of cancer cell invasion of tissue: the role of the urokinase plasminogen activation system. Math Models Methods Appl Sci 15(11):1685–1734
94. Del Monte U (2009) Does the cell number 10(9) still really fit one gram of tumor tissue? Cell Cycle 8(3):505–506
95. Mochan E and Keler T (1984) Plasmin degradation of cartilage proteoglycan. Biochim. Biophys. Acta 800(3):312–315
96. Haessler U, Pisano M, Wu M, Swartz MA (2011) Dendritic cell chemotaxis in 3D under defined chemokine gradients reveals differential response to ligands CCL21 and CCL19. Proc Natl Acad Sci USA 108(14):5614–5619
97. Stacker SA, Stenvers K, Caesar C, Vitali A, Domagala T, Nice E, Roufail S, Simpson RJ, Moritz R, Karpanen T, Alitalo K, Achen MG (1999) Biosynthesis of vascular endothelial growth factor-D involves proteolytic processing which generates non-covalent homodimers. J Biol Chem 274(45):32127–32136
98. Yonemura Y, Endo Y, Tabata K, Kawamura T, Yun HY, Bandou E, Sasaki T, Miura M (2005) Role of VEGF-C and VEGF-D in lymphangiogenesis in gastric cancer. Int J Clin Oncol 10(5):318–327
99. Issa A, Le TX, Shoushtari AN, Shields JD, Swartz MA (2009) Vascular endothelial growth factor-C and C-C chemokine receptor 7 in tumor cell-lymphatic cross-talk promote invasive phenotype. Cancer Res 69(1):349–357
100. Tissot JD, Schneider P, Hauert J, Ruegg M, Kruithof EK, Bachmann F (1982) Isolation from human plasma of a plasminogen activator identical to urinary high molecular weight urokinase. J Clin Invest 70(6):1320–1323
101. Wun TC, Schleuning WD, Reich E (1982) Isolation and characterization of urokinase from human plasma. J Biol Chem 257(6):3276–3283
102. Vassalli JD, Baccino D, Belin D (1985) A cellular binding site for the Mr 55,000 form of the human plasminogen activator, urokinase. J Cell Biol 100(1):86–92
103. Markus G, Camiolo SM, Kohga S, Madeja JM, Mittelman A (1983) Plasminogen activator secretion of human tumors in short-term organ culture, including a comparison of primary and metastatic colon tumors. Cancer Res 43(11):5517–5525
104. Barlow GH (1976) Proteolytic enzymes, part B (methods in enzymology). Academic Press, New York
105. Ellis V (1996) Functional analysis of the cellular receptor for urokinase in plasminogen activation. Receptor binding has no influence on the zymogenic nature of pro-urokinase. J Biol Chem 271(25):14779–14784
106. Wu HL, Shi GY, and Bender ML (1987) Preparation and purification of microplasmin. Proc Natl Acad Sci USA 84(23):8292–8295
107. Sherratt JA, Murray JD (1990) Models of epidermal wound healing. Proc Biol Sci 241(1300):29–36
108. Stokes CL, Lauffenburger DA, Williams SK (1991) Migration of individual microvessel endothelial cells: stochastic model and parameter measurement. J Cell Sci 99(Pt 2):419–430
109. Robbins KC, Summaria L, Elwyn D, Barlow GH (1965) Further studies on the purification and characterization of human plasminogen and plasmin. J Biol Chem 240:541–550
110. Fleury ME, Boardman KC, Swartz MA (2006) Autologous morphogen gradients by subtle interstitial flow and matrix interactions. Biophys J 91(1):113–121
111. Chaplain MA (1995) The mathematical modelling of tumour angiogenesis and invasion. Acta Biotheor 43(4):387–402
112. Chaplain MAJ, Giles SM, Sleeman BD, Jarvis RJ (1995) A mathematical analysis of a

model for tumor angiogenesis. J Math Biol 33: 744–770
113. White WF, Barlow GH, Mozen MM (1966) The isolation and characterization of plasminogen activators (urokinase) from human urine. Biochemistry 5(7):2160–2169
114. Shin Y, Han S, Jeon JS, Yamamoto K, Zervantonakis IK, Sudo R, Kamm RD, Chung S (2012) Microfluidic assay for simultaneous culture of multiple cell types on surfaces or within hydrogels. Nat Protoc 7(7):1247–1259
115. Yu W, Kim J, Ossowski L (1997) Reduction in surface urokinase receptor forces malignant cells into a protracted state of dormancy. J Cell Biol 137(3):767–777
116. Stokes CL, Lauffenburger DA (1991) Analysis of the roles of microvessel endothelial cell random motility and chemotaxis in angiogenesis. J Theor Biol 152(3):377–403
117. Chaplain MAJ (1996) Avascular growth, angiogenesis and vascular growth in solid ttumour: the mathematical modelling of the stages of tumour development. Math Comput Model 23(6):47–87
118. Leak LV, Jones M (1993) Lymphatic endothelium isolation, characterization and long-term culture. Anat Rec 236(4):641–652
119. Nguyen VP, Chen SH, Trinh J, Kim H, Coomber BL, Dumont DJ (2007) Differential response of lymphatic, venous and arterial endothelial cells to angiopoietin-1 and angiopoietin-2. BMC Cell Biol 8:10
120. Park JE, Keller GA, Ferrara N (1993) The vascular endothelial growth factor (VEGF) isoforms: differential deposition into the subepithelial extracellular matrix and bioactivity of extracellular matrix-bound VEGF. Mol Biol Cell 4(12):1317–1326
121. Weich HA, Bando H, Brokelmann M, Baumann P, Toi M, Barleon B, Alitalo K, Sipos B, Sleeman J (2004) Quantification of vascular endothelial growth factor-C (VEGF-C) by a novel ELISA. J Immunol Methods 285(2):145–155
122. Bocci G, Fasciani A, Danesi R, Viacava P, Genazzani AR, Del Tacca M (2001) In-vitro evidence of autocrine secretion of vascular endothelial growth factor by endothelial cells from human placental blood vessels. Mol Hum Reprod 7(8):771–777
123. Bianchi A, Painter KJ, Sherratt JA (2015) A mathematical model for lymphangiogenesis in normal and diabetic wounds. J Theor Biol 383:61–86
124. Imoukhuede PI, Dokun AO, Annex BH, Popel AS (2013) Endothelial cell-by-cell profiling reveals the temporal dynamics of VEGFR1 and VEGFR2 membrane localization after murine hindlimb ischemia. Am J Physiol Heart Circ Physiol 304(8):H1085–1093
125. Baldwin ME, Catimel B, Nice EC, Roufail S, Hall NE, Stenvers KL, Karkkainen MJ, Alitalo K, Stacker SA, and Achen MG (2001) The specificity of receptor binding by vascular endothelial growth factor-d is different in mouse and man. J Biol Chem 276(22):19166–19171
126. Bando H, Brokelmann M, Toi M, Alitalo K, Sleeman JP, Sipos B, Grone HJ, Weich HA (2004) Immunodetection and quantification of vascular endothelial growth factor receptor-3 in human malignant tumor tissues. Int J Cancer 111(2):184–191
127. Wang Y, Nakayama M, Pitulescu ME, Schmidt TS, Bochenek ML, Sakakibara A, Adams S, Davy A, Deutsch U, Luthi U, Barberis A, Benjamin LE, Makinen T, Nobes CD, Adams RH (2010) Ephrin-B2 controls VEGF-induced angiogenesis and lymphangiogenesis. Nature 465(7297):483–486
128. Mandriota SJ, Jussila L, Jeltsch M, Compagni A, Baetens D, Prevo R, Banerji S, Huarte J, Montesano R, Jackson DG, Orci L, Alitalo K, Christofori G, Pepper MS (2001) Vascular endothelial growth factor-C-mediated lymphangiogenesis promotes tumour metastasis. EMBO J 20(4):672–682
129. Enholm B, Paavonen K, Ristimaki A, Kumar V, Gunji Y, Klefstrom J, Kivinen L, Laiho M, Olofsson B, Joukov V, Eriksson U, Alitalo K (1997) Comparison of VEGF, VEGF-B, VEGF-C and Ang-1 mRNA regulation by serum, growth factors, oncoproteins and hypoxia. Oncogene 14(20):2475–2483
130. Ristimaki A, Narko K, Enholm B, Joukov V, Alitalo K (1998) Proinflammatory cytokines regulate expression of the lymphatic endothelial mitogen vascular endothelial growth factor-C. J Biol Chem 273(14):8413–8418
131. Atkins P, De Paula J (2006) Physical chemistry, 8th edn. Oxford University Press, Oxford
132. Estreicher A, Muhlhauser J, Carpentier JL, Orci L, Vassalli JD (1990) The receptor for urokinase type plasminogen activator polarizes expression of the protease to the leading edge of migrating monocytes and promotes degradation of enzyme inhibitor complexes. J Cell Biol 111(2):783–792
133. Stoppelli MP, Corti A, Soffientini A, Cassani G, Blasi F, Assoian RK (1985) Differentiation-enhanced binding of the amino-terminal fragment of human urokinase plasminogen activator to a specific receptor on U937 monocytes. Proc Natl Acad Sci USA 82(15):4939–4943
134. Bajpai A, Baker JB (1985) Cryptic urokinase binding sites on human foreskin fibroblasts. Biochem Biophys Res Commun 133(2):475–482
135. Barnathan ES, Kuo A, Rosenfeld L, Kariko K, Leski M, Robbiati F, Nolli ML, Henkin J, Cines DB (1990) Interaction of single-chain urokinase-type plasminogen activator with human endothelial cells. J Biol Chem 265(5):2865–2872
136. Sillaber C, Baghestanian M, Hofbauer R, Virgolini I, Bankl HC, Fureder W, Agis H, Willheim M, Leimer M, Scheiner O, Binder BR, Kiener HP, Bevec D, Fritsch G, Majdic O, Kress HG, Gadner H, Lechner K, Valent P (1997) Molecular and functional charac-

terization of the urokinase receptor on human mast cells. J Biol Chem 272(12):7824–7832
137. Olson D, Pollanen J, Høyer-Hansen G, Rønne E, Sakaguchi K, Wun TC, Appella E, Danø K, Blasi F (1992) Internalization of the urokinase-plasminogen activator inhibitor type-1 complex is mediated by the urokinase receptor. J Biol Chem 267(13):9129–9133
138. Mignatti P, Mazzieri R, Rifkin DB (1991) Expression of the urokinase receptor in vascular endothelial cells is stimulated by basic fibroblast growth factor. J Cell Biol 113(5):1193–1201
139. Grover WH, Bryan AK, Diez-Silva M, Suresh S, Higgins JM, Manalis SR (2011) Measuring single-cell density. Proc Natl Acad Sci USA 108(27):10992–10996
140. Camiolo SM, Markus G, Englander LS, Siuta MR, Hobika GH, Kohga S (1984) Plasminogen activator content and secretion in explants of neoplastic and benign human prostate tissues. Cancer Res 44(1):311–318
141. Harvey SR, Lawrence DD, Madeja JM, Abbey SJ, Markus G (1988) Secretion of plasminogen activators by human colorectal and gastric tumor explants. Clin Exp Metastasis 6(6):431–450
142. Werb Z, Bainton DF, Jones PA (1980) Degradation of connective tissue matrices by macrophages. III. Morphological and biochemical studies on extracellular, pericellular, and intracellular events in matrix proteolysis by macrophages in culture. J Exp Med 152(6):1537–1553
143. Allen LE, Dubeau L, Alvarez O, Jones PA (1990) Rapid degradation of extracellular matrix proteins by normal human uroepithelial cells. Cancer Res 50(6):1897–1904
144. Jones PA, Werb Z (1980) Degradation of connective tissue matrices by macrophages. II. Influence of matrix composition on proteolysis of glycoproteins, elastin, and collagen by macrophages in culture. J Exp Med 152(6):1527–1536
145. Pins GD, Collins-Pavao ME, Van De Water L, Yarmush ML, Morgan JR (2000) Plasmin triggers rapid contraction and degradation of fibroblast-populated collagen lattices. J Invest Dermatol 114(4):647–653
146. Jones PA, Scott-Burden T, Gevers W (1979) Glycoprotein, elastin, and collagen secretion by rat smooth muscle cells. Proc Natl Acad Sci USA 76(1):353–357
147. Miyashita C, Wenzel E, Heiden M (1988) Plasminogen: a brief introduction into its biochemistry and function. Haemostasis 18(1):7–13
148. Skeel RD, Berzins M (1990) A method for the spatial discretization of parabolic equations in one space variable. SIAM J Sci Stat Comput 11(1):1–32
149. Kramer RH, Bensch KG, Wong J (1986) Invasion of reconstituted basement membrane matrix by metastatic human tumor cells. Cancer Res 46(4 Pt 2):1980–1989
150. Fidler IJ (1978) Tumor heterogeneity and the biology of cancer invasion and metastasis. Cancer Res 38(9):2651–2660
151. Friedl P, Noble PB, Walton PA, Laird DW, Chauvin PJ, Tabah RJ, Black M, Zanker KS (1995) Migration of coordinated cell clusters in mesenchymal and epithelial cancer explants in vitro. Cancer Res 55(20):4557–4560
152. Friedl P, Brocker EB (2000) The biology of cell locomotion within three-dimensional extracellular matrix. Cell Mol Life Sci 57(1):41–64
153. Aceto N, Bardia A, Miyamoto DT, Donaldson MC, Wittner BS, Spencer JA, Yu M, Pely A, Engstrom A, Zhu H, Brannigan BW, Kapur R, Stott SL, Shioda T, Ramaswamy S, Ting DT, Lin CP, Toner M, Haber DA, Maheswaran S (2014) Circulating tumor cell clusters are oligoclonal precursors of breast cancer metastasis. Cell 158(5):1110–1122
154. Kolega J (1981) The movement of cell clusters in vitro: morphology and directionality. J Cell Sci 49:15–32
155. Liotta LA, Saidel MG, Kleinerman J (1976) The significance of hematogenous tumor cell clumps in the metastatic process. Cancer Res 36(3):889–894
156. Menashi S, Dehem M, Souliac I, Legrand Y, Fridman R (1998) Density-dependent regulation of cell-surface association of matrix metalloproteinase-2 (MMP-2) in breast-carcinoma cells. Int J Cancer 75(2):259–265

Positive Feedback Loops Between Inflammatory, Bone and Cancer Cells During Metastatic Niche Construction

7

Ardeshir Kianercy and Kenneth J. Pienta

Abstract

Bone, which includes several cell populations and numerous cytokines and chemokines that provide cell-cell signaling, is a common destination for many cancer metastases. Bone metastasis skews this signaling to develop vicious cycles between immune, bone and cancer populations that lead to abnormal bone remodeling during cancer niche construction. Temporal models utilize positive feedback systems as an integrative tool providing insights into the rate-limiting processes that determine multiple stages of the bone metastasis. We develop a logical-transient-threshold framework by linking temporal responses of the cancer, bone and immune systems through macrophages during ecological niche construction of cancer in host bone.

Keywords

Bone metastasis • Cancer ecology • Feedback systems • Tumor vicious cycle

7.1 Introduction

The three most common neoplasms in humans – breast, prostate, and lung cancer are often associated with bone metastases and remodeling of the skeleton to a tumor nourishing ecology (Fig. 7.1). In the U.S., the estimated number of deaths due to prostate cancer in 2015 is 28,000 and majority of patients die of metastatic bone involvement with a 5-year survival rate of approximately 28 % [1].

A. Kianercy
Brady Urological Institute, Johns Hopkins Hospital, Baltimore, MD, USA

Department of Molecular Physiology and Biophysics, Vanderbilt University, Nashville, TN, USA
e-mail: kianercy@usc.edu

K.J. Pienta (✉)
Brady Urological Institute, Johns Hopkins Hospital, Baltimore, MD, USA
e-mail: kpienta1@jhmi.edu

K.A. Rejniak (eds.), *Systems Biology of Tumor Microenvironment*, Advances in Experimental Medicine and Biology 936, DOI 10.1007/978-3-319-42023-3_7

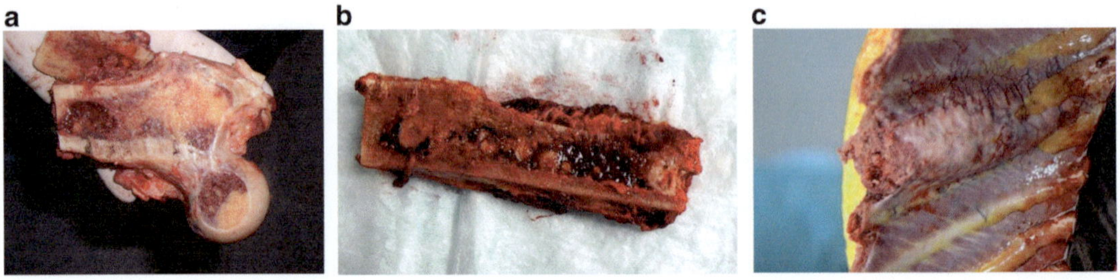

Fig. 7.1 Typical metastatic prostate cancer bone lesions. (**a**) Osteoblastic (abnormal bone forming) lesions in a hip. (**b**) Multiple lesions in a humerus. (**c**) Metastatic growth replaced a rib

In normal bone, the bone resorption rate, and the bone formation rate are all relatively constant. The bone remodeling process maintains the integrity of the bone through temporary balanced anatomic structures called basic multicellular units (BMUs) consisting of osteoblasts (bone-forming) and osteoclast (bone-lysing) cells [2]. Several positive cycles between the immune, BMU and cancer system can change this balance.

Cancer cells secrete factors in their microenvironment that can cause an inflammatory immune system response, thereby utilizing inflammation for its growth and irreversibly shifting the bone's state [3]. Thus, inflammation provides supporting conditions for enhancing tumor growth. The inflammatory tumor mass comprises a number of different cell types. It includes granulocytes, lymphoid cells, mast cells, natural killer cells, dendritic cells and macrophages. Macrophages, the most abundant inflammatory cell type, play a significant regulatory role in inflammation and cancer. To utilize inflammation for its growth, cancer cell secretes factors that shift macrophage polarization from M1-like into M2-like type tumor-associated macrophages that subsequently promote tumor growth, and progression [4, 5].

Furthermore, certain types of solid tumors metastasize to bone and recruit osteoclast and osteoblast cells which secret growth factors enhancing tumor growth – resulting in positive cycles [6].

In ecological systems, a niche describes the organism response to the available resources as well as the competition with other species for those resources, usually in a defined space. Resource limitation and competition modulate the biological and physical ecosystem components. A successful bone metastasis niche construction is the result of driven cancer disturbance of the cell-cell network signaling in the bone microenvironment. We demonstrate a `vicious triangle` with positive feedback loops (vicious cycles) between Macrophages, BMUs and cancer cells supporting the metastatic bone niche construction.

Mathematical case studies have defined inflammation-associated bone remodeling homeostasis [7], myeloma affects on bone populations [8], bone-immune system [9], and bone-cancer dynamics [10]. Other studies have introduced frameworks that can capture inflammatory control mechanisms [11], dynamic interactions between the immune system and cancer [11] and dynamic interactions between the immune system and cancer [9, 12].

We utilize a modularization (functional block) framework to integrate different temporal modelings, simplifying cells gene regulatory system and cell-cell networks into conceptually manageable functional modules [13].

The mammalian gene regulatory network temporal responses to environmental stimuli are either impulse, state transitions or oscillatory responses. Thus, we adopt a second order system as one of the simplest dynamics that can capture these gene regulatory transient responses [14].

In addition, the tumour microenvironment progresses through rate-limiting stages that lead to several phenotypic transitions [15]. There is

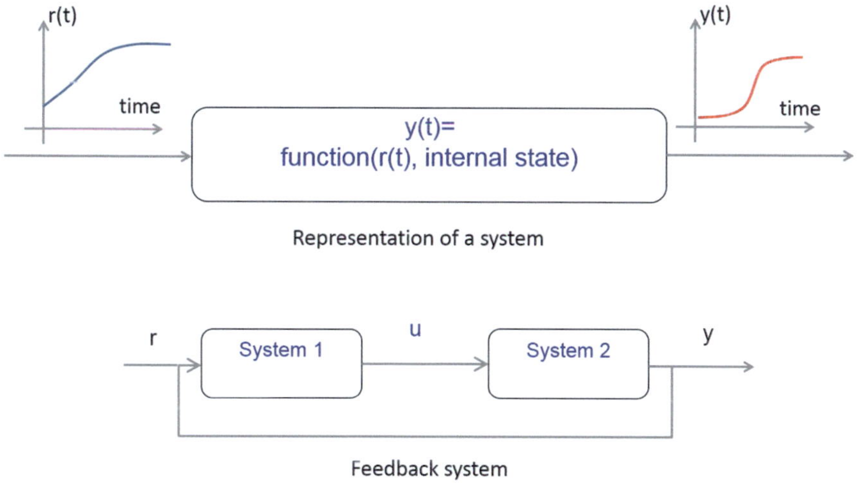

Fig. 7.2 Block diagrams helps to represent the network of interconnected dynamical systems. (*Top*) Information cascades from input r, is processed by a dynamical system toward the final system output y over time (t). (*Down*) Feedback system: System 1 output *u* is an input for system 2 while system 2 output *y* becomes an input of system 1. *r* is an environmental input signal. The feedback signal can be either stimulatory (positive) or inhibitory (negative)

experimental evidence of phenotypic transitions in cancer [16, 17], macrophage [18, 19] and bone [20]. This evidence suggests that a threshold module that takes the transient response as an input is in place to determine the cellular phenotypic switching. Also, it is suggested, that cell decision-making processes may go through logical modules to integrate the multiple inputs of stimuli from the environment [21].

We reconcile these three common functional modules, providing a systematic method for dynamical cell-cell networks, describe the rate-limiting processes during bone metastases, and study positive cycles between inflammatory, bone remodeling and cancer populations in the metastatic bone.

7.2 Cellular Systems in Tumor Inflammation-Associated Bone Remodeling

A difficulty in complex system modeling is interconnecting systems from many different domains. In metastatic bone niche construction, the system, at a minimum, includes immune cell populations, bone remodeling units, and cancer cells.

Biological systems make use of feedback in different scales ranging from molecules, cells, organisms and ecosystems [22]. Feedback mechanisms in biological systems were first defined by Walter Cannon to describe homeostasis in physiological variables within a predefined range [23]. The term *feedback* refers to two or more dynamical systems cross talking such that their dynamics are coupled. Figure 7.2 illustrates block diagram of a simple feedback system (closed loop).

A feedback system can contain negative or positive loops. Negative feedback can reduce noise and increase resistance to disturbances inside the feedback loop, but it can also bring oscillations if the disturbance is too strong. Positive feedback, on the other hand, increases the system's sensitivity to the disturbances and may also lead to bi-stability or a sudden dramatic state switch [24]. In general, positive cycles (loops) can lead to alternative stable state. Thus, we focus on the positive feedback loops to describe the state switching within metastasis bone in favor of the cancer bone niche construction.

7.2.1 Bone Basic Multi-cellular Unit (BMU)

On a molecular level, the balance of bone formation and resorption is coordinated by different cell types: osteoblasts, osteocytes, adipocytes, chondrocytes, and fibroblasts. The bone cells communication occurs through cytokines, ligand-receptor interactions and other cell-cell interactions. However, simple models of bone homeostasis focus on only two specialized cells, osteoblast and osteoclast cells as a Basic Multi-cellular Unit (BMU) that coordinate bone formation and resorption through RANK/RANKL/OPG signaling pathway.

Osteoclasts are derived from the monocyte cell linage in the local bone ecosystem and cytokine milieu. Osteoclast differentiation relies on the presence of Macrophage Colony-Stimulating Factor (M-CSF) and Receptor for Activation of Nuclear factor Kappa B Ligand (RANKL). On the other hand, osteoblasts derive from mesenchymal stem cells, act as both negative and positive regulators of osteoclast differentiation by adapting their expression of RANKL and OPG.

RANK is a protein expressed by osteoclasts and is a receptor for RANKL, a protein produced by osteoblasts. RANK/RANKL signaling triggers osteoclast differentiation, proliferation and thus, the resorption phase during bone remodeling. Osteoprotegerin (OPG) is a decoy receptor for RANKL, since OPG is also a receptor for RANKL. Mature osteoblasts express OPG that binds to RANKL, thus inhibiting osteoclast production. When RANKL expression is high, osteoprotegerin levels are low and vice versa [7,25,26]. The balance between OPG and RANKL is critical for normal osteoclast differentiation. Local factors, such as cancer and immune cytokines influence the ratio RANKL/OPG.

7.3 Vicious Cycles Within the Bone Metastasis Microenvironment

The metastatic niche model suggests that a microenvironment (pre-metastatic niche) must evolve for tumor cells to be able to proliferate at the secondary site (metastatic niche) [27, 28]. Several reinforcing cycles between tumor cells catalyze this metastatic niche construction in bone.

Patients with bone metastases can develop two types of lesions: osteolytic or osteoblastic. During osteolytic (bone lysing) metastases, typical for myeloma, high osteoclast activation results from direct and indirect actions by metastatic cancer cells. The clearest examples of cancer niche constructions are osteolytic bone metastasis of breast cancer [28]. On other hand, metastases from cancers such as prostate cancer tend to be predominantly osteoblastic (bone forming).

In both cases, usually cancer, bone marrow and immune systems can establish positive cycles supporting cancer growth and bone destruction [7, 25, 29–31]. Here, we describe three important types of these positive vicious cycles during bone metastases (Fig. 7.3).

7.3.1 Bone-Cancer Vicious Cycle

The imbalance between the activities of osteoblasts and osteoclasts determines the phenotype of metastatic bone lesions. In general, cancer bone metastases demonstrate mixed bone lesions. For instance prostate cancer metastases in bone tend to be more osteoblastic rather than osteolytic, while breast bone metastases predominantly reside in a niche that exhibits features of osteoblast and later becomes osteolytic [33, 34].

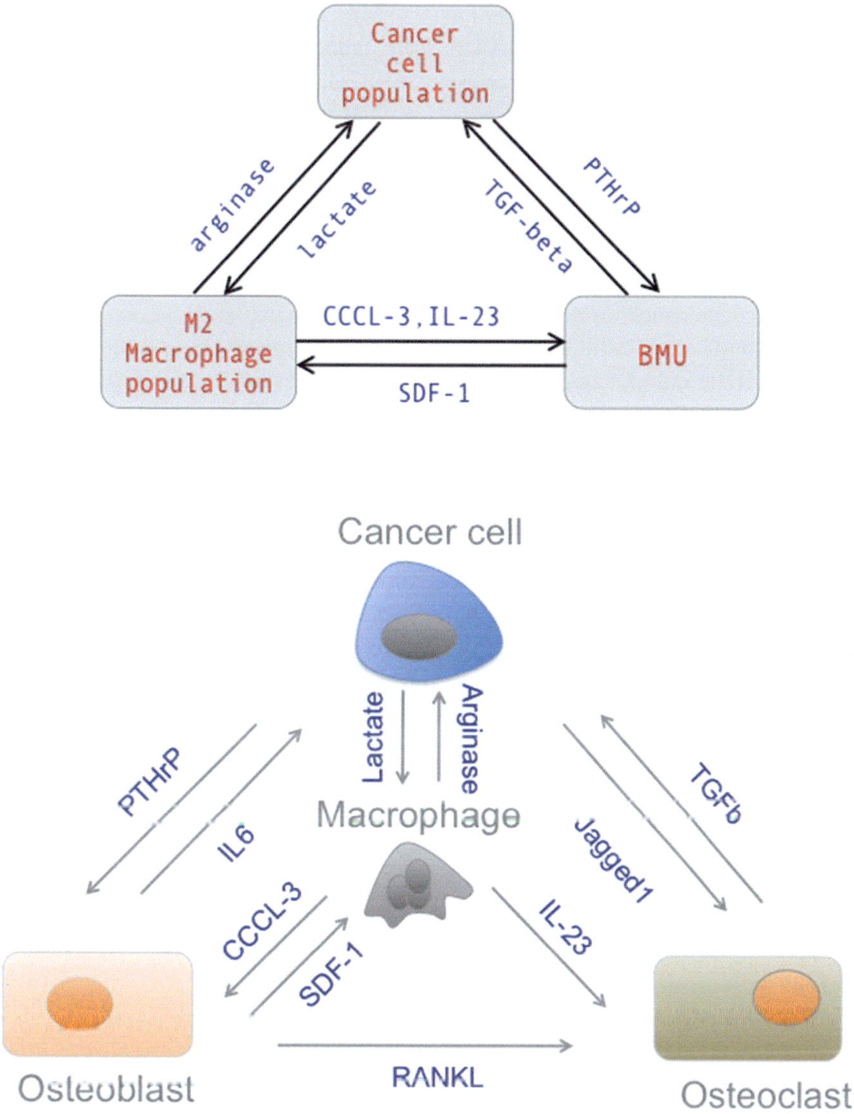

Fig. 7.3 (*Top*) A simple schematic of vicious cycles in metastasis bone between immune system and BMU (Bone basic multi-cellular units). The vicious cycles build up a vicious triangle which support cancer cells, M2-like Macrophages and abnormal bone remodeling. (*Bottom*) A simple expanded schematic of cell-cell network in metastatic bone. The balances between osteoblast and osteoclast shifts through caner and macrophage signals. The osteoclast-mediated bone matrix resorption releasesTGF-β and IGF1, which in turn upregulate the expression of PTHrP, Jagged1 to create a vicious cycle of bone destruction. Physical factors within the bone microenvironment such as hypoxia, low pH, and high extracellular calcium concentration, may also enahnce this vicious triangle (For more comprehensive details [28, 32])

In osteolytic metastases, cytokines or growth factors secreted by invading cancer cells such as parathyroid hormone-related protein (PTHrP) act to stimulate osteoblasts to produce more receptor activator of RANKL and less OPG. An increase in serum RANKL/OPG ratio leads to a higher osteolytic activity. Activated osteoclasts degrade bone matrix, releasing growth factors such as transforming growth factor beta (*TGFβ*) and insulin-like growth factor (IGF). These molecules, in return, stimulate further cancer cell growth thus establishing a positive feedback loop.

This series of events provides an explanation of the osteolytic stage of breast and prostate cancer metastasis in bone; that is, an increase in osteoclast activation leads to excess bone breakdown and further stimulation of cancer cells [28, 30, 31, 35].

In osteoblastic metastases, the osteoblast population expands more as compared to the osteoclast population, via stimulation of PTHrP [36]. The signature of prostate cancer bone metastasis is the clear presence of osteoblastic lesions with an underlying osteolytic component. In fact, in the early stage of prostate cancer bone metastases, osteoclast activity increases and in the subsequent stage, osteoblast cells are stimulated while osteoclast cells are less active [3].

7.3.2 Cancer-Macrophage Vicious Cycle

Among different immune cells in metastatic bone, one of the major players is the macrophage. For example in metastatic breast and prostate cancers, macrophages may account for up to 50 % of the tumor cell mass. Cancer cells in the tumor microenvironment recruit and stimulate macrophage polarization to M2-like macrophages that contribute both to tumor development and bone remodeling [37, 38].

Macrophages differentiate in tissues from monocytes originating in the bone marrow and its phenotype can be altered according to their resident microenvironment. Macrophage phenotypes lie within a phenotypic spectrum with two extreme ends M1 and M2. The macrophage phenotypic state is a key element that links inflammation to cancer.

Activators such as interferonγ and lipopolysaccharide promote M1, or classically activated macrophages. They are characterized by high IL-23 and IL-12 production, high capacity for antigen presentation and eventual T cells activation to Th1 responses. M1-like macrophage mediate pro-inflammatory, and cytotoxicity against tumor cells is a key feature of M1-like macrophages, mediated by the release of TNF, nitric oxide and reactive oxygen intermediates.

M2, or alternatively activated macrophages, are induced by IL-4, IL-10 and or IL-13. The main characteristics of M2 macrophages are wound healing, tissue remodeling, suppression of inflammatory responses, and stimulation of Th2 responses [37, 39]. The phenotype and the function of differentiated mature tumor associated macrophges resembles the M2-like, rather than the M1-like phenotype.

Experiments suggest various tumour cells, including breast and prostate cancers, secrete increased levels of Monocyte Chemo-attractant Protein 1(MCP-1). MCP-1 over-expression in tumor cells increases macrophage infiltration into the bone. In return, the secretion of IL-1 and TNFα by macrophages stimulates the production of IL-6 by tumor cells. Many cancer related processes including tumor cell proliferation express IL-6 [39, 40].

Furthermore, the lactic acid produced by tumour cells, as a by-product of glycolysis, is another factor which induce the M2-like polarization of tumour-associated macrophages. In return, the lactate-induced expression of arginase-1 by macrophages has an important role in tumour growth [41].

7.3.3 Macrophage-Bone Vicious Cycle

Osteoimmunology elucidates inflammation-associated changes in bone homeostasis by looking at osteoclasts/osteoblast and immune cells crosstalk [7, 26]. This crosstalk can change to vicious cycles in bone metastases as both osteoclasts and osteoblasts can manipulate the immune system toward cancer niche construction.

Macrophages can promote both osteolytic and osteoblastic bone niches. Studies have identified that Human Macrophage Inflammatory Protein-1 α (hMIP-1α), also termed CCL3, produced by macrophages, may be responsible for the increase of osteoblast population activity in bone marrow [42]. Macrophages can also support osteoclytic activity by secreting IL-23 that promotes osteoclastic formation [43]. In return, macrophages exposed to Stromal cell-Derived Factor 1 (SDF-1 also known as CXCL12), expressed by osteoblasts demonstrate a M2-like phenotype, for instance during prostate carcinoma progression [44, 45].

In conclusion, cancer, BMU, and macrophage populations build a triangle. We term this triangle a `vicious triangle` that promotes cancer proliferation, anti-inflammatory macrophages and abnormal bone remodeling simultaneously (Fig. 7.3).

7.4 Logical-Transient-Threshold Dynamics

Understanding the dynamics of cellular responses requires collecting high-quality time series data of cell information transmission, processing and decision making. The development of fluorescent labeling and time-lapse technology improved our ability to quantify the dynamics of biological responses in single cells providing rich insights about cellular signaling pathways and cell-cell network dynamics.

Cell signaling dynamics studies the shape of signals curve through the amplitude, frequency, amplitude, duration, or state modification over time. The time scale of the dynamics can change from seconds, in intracellular events, to days, for observable phenotypes such as cell morphology or expression of cell-surface markers [46].

Functional module or blocks representation provide a simple and intuitive picture without describing the detailed biochemistry of transmitting information in cells and between cells, providing a dynamical description, ranging from a cell dynamics to the average dynamical behavior of a population [47].

The typical temporal patterns of eukaryote cells gene transcription response to environmental stimuli can be placed into three classes: Impulse (transient), Sustained (switching between states) and Oscillating (cyclic) patterns.

In fact, impulse transient patterns are prevalent in responses to environmental changes in all organisms from bacteria to mammals. In addition, some transient responses induce strong gene expression necessary for shifting the state to a long term sustained cell's state [14, 46, 48].

For example, the tumor suppressor p53 shows both stimulus and amplitude dependent dynamics. Also, a combination of transient and state switching was observed during Phorbol Myristate Acetate (PMA) induced differentiation of myelomonocytic leukemia cells. This state switching over time has been observed in many other mammalian populations, including responses to growth, pathogens and stress signals (Ref. [14, 46, 48] for comprehensive review). To capture these gene transcription transient behaviors, we suggest a second order dynamical system module.[1]

In addition, hysteresis is part of the cell dynamical pattern when cell state or phenotype not only depend on its current inputs abut also depends on current state but also depend on an internal state that is formed based on the past input signals. In other words, cell hysteresis arises because a cell's biological history affects the value of its current internal state. Studies shows that complex nonlinear gene regulatory networks demonstrate a threshold and hysteresis that lead to phenotypic switching [49]. Thus, sudden phenotypic transition over time can model in a step-like hysteresis module.

Furthermore, eukaryotic stress response pathways contain redundant components in the sensory level that make logical operation an easy module for integrating those multiple inputs. In our case, the output of each cells can be translated into a bit like (high-down, one-zero, ...) output. Thus, we suggest these multiple bit-like inputs integration occurs in biological modules similar to logical operators such as AND, OR operators (Fig. 7.4).

A simple example of a cell-cell network of the vicious triangle is shown in Fig. 7.3. During the bone niche construction, in the simplest model, populations obtain two states; cancer cells obtain two states, macrophages polarize to either M1 or

[1] A second order behavior of variable x under external stimuli $f(t)$ is given by a second order linear differential system as $\frac{d^2x}{dt^2} + a_1\frac{dx}{dt} + a_2x = f(t)$.

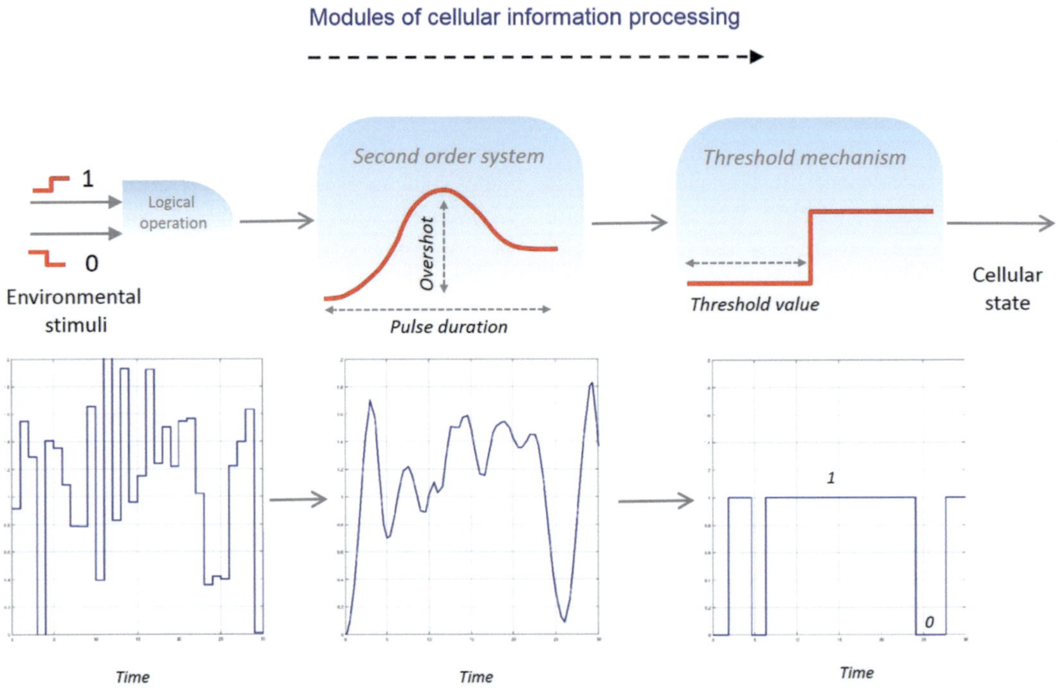

Fig. 7.4 Logical-Transient-Threshold model: A simple modularization of the cellular information processing, thereby leading phenotypic switching (*Top*) A simple second order linear differential equation creates the common cellular transient response patterns. Also, threshold mechanism sensitive to dynamical characteristics, such as transient overshot or pulse duration, switches the cells state to an alternative state or phenotype. The logical mechanisms of cells handles the multiple inputs from environment such as signals form other cells. (*Bottom*) A Simulink simulation of Logical-Transient-Threshold response to noisy step input. Each graph correspond to the block above it. A logical operator, here an OR, receives two noisy signals and produces an output signal (the first graph). This signal, a step like signal with a Gaussian noise $\sigma = 0.3$ is an input to a second order system $\frac{d^2x}{dt^2} + \frac{dx}{dt} + 0.7x$, sending a temporal profile to a hysteresis function that switches on (i.e. change of state to 1) at 1 and off (i.e. change of state to 0) at 0.8 of $x(t)$ value that is the output of the second order system

M2 and bone remodeling is either osteolytic or osteoblastic.

An example of the logical-transient-threshold model simulation for the three populations is demonstrated in Fig. 7.4. It shows the temporal profiles, sudden changes of cells phenotypic switching and how some rate limiting processes (based on threshold criteria, type of the logical operations on signals or the speed of transient response) lead to distinctive stages of tumor.

There are multiple steps during cancer metastatic niche construction. As cancer cells evolve through these intermediate steps, the slowest step, i.e. rate-limiting process, determines the rate of appearance of the phenotypic shifts.

The Logical-Transient-Threshold dynamics provides a simple tool to consider the rate-limiting factors during the cancer niche construction in tumor. In this model, rate-limiting factors are determined by cell's metabolic dynamics (second order dynamical system), cell's integration logic of the stimuli (logical operation) and cells' tolerance to these stimuli (threshold mechanism).

It can be seen, that even in a simple model, one can capture interesting phenomena. For example in Fig. 7.5, the number of cancer cells at the beginning at state 1 (for instance high proliferating) while the osteoclast cells proliferate (osteolytic metastatic stage). Later, the number of osteoclast cells decreases, and the cancer cells, after a delay, goes to state 2 (more proliferation). The population of M2-like Macrophages increases in the middle of the bone niche construction.

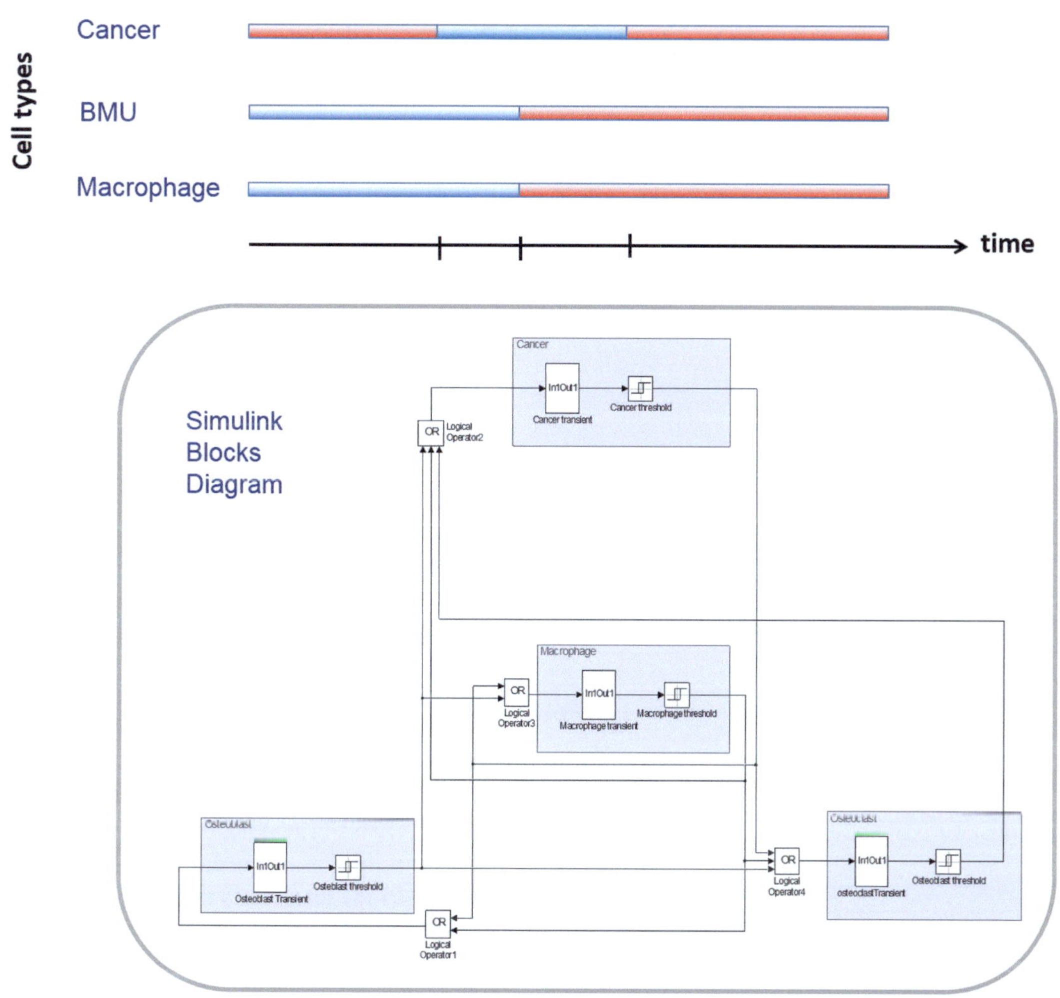

Fig. 7.5 Metastatic bone phenotypic changes over time (*Top*) An example of Simulink simulation results Transient-Threshold-logical provides an insight of rate-limiting steps during multistage carcinogenesis. The lines color change overtime as an indication of the cell state switching. Each population obtain two states (*red* or *blues*). Cancer have two states, bone is either in osteolytic or osteoblastic state, and macrophage polarization are either in M1-Like ot M2-like. (*Bottom*) The Simulink functional blocks networks composed of three logical operator, three transient response and three threshold mechanisms

7.4.1 Modules' Parameters Estimations

Logical – transient – threshold modules help to systematically design experiments for characterizing parameters of cellular transient behavior, criteria for transitions and types of the biological signal integration.

For instance, a second order gene regulatory transient module that undergoes a cyclic input, such as sinusoidal signal, can be fully characterized by utilizing control systems frequency response methods.

In addition, the threshold criteria of the cells state transition can be tested, for instance, by the genes response overshot, maximum value of the biological pulse, duration of the gene transient response, or mixture of the two (by pulse integration). Finally, experimental combination of multiple stimuli or co-culturing cells involved in tumor microenvironment can help to study the logical integration of multiple input signals.

Additional Temporal Modules

Nonlinear cellular growth: During the cancer initial adaptation in bone, there is little evidence of an increase in cell number as it is a period during which the cell attaches to the bone substrate and it can refer as lag phase of the population growth. However, to consider caner cell growth overtime, a growth module should be added to the model.

Differential equations from the simple one-parameter exponential growth model to more advanced models can capture the nonlinear growth of a cancer population [50]. As an example, the logistic (or Pearl-Verhulst) equation has successfully been applied to a large number of biologic phenomena, ranging from tumor populations to bacterial growth:

$$\frac{dC}{dt} = rC(1 - \frac{C}{K}) \tag{7.1}$$

In this equation, the dynamics of a cancer population C are described with an intrinsic growth rate r, and tumor microenvironment carrying capacity K.

Tumor macrophage recruiting dynamics: Cancer cells in the tumor microenvironment recruit macrophages in the metastatic bone by highly expressing a number of monocyte chemoattractants, including Chemo-attractant Protein 1(MCP-1), chemokine (C-C motif) ligands CCL2, CCL3, CCL4, CCL8 and CCL5 to recruit macrophages that increase in number and then differentiation inside the bone [5]. To model this macrophage recruiting mechanism, we can use dynamical equation

$$\frac{dM}{dt} = \frac{M_{max}C}{k + C} \tag{7.2}$$

where M, and C are macrophage and cancer cells intensity, receptively. k is a constant, and M_{max} is the maximum capacity of the bone for accepting macrophages.

7.5 Conclusion

The development of metastasis is a multistage process with several distinctive steps. Here, we propose a systematic method to find the rate-limiting steps that leads to phenotypic state transition during bone metastases, suggesting a logical-transient-threshold model by utilizing only a few cell decision modules.

We suggest a metastatic bone ecosystem with three components: the immune, cancer cell populations, and BMUs (osteoblasts and osteoclasts) system and highlight positive feedback systems role in the maintenance of the tumor bone metastases. In a simplified model, cells adopt only two states (M1/M2, osteolytic/osteoblastic, stage1/stage2). The model can be optimized together with experimental studies for hypothesis driven experimental studies in cell-cell signaling studies.

The logical-transient-threshold framework can be expanded for more complex tumor cell-cell networks. For instance, experimental data show that infiltrating tumor cells need to induce periostin (POSTN), a component of the extracellular matrix, that is expressed by fibroblasts in the stroma of the metastases target organ to initiate cancer colonization. Thus, stroma population can support metastatic colonization and may be added to the model [51]. Also, the number of the possible phenotypic states can increase above two choices; leading to even more complex, but potentially insightful models.

The bone marrow serves as a reservoir for tumor cells that can resist chemotherapeutic stress and predictive models may suggest new strategies for therapy. Our simple model reveals the potential importance of temporal profiling and timing of the cancer therapy. It proposes, that a successful cancer therapy should consider what threshold or transient responses may lead to the abnormal bone remodeling during cancer niche construction.

Acknowledgements NCI Grant numbers: U54CA143803, CA163124, CA093900, CA143055 supported this study.

References

1. Surveillance, epidemiology, and end results program-turning cancer data into discovery. http://seer.cancer.gov/statfacts/html/prost.html (2014)
2. Sims NA, John Martin T (2014) Coupling the activities of bone formation and resorption: a multitude of signals within the basic multicellular unit. BoneKEy Rep 3:481
3. Roato I (2013) Interaction among cells of bone, immune system, and solid tumors leads to bone metastases. Clin Dev Immunol 2013:1–7
4. Sica A, Larghi P, Mancino A, Rubino L, Porta C, Totaro MG, Rimoldi M, Biswas SK, Allavena P, Mantovani A (2008) Macrophage polarization in tumour progression. Semin Cancer Biol 18(5):349–355, Elsevier
5. Zarif JC, Taichman RS, Pienta KJ (2014) Tam macrophages promote growth and metastasis within the cancer ecosystem. OncoImmunology 3(7):e941734
6. Weilbaecher KN, Guise TA, McCauley LK (2011) Cancer to bone: a fatal attraction. Nat Rev Cancer 11(6):411–425
7. Liò P, Paoletti N, Moni MA, Atwell K, Merelli E, Viceconti M (2012) Modelling osteomyelitis. BMC Bioinform 13(Suppl 14):S12
8. Ayati BP, Edwards CM, Webb GF, Wikswo JP (2010) Research a mathematical model of bone remodeling dynamics for normal bone cell populations and myeloma bone disease. Biol Direct 5:1–17
9. Eftimie R, Bramson JL, Earn DJD (2011) Interactions between the immune system and cancer: a brief review of non-spatial mathematical models. Bull Math Biol 73(1):2–32
10. Ryser MD, Qu Y, Komarova SV (2012) Osteoprotegerin in bone metastases: mathematical solution to the puzzle. PLOS Comput Biol 8(10):e1002703
11. Kotas ME, Medzhitov R (2015) Homeostasis, inflammation, and disease susceptibility. Cell 160(5):816–827
12. Eladdadi A, Kim P, Mallet D (2014) Mathematical models of tumor-immune system dynamics, vol 107. Springer, New York
13. Papin JA, Reed JL, Palsson BO (2004) Hierarchical thinking in network biology: the unbiased modularization of biochemical networks. Trends Biochem Sci 29(12):641–647
14. Yosef N, Regev A (2011) Impulse control: temporal dynamics in gene transcription. Cell 144(6):886–896
15. Barcellos-Hoff MH, Lyden D, Wang TC (2013) The evolution of the cancer niche during multistage carcinogenesis. Nat Rev Cancer 13(7):511–518
16. Chaffer CL, Weinberg RA (2011) A perspective on cancer cell metastasis. Science 331(6024):1559–1564
17. Gupta PB, Fillmore CM, Jiang G, Shapira SD, Tao K, Kuperwasser C, Lander ES (2011) Stochastic state transitions give rise to phenotypic equilibrium in populations of cancer cells. Cell 146(4):633–644
18. Liddiard K, Taylor PR (2015) Understanding local macrophage phenotypes in disease: shape-shifting macrophages. Nat Med 21(2):119–120
19. Bronte V, Murray PJ (2015) Understanding local macrophage phenotypes in disease: modulating macrophage function to treat cancer. Nat Med 21(2):117–119
20. Guise TA, Mohammad KS, Clines G, Stebbins EG, Wong DH, Higgins LS, Vessella R, Corey E, Padalecki S, Suva L et al (2006) Basic mechanisms responsible for osteolytic and osteoblastic bone metastases. Clin Cancer Res 12(20):6213s–6216s
21. Morris MK, Saez-Rodriguez J, Sorger PK, Lauffenburger DA (2010) Logic-based models for the analysis of cell signaling networks. Biochemistry 49(15):3216–3224
22. Aström KJ, Murray RM (2010) Feedback systems: an introduction for scientists and engineers. Princeton university press, Princeton
23. Cannon WB (1929) Organization for physiological homeostasis. Physiol Rev 9(3):399–431
24. Kholodenko BN (2006) Cell-signalling dynamics in time and space. Nat Rev Mol Cell Biol 7(3):165–176
25. Guise TA, Mundy GR (1998) Cancer and bone 1. Endocr Rev 19(1):18–54
26. Scholtysek C, Kronke G, Schett G (2012) Inflammation-associated changes in bone homeostasis. Inflamm Allergy-Drug Targets (Former Curr Drug Targets-Inflamm Allergy) 11(3):188–195
27. Psaila B, Lyden D (2009) The metastatic niche: adapting the foreign soil. Nat Rev Cancer 9(4):285–293
28. Oskarsson T, Batlle E, Massagué J (2014) Metastatic stem cells: sources, niches, and vital pathways. Cell Stem Cell 14(3):306–321
29. Bussard KM, Gay CV, Mastro AM (2008) The bone microenvironment in metastasis; what is special about bone? Cancer Metastasis Rev 27(1):41–55
30. Sosnoski DM, Krishnan V, Kraemer WJ, Dunn-Lewis C, Mastro AM (2012) Changes in cytokines of the bone microenvironment during breast cancer metastasis. Int J Breast Cancer 2012:1–9
31. Zheng Y, Zhou H, Dunstan CR, Sutherland RL, Seibel MJ (2013) The role of the bone microenvironment in skeletal metastasis. J Bone Oncol 2(1):47–57
32. Kingsley LA, Fournier PGJ, Chirgwin JM, Guise TA (2007) Molecular biology of bone metastasis. Mol Cancer Ther 6(10):2609–2617
33. Logothetis CJ, Lin S-H (2005) Osteoblasts in prostate cancer metastasis to bone. Nat Rev Cancer 5(1):21–28
34. Wang H, Yu C, Gao X, Welte T, Muscarella AM, Tian L, Zhao H, Zhao Z, Du S, Tao J et al (2015) The osteogenic niche promotes early-stage bone colonization of disseminated breast cancer cells. Cancer Cell 27(2):193–210
35. Roodman GD (2001) Biology of osteoclast activation in cancer. J Clin Oncol 19(15):3562–3571
36. Walsh MC, Kim N, Kadono Y, Rho J, Lee SY, Lorenzo J, Choi Y (2006) Osteoimmunology: in-

terplay between the immune system and bone metabolism. Annu Rev Immunol 24:33–63
37. Allavena P, Sica A, Solinas G, Porta C, Mantovani A (2008) The inflammatory micro-environment in tumor progression: the role of tumor-associated macrophages. Crit Rev Oncol/Hematol 66(1):1–9
38. Vasiliadou I, Holen I The role of macrophages in bone metastasis. J Bone Oncol 2(4):158–166 (2013)
39. Biswas SK, Mantovani A (2010) Macrophage plasticity and interaction with lymphocyte subsets: cancer as a paradigm. Nat Immunol 11(10):889–896
40. Kim SW, Kim JS, Papadopoulos J, Choi HJ, He J, Maya M, Langley RR, Fan D, Fidler IJ, Kim S-J Consistent interactions between tumor cell il-6 and macrophage tnf-α enhance the growth of human prostate cancer cells in the bone of nude mouse. Int Immunopharmacol 11(7):862–872 (2011)
41. Colegio OR, Chu N-Q, Szabo AL, Chu T, Rhebergen AM, Jairam V, Cyrus N, Brokowski CE, Eisenbarth SC, Phillips GM et al (2014) Functional polarization of tumour-associated macrophages by tumour-derived lactic acid. Nature 513:559–563
42. Han J-H, Choi SJ, Kurihara N, Koide M, Oba Y, Roodman GD (2001) Macrophage inflammatory protein-1α is an osteoclastogenic factor in myeloma that is independent of receptor activator of nuclear factor κb ligand. Blood 97(11):3349–3353
43. Yago T, Nanke Y, Kawamoto M, Furuya T, Kobashigawa T, Kamatani N, Kotake S (2007) Il-23 induces human osteoclastogenesis via il-17 in vitro, and anti-il-23 antibody attenuates collagen-induced arthritis in rats. Arthritis Res Ther 9(5):R96
44. Kollet O, Dar A, Lapidot T (2007) The multiple roles of osteoclasts in host defense: bone remodeling and hematopoietic stem cell mobilization. Annu Rev Immunol 25:51–69
45. Comito G, Giannoni E, Segura CP, Barcellos-de Souza P, Raspollini MR, aroni G, Lanciotti M, Serni S, Chiarugi P (2014) Cancer-associated fibroblasts and m2-polarized macrophages synergize during prostate carcinoma progression. Oncogene 33(19):2423–2431
46. Purvis JE, Lahav G Encoding and decoding cellular information through signaling dynamics. Cell 152(5):945–956 (2013)
47. Behar M, Hoffmann A (2010) Understanding the temporal codes of intra-cellular signals. Curr Opin Genet Dev 20(6):684–693
48. Davis DM, Purvis JE (2015) Computational analysis of signaling patterns in single cells. Semin Cell Dev Biol 37:35–43, Elsevier
49. Thomas P, Popović N, Grima R (2014) Phenotypic switching in gene regulatory networks. Proc Natl Acad Sci 111(19):6994–6999
50. Gerlee P (2013) The model muddle: in search of tumor growth laws. Cancer Res 73(8):2407–2411
51. Malanchi I, Santamaria-Martínez A, Susanto E, Peng H, Lehr H-A, Delaloye J-F, Huelsken J (2012) Interactions between cancer stem cells and their niche govern metastatic colonization. Nature 481(7379):85–89

Microenvironmental Niches and Sanctuaries: A Route to Acquired Resistance

8

Judith Pérez-Velázquez, Jana L. Gevertz, Aleksandra Karolak, and Katarzyna A. Rejniak

Abstract

A tumor vasculature that is functionally abnormal results in irregular gradients of metabolites and drugs within the tumor tissue. Recently, significant efforts have been committed to experimentally examine how cellular response to anti-cancer treatments varies based on the environment in which the cells are grown. In vitro studies point to specific conditions in which tumor cells can remain dormant and survive the treatment. In vivo results suggest that cells can escape the effects of drug therapy in tissue regions that are poorly penetrated by the drugs. Better understanding how the tumor microenvironments influence the emergence of drug resistance in both primary and metastatic tumors may improve drug development and the design of more effective therapeutic protocols. This chapter presents a hybrid agent-based model of the growth of tumor micrometastases and explores how microenvironmental factors can contribute to the development of acquired resistance in response to a DNA damaging drug. The specific microenvironments of interest in this work are tumor hypoxic niches and tumor normoxic sanctuaries with poor drug penetration. We aim to quantify how spatial constraints of limited drug transport and quiescent cell survival contribute to the development of drug resistant tumors.

J. Pérez-Velázquez (✉)
Mathematical Modeling of Biological Systems, Centre for Mathematical Science, Technical University of Munich, Garching, Germany
e-mail: cerit@ma.tum.de

J.L. Gevertz
Department of Mathematics and Statistics, The College of New Jersey, Ewing, NJ, USA
e-mail: gevertz@tcnj.edu

A. Karolak
Integrated Mathematical Oncology, H. Lee Moffitt Cancer Center & Research Institute, Tampa, FL, USA
e-mail: Aleksandra.Karolak@moffitt.org

K.A. Rejniak
Integrated Mathematical Oncology, H. Lee Moffitt Cancer Center & Research Institute, Tampa, FL, USA

Department of Oncologic Sciences, College of Medicine, University of South Florida, Tampa, FL, USA
e-mail: Kasia.Rejniak@moffitt.org

K.A. Rejniak (eds.), *Systems Biology of Tumor Microenvironment*, Advances in Experimental Medicine and Biology 936, DOI 10.1007/978-3-319-42023-3_8

Keywords
Anticancer drug resistance • Hypoxic niche • Drug sanctuary • Tumor microenvironment • Agent-based model

8.1 Introduction

There is an ongoing discussion about the origin of anti-cancer drug resistance [1, 13, 25, 47]. Clinically, drug resistance is defined as a reduced effectiveness of treatment during or after the course of therapy. As a result, a cancer can progress even months or years after treatment, leading to tumor recurrence. Currently, no pretreatment tests or analyses can predict whether or not a particular patient will develop resistance.

Further, the contribution of two potential modes of resistance – pre-existing and acquired – is also challenging to determine. The term pre-existing resistance refers to the fact that, before the therapy is applied, the tumor may harbor some cells or cell clones that will not respond to the treatment. These cells are passively selected by the therapy, expand and overtake the population of non-resistant cells. In contrast, acquired resistance means that all cells are initially sensitive to the treatment, and that mechanisms of resistance develop during the therapy. The cells will then become progressively more tolerant to the drug as a result of treatment, and the end result is similar to what occurs in the case of pre-existing resistance: the population of non-resistant cells is overtaken by resistant ones (see Fig. 8.1).

In this chapter we will concentrate on the case of drug resistance that is induced by treatment (acquired resistance). Many experimental results support the idea that cancer cells may harbor the potential to acquire a variety of drug-resistance mechanisms activated in response to therapy [23, 50, 57, 60]. In the in vitro experiment described in [57], a lung cancer cell line was exposed to a targeted therapy, and a small population of cells that remained alive was subsequently cultured in the presence of that drug for a year. This created a handful of "persister" cells that survived a prolonged drug exposure with almost no population growth. These cells attained protection from lethal drug levels and, finally, gained ability to proliferate in spite of the drug. The extensive screening of drug-tolerant cell colonies that arose from persister cells revealed that they exhibit diverse resistance mechanisms. This experiments showed that heterogeneous resistance mechanisms can emerge in cells derived from the same cell line and exposed to identical selective pressures. Another example is DNA damage caused by anticancer methylating agents that is typically processed by methyltransferases or directed for mismatch repair (MMR). Progressive inactivation of MMR by reducing or inhibiting the expression of repair proteins, for example under hypoxic conditions, leads to increased tumor cell tolerance to this intended damage [28]. Similarly, in the case of DNA-crosslinking cisplatin, the tumor cell may bypass cisplatin-DNA adducts by translesion synthesis polymerases [14, 65].

The focus in this work is on a DNA damaging drug. The mechanism through which cancer cells could acquire resistance to such a drug arise from the natural tendency of cells to preserve their genomic integrity – even normal cells continuously fight against endogenous metabolic products and exogenous damaging compounds through various repair mechanisms. Agents damaging DNA on the level of a single base, a dinucleotide, strand crosslinks, strand breaks, or alkylating agents all activate the response from DNA damage repair machinery (DDR) [10]. DDR activates various mechanisms, such as nucleotide excision repair, base excision repair or MMR. These repair mechanisms work individually or in parallel to maintain genomic stability. Under normal conditions they should lead the cell to arrest in one of the cell cycle checkpoints that will resume only after the damage is repaired or, they should result in a permanent cell-cycle arrest and cell apoptosis if the damage is unrepairable [6]. The cytotoxic agents used in many anticancer therapies are designed to target various DNA sites with the ultimate goal to overcome repair mechanisms

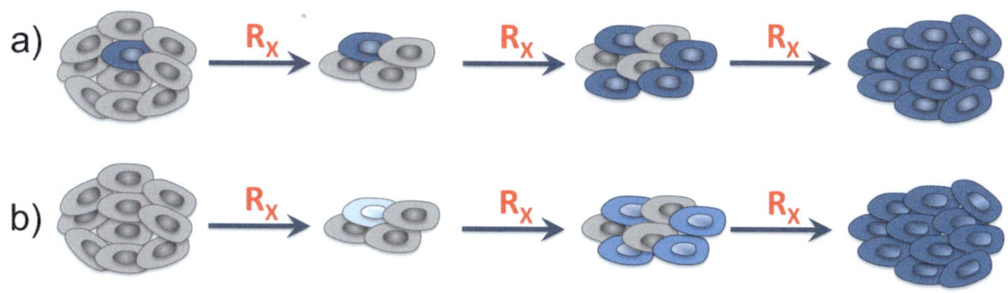

Fig. 8.1 Schematics of the development of pre-existing and acquired drug resistance in tumor tissue under continuously administered treatment. (**a**) In the pre-existing resistance case, the resistant cells are selected for and overgrow the sensitive cells. (**b**) In the acquired resistance case, the cells progressively acquire resistance as a result of being exposed to the drug. The final outcome in both cases is a resistant tumor, which makes it difficult to determine what mechanism(s) of resistance a given tumor possesses

and lead to cell death. Nonetheless, as a part of a natural cell response to damage, cells try to escape the arrest and maintain genomic stability. Here, in addition to transcription-related DDR pathways, the replication-related DNA-damage tolerance (DDT) mechanisms are activated leaving unrepaired or misrepaired DNA modifications to be repaired later and the cell proceeds with its cell cycle [9, 19]. This increases cell tolerance to accumulation of DNA damage and, although highly disadvantageous for therapeutic outcomes, such effect is commonly observed in anticancer therapies.

8.2 Microenvironmental Niches and Sanctuaries

It has been postulated that tumor cells can survive in dormant niches in a growth-arrested state for many years before they re-enter proliferation [17]. Tumor niches are specialized regions in tumor microenvironment that can provide required factors for tumor cell survival, and may be permissive either for tumor initiation, growth or progression to metastasis. Several kinds of microenvironmental niches have been identified and are subject of ongoing research, among these are pre-cancerous, pre-metastatic or stem-cell niches [3, 5, 18, 27, 38, 39, 49, 51]. Such regions allow the tumor cells to survive and maintain a quiescent state while regulating their metabolic needs and immune cell surveillance. The niche conditions can also influence specific changes in gene expression levels and can induce DNA mutations or cell differentiation mechanisms that may lead to tumor cell invasion and metastatic spread [8, 20, 21, 40].

Tumor tissues may also contain regions that are poorly penetrated by therapeutics, forming what are termed drug-limited or pharmacologic sanctuaries [11, 56]. One of the main sanctuary sites in our body is the central nervous system [48, 56]. Since the blood-brain barrier (BBB) prevents tumor cells from drug exposure, the cells are able to escape the effects of drug therapy and maintain fast growth forming brain metastases. Beyond this organ-level sanctuary, the non-uniform distribution of drugs within a specific tissue can form tissue-level sanctuaries.

Multiple factors can contribute to the formation of these tissue-level sanctuaries. For instance, drug interstitial transport depends not only on physicochemical properties of the drug, but also on the structure of the extracellular matrix that may hinder drug transport, for example in regions of higher fiber density or cross-linking. Cellular architecture of the tissue may also prohibit free diffusion of drug molecules. Moreover, drug internalization may vary between individual cells. While measurements of drug concentrations in blood plasma are easy to obtain, the distribution within a tissue compartment is more difficult to quantify. Recent developments using in vivo imaging techniques now allow for the visualization of heterogeneous drug distributions within tissues, as well as the visual assessment of cells' response to fluorescently labeled drugs or fluorescent non-therapeutic imaging agents [43, 55, 63, 64].

In an effort to better understand the impact of drug sanctuaries, attempts have been made to quantify tumor cell response in various conditions. A comprehensive set of in vitro experiments looked at three different tumor cell lines grown in 54 homogeneous microenvironments that differed from one another in the levels of glucose and oxygen, and with diverse concentrations of the clinically used drug erlotinib [44]. Quantitative time series data were collected and subsequently used to calibrate a non-spatial stochastic branching model containing populations of drug-sensitive and drug-resistant cells. Simulations of this experimentally-calibrated model showed that the tumor microenvironment has a strong influence on tumor evolutionary dynamics under the pulsed drug administration schedule. In particular, the computational model demonstrated that under heterogeneous microenvironmental conditions there was no reduction in tumor burden. On the other hand, the same tumors in homogeneous conditions responded positively to the therapy.

Therefore, heterogeneities in the tumor microenvironment, such as tumor-associated niches and sanctuaries, may play a crucial role in tumor promotion, survival, progression, and response to therapies. Can these niches (formed due to actions of drug and/or stromal cells) and/or sanctuaries (formed as a result of limited drug penetration) that emerge within the tumor microenvironment promote anti-cancer drug resistance by enabling drug-induced tolerance? In this chapter we will address this question by studying tumor response to a simulated DNA damaging agent using a spatial agent-based model with explicit heterogeneous tissue morphology.

8.3 The Mathematical Model of the Tumor and Its Microenvironment

In order to create a heterogeneous tumor microenvironment, we consider a small patch of tissue with four non-evolving blood vessels placed inside the tissue in an irregular pattern as shown in Fig. 8.2. These vessels supply both oxygen and drug that subsequently diffuse through the domain and are absorbed by tumor cells. Additionally, we assume that the drug is subjected to decay due to its half-life. Tumor cells are modeled as individual entities whose behavior is modulated by both the properties inherited from their mother cells (the age at which the cells can divide, the initial tolerance to DNA damage, the amounts of accumulated drug and DNA damage), and by their immediate environment

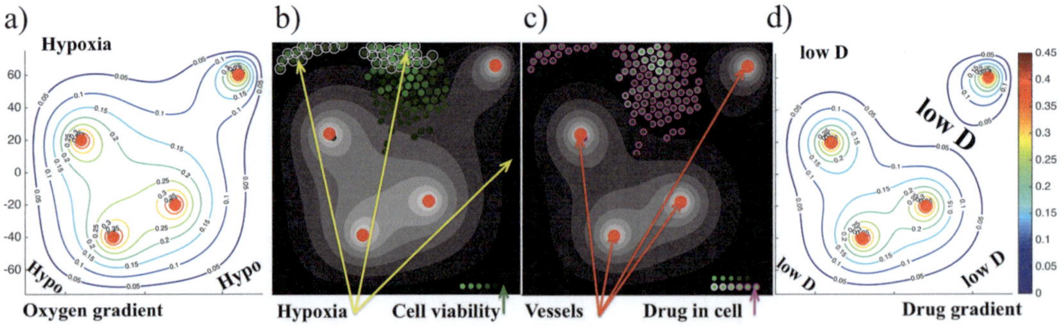

Fig. 8.2 Schematics of the tumor microenvironment with metabolic and drug gradients. (**a**) Oxygen supplied by four vessels (*red circles*) creates a gradient (*color-coded contours lines*) due to its diffusion through the tissue and cellular uptake – the levels of low oxygen (hypoxia) are indicated; (**b**) the corresponding oxygen gradient (*shades of gray*) with hypoxic cells circled and cell viability levels (based on how much cell's DNA has been damaged) are represented by shades of green; (**c**) the corresponding drug gradient (*shades of gray*) with intracellular drug accumulation represented by shades of purple; (**d**) color-coded contours of drug gradient – the levels of low drug concentration are indicated by low D

(the levels of extracellular oxygen and drug, and their interactions with the neighboring cells). For example, cell division and cell relocation can be suppressed due to lack of free space and cellular overcrowding. Another example is that cells can become dormant if they move to a tissue region with hypoxic levels of oxygen. A cell's initial viability is also inherited from its mother cell because both the DNA damage and the cell's tolerance to DNA damage are passed from a mother to both daughter cells. This model setup enables tracing of both cells' ancestry and how the cell properties evolve upon treatment.

A distorted tumor vasculature such as the one we are considering results in irregular gradients of metabolites. This includes significant regions of low oxygen content (hypoxia), which is often observed in animal tumors and in biomedical images used in the clinic for diagnostic purposes [31, 62]. In our model, we placed blood vessels in order to create regions that contain no more than 5 % of the oxygen level found in the most oxygenated areas near the blood vessels (Fig. 8.2a). Since both oxygen and drug are supplied by the same vascular system, it may seem that the resulting gradients will also be similar. However, the actual levels of oxygen and drug at the same spatial location may be different since the interstitial transport of oxygen and drug depend not only on their vascular influx and interstitial diffusion, but also on cellular uptake (which may be different), and the decay in the case of the drug.

Figures 8.2a, d show contours of both extracellular gradients in the same scale. These figures indicate that the regions near the domain boundaries are hypoxic and contain low concentrations of drugs (we call such regions the hypoxic niches), and that the region in between the vessel in the top-right corner and three other vessels have normoxic levels of oxygen but low levels of the drug (we call such tissue regions the sanctuaries). Figures 8.2b, c show cellular response to the gradients in these specific tissue regions (grey-scale gradients in these figures are identical to the contours shown in Figs. 8.2a, d, respectively). In Fig. 8.2b the hypoxic cells are circled, and the gradient of oxygen from Fig. 8.2a is shown in the background. The same cells in Fig. 8.2c are color-coded according to the accumulated levels of drug, and the extracellular gradient of the drug from Fig. 8.2d is shown in the background. The cells that were recently born are colored in light purple, since they inherited half of the drug accumulated in the mother cell. Upon exposure to the drug, cells accumulate damage and become less viable, which is color-coded in Fig. 8.2b – the lighter the color the more viable the cell is. Note that the clusters of cells in the hypoxic areas are more viable, while the cells near the vasculature are close to reaching the DNA damage tolerance level that will result in cell death.

Mathematically, our modeling approach combines the off-lattice agent-based model of individual cells with physical cell-cell interactions, and the continuous description of extracellular nutrients and metabolites, as well as stiffness/viscous properties of the extracellular matrix. The temporal and spatial changes in oxygen ξ and drug γ concentrations in the interstitial space are given in Eqs. (8.1) and (8.2). These metabolites are supplied from the vessels with the supply rates S_ξ and $S_\gamma(t)$, respectively. They diffuse through the extracellular space with diffusion coefficients D_ξ and D_γ. They are absorbed by the cells with uptake rates ρ_ξ and ρ_γ. Additionally, drug decays with decay rate d_γ. Each cell $C_k(t)$ is defined by its position $C_k^{(X,Y)}(t)$, and is characterized by several properties, such as current cell age $C_k^{age}(t)$, and cell maturation age at which the cell is ready to divide C_k^{mat}. Moreover, cell behavior depends also on the level of sensed oxygen $C_k^{\xi}(t)$, the amount of accumulated drug $C_k^{\gamma}(t)$, the duration of cell exposure to high drug concentration $C_k^{exp}(t)$, the level of accumulated damage $C_k^{dam}(t)$, and the level of damage that the cell can withstand without committing to death (also called the level of tolerated damage or death threshold) $C_k^{death}(t)$. The quantity of oxygen taken up by the cell is described in Eq. (8.3), and that of drug in Eq. (8.4). Both of these metabolites are obtained from the local cell neighborhood χC_k, as defined in Eq. (8.5).

Equations 1 present the continuous description of the interstitial kinetics of oxygen (8.1) and drug (8.2); cellular uptake of the oxygen (8.3) and drug (8.4); definition of the cellular C_k and vascular V_j neighborhoods with radii R_C and R_V for cellular uptake and vascular supply, respectively (8.5).

$$\frac{\partial \xi(\mathbf{x},t)}{\partial t} = \underbrace{\mathcal{D}_\xi \Delta \xi(\mathbf{x},t)}_{diffusion} - \underbrace{\min\left(\xi(\mathbf{x},t), \rho_\xi \sum_k \chi_{C_k}(\mathbf{x},t)\right)}_{uptake\ by\ the\ cells} + \underbrace{S_\xi \sum_j \chi_{V_j}(\mathbf{x},t)}_{supply}, \tag{8.1}$$

$$\frac{\partial \gamma(\mathbf{x},t)}{\partial t} = \underbrace{\mathcal{D}_\gamma \Delta \gamma(\mathbf{x},t)}_{diffusion} - \underbrace{d_\gamma \gamma(\mathbf{x},t)}_{decay} - \underbrace{\min\left(\gamma(\mathbf{x},t), \rho_\gamma \sum_k \chi_{C_k}(\mathbf{x},t)\right)}_{uptake\ by\ the\ cells} + \underbrace{S_\gamma(t) \sum_j \chi_{V_j}(\mathbf{x},t)}_{supply}, \tag{8.2}$$

$$C_k^\xi(t+\Delta t) = \underbrace{\sum_{\mathbf{x}} \xi(\mathbf{x},t)}_{sensed\ \&\ used}, \quad \text{where } \mathbf{x} \in \chi_{C_k}(\mathbf{x},t), \tag{8.3}$$

$$C_k^\gamma(t+\Delta t) = C_k^\gamma(t) + \left[\max\left(0, \sum_{\mathbf{x}} \underbrace{\min\left(\gamma(\mathbf{x},t), \rho_\gamma\right)}_{uptake} - \underbrace{d_\gamma C_k^\gamma(t)}_{decay}\right)\right]\Delta t, \quad \mathbf{x} \in \chi_{C_k}(\mathbf{x},t), \tag{8.4}$$

$$\chi_{C_k}(\mathbf{x},t) = \begin{cases} 1 \text{ if } \left\|\mathbf{x} - C_k^{(X,Y)}(t)\right\| < R_C \\ 0 \text{ otherwise,} \end{cases} \quad \chi_{V_j}(\mathbf{x},t) = \begin{cases} 1 \text{ if } \left\|\mathbf{x} - V_j^{(X,Y)}(t)\right\| < R_V \\ 0 \text{ otherwise.} \end{cases} \tag{8.5}$$

The drug-induced damage depends on the current increase in drug consumed by the cell (drug uptake minus drug decay) and on damage repair, that is proportional to the current damage with rate p (Eq. (8.6)). The damage level tolerated by the cells increases when the cell is exposed to high enough drug concentrations γ_{exp} for a prolonged time t_{exp} (Eqs. (8.7) and (8.8)). When the cell $C_k(t)$ divides, one of its daughter cells takes the coordinates of its mother: $C_{k_1}{}^{(X,Y)}(t) = C_k{}^{(X,Y)}(t)$, whereas the second is placed randomly near the mother cell: $C_{k_2}{}^{(X,Y)}(t) = C_k{}^{(X,Y)}(t) + R_C$ $(cos\theta,\ sin\theta)$, where R_C is the cell radius. The current age of both daughter cells is set to 0, and the cell maturation age is inherited with a small noise term. The cell damage, damage tolerance and the current drug exposure time are all inherited from the mother cell. On the other hand, the level of accumulated drug is split in half between both daughter cells. Finally, the level of sensed oxygen is determined independently for each cell based on the oxygen concentration in the cell's vicinity.

In order to keep the neighboring cells from overlapping, repulsive forces are applied as described in Eq. (8.9). These overdamped springs (ν is a damping coefficient) allow the whole multicellular system to return to equilibrium without oscillations. The repulsive forces with spring stiffness F_{spr} are used to move the cells apart until they reach the distance equal to cell diameter ($2R_C$) as defined in Eq. (8.10) for two neighboring cells, and in Eq. (8.11) for multiple cells. For simplicity, we do not include other microenvironmental components, such as stromal cells and other metabolites. More details of the model, its implementation, and the parameter self-calibration can be found in [16].

Equations 2 define accumulation of cell damage (8.6), increase in cell tolerance to damage (8.7), and cell exposure time to the drug (8.8); mechanical equations of cell movement (8.9) determined by repulsive forces between two (8.10) and multiple (8.11) neighboring cells.

$$C_k^{dam}(t+\Delta t) = C_k^{dam}(t) + \left[\max\left(0, \underbrace{\sum_{\mathbf{x}} \min\left(\gamma(\mathbf{x},t), \rho_\gamma\right)}_{uptake} - \underbrace{d_\gamma C_k^\gamma(t)}_{decay}\right) - \underbrace{pC_k^{\mathrm{dam}}(t)}_{repair}\right]\Delta t, \quad (8.6)$$

$$C_k^{death}(t+\Delta t) = \begin{cases} C_k^{death}(t) + \Delta_{death}\,\Delta\, t \text{ if } C_k^{exp}(t) > t_{exp} \\ C_k^{death}(t) \quad \text{otherwise,} \end{cases} \quad (8.7)$$

$$C_k^{exp}(t+\Delta t) = \begin{cases} C_k^{exp}(t) + \Delta t \text{ if } C_k^\gamma(t) > \gamma_{exp} \\ 0 \quad \text{otherwise,} \end{cases} \quad (8.8)$$

$$F_i = -\nu \frac{d\mathbf{X}_i}{dt} \quad \text{and} \quad \mathbf{X}_i(t+\Delta t) = \mathbf{X}_i(t) - \frac{1}{\nu} F_i \Delta t. \quad (8.9)$$

$$f_{i,j} = \begin{cases} \mathcal{F}_{spr}\left(2R_C - \|\mathbf{X}_i - \mathbf{X}_j\|\right) \frac{\mathbf{X}_i - \mathbf{X}_j}{\|\mathbf{X}_i - \mathbf{X}_j\|} \text{ if } \|\mathbf{X}_i - \mathbf{X}_j\| < 2R_C \\ 0 \quad \text{otherwise,} \end{cases} \quad (8.10)$$

$$F_i = \underbrace{\mathcal{F}_{spr}\left(2R_C - \|\mathbf{X}_i - \mathbf{X}_{j1}\|\right) \frac{\mathbf{X}_i - \mathbf{X}_{j1}}{\|\mathbf{X}_i - \mathbf{X}_{j1}\|}}_{f_{i,j1}} + \cdots + \underbrace{\mathcal{F}_{spr}\left(2R_C - \|\mathbf{X}_i - \mathbf{X}_{jM}\|\right) \frac{\mathbf{X}_i - \mathbf{X}_{jM}}{\|\mathbf{X}_i - \mathbf{X}_{jM}\|}}_{f_{i,jM}}, \quad (8.11)$$

8.4 Non-resistant Tumor Dynamics Under Treatment

In the model described here, the drug is supplied continuously from four blood vessels inside the tumor. We consider a small cluster of tumor cells, as could arise in a micrometastasis. In the case of non-resistant cells, cell tolerance to the drug-induced damage does not change in time. Thus continuous drug exposure will eventually cause each cell to accumulate a damage level in excess of the threshold, which will result in tumor eradication. The time course of tumor progression in this example is shown in Fig. 8.3.

At the initiation of the simulation there is no damage, and there is a uniform (as indicated by a standard deviation of zero) damage tolerance level of 0.5, as shown in Fig. 8.3a. The initial 65 cells shown in Fig. 8.3b proliferate

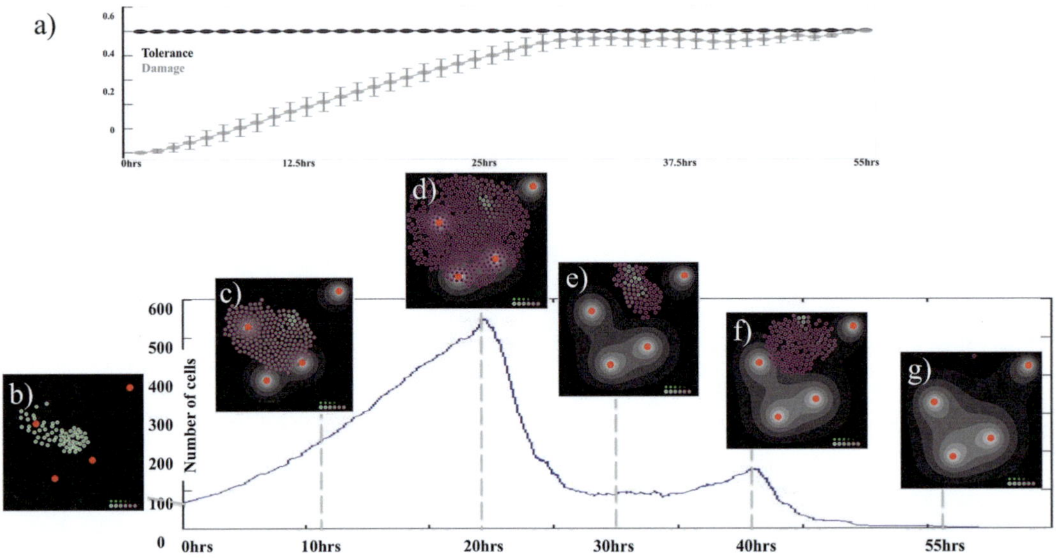

Fig. 8.3 Time course of non-resistant tumor progression. The bottom curve indicates the total number of cells over time. The left-top inset (**a**) shows the time course in both accumulated damage (*grey*) and damage tolerance (*black*) averaged over all tumor cells at a given time (standard deviation also shown); note, that the averaged damage tolerance is constant for non-resistant cells. The insets (**b**–**g**) show tumor morphology at the specific time points with the internally accumulated drug (*shades of purple*)

extensively while the absorbed drug leads to the accumulation of drug-induced damage – note the steady increase in the average damage levels in Fig. 8.3a. Cells primarily located near the vessels quickly accumulate damage levels above their tolerance level, and as a result die. Dying cells give the remaining cells space to divide, and the net result is steady tumor growth during the first 20 simulated hours (Figs. 8.3c, d), with the tumor reaching its maximal size of 528 cells after 20 h (Fig. 8.3d). The newly divided cells are indicated by a light purple color, since daughter cells inherit half of the drug accumulated by the mother cell.

Most of the newly divided cells are located on the rim of the tumor cluster. However, during the 20 h of drug exposure, many cells have accumulated such high levels of damage that they rapidly died after that time (Fig. 8.3e). At 30 h, a small cluster of surviving cells is located in the area of low drug content (the tissue sanctuary) that allows for a short rebound in tumor growth between 30 and 40 h (Fig. 8.3f). During this time, the DNA damage averaged over the total number of cells is very close to the value of cell tolerance to DNA damage shown in Fig. 8.3a (note, that for the non-resistant tumor the tolerance to DNA damage is identical for all cells and is constant in time). However, the accumulated damage of the 153 surviving cells after 40 h is very high and the whole tumor is eradicated within 55 h (Fig. 8.3g). In Fig. 8.3a this is confirmed by the close proximity of both averaged curves over this time with, finally, the damage curve intersecting the damage tolerance curve at the time when all tumor cells die.

8.5 Tumor Dynamics with Acquired Resistance Under the Treatment

In contrast to the non-resistant case in which tumor cells have a constant tolerance, in the case of acquired resistance the tolerance of individual cells to DNA damage can increase as a result of prolonged drug exposure. The cells can still die when they are damaged beyond the tolerance threshold, but cell fate depends on whether or not the cell tolerance level grows faster than the level of cell damage. The time course of tumor

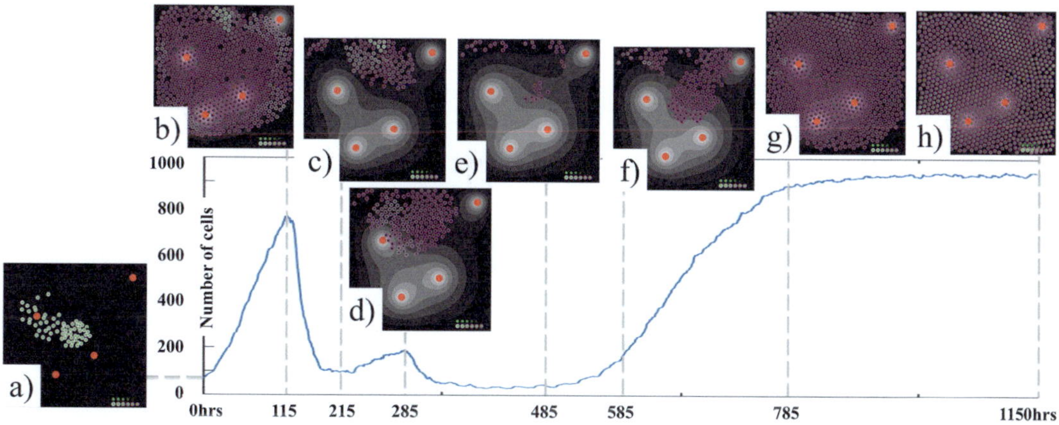

Fig. 8.4 Time course of progression of a tumor with acquired resistance. The curve indicates the total number of cells over time. The insets (**a–h**) show tumor morphology at the specific time points with the internally accumulated drug (*shades of purple*)

progression when the individual cells can acquire resistance to the drug (at rate 9.5×10^{-5} per iteration) is shown in Fig. 8.4. The initial 65 cells seen in Fig. 8.4a proliferate extensively, reaching about 800 cells in 115 h as shown in Fig. 8.4b. Note that the non-resistant tumor was completely eradicated in less than half of this time. This indicates that cell tolerance levels, at least in some cells, have increased above the level of non-resistant cells leading to a tenfold expansion of the tumor.

Although tolerance levels are increasing in many cells, it has not necessarily increased fast enough for a subpopulation of the cells to evade death. In fact, a large subpopulation of tumor cells has died, and the tumor shrunk to about 100 cells in 215 h, as shown in Fig. 8.4c. In the next period of time it is evident that a competition between increasing cell tolerance to the drug induced damage and increasing cell DNA damage due to the drug exposure takes place. This is evidenced by fluctuations in tumor size. The tumor grows to about 200 cells in next 70 h (shown in Fig. 8.4d), and then rapidly retreats to about 30 cells. This small cell population is located in the hypoxic niches, in which drug concentrations are negligible. As a result, no significant damage is induced in these dormant cells, thus the cells can continue to repair DNA damage. With lower cell damage and increased tolerance, these cells expand fivefold in a short amount of time when the local oxygenation changes slightly (Figs. 8.4e, f). Since the damage and tolerance properties are passed to the daughter cells, the tumor is able to avoid complete eradication. While individual cells can still die, the tumor as a whole overtakes the available tissue space – it reaches about 950 cells (Fig. 8.4g), and remains at that size for the rest of the simulation as shown in Fig. 8.4h.

8.6 Development of a Drug Resistant Tumor

The lack of tumor shrinkage at the end of the simulation shown in Fig. 8.4 does not directly imply that the tumor became resistant. However, here we prove that the tumor in Fig. 8.4 is resistant to the simulated DNA damaging agent by analyzing tumor cell viability. In our model cell viability is defined as how close the level of accumulated damage is to reaching the level of damage that the cell can tolerate (damage tolerance minus damage accumulated). In case of the non-resistant tumor shown in Fig. 8.3, the tolerance level was the same for every cell and constant during the entire simulation. Thus, with the increase of drug-induced damage, cell viability was continually decreasing, eventually leading to death of the whole tumor.

In the case of the resistance-acquiring tumor shown in Fig. 8.4, cell tolerance to damage can be

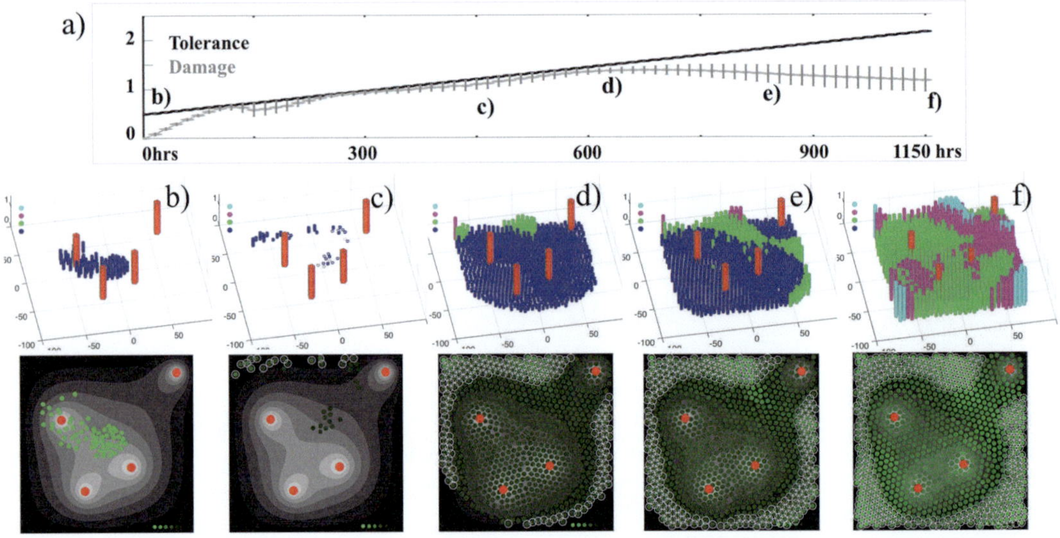

Fig. 8.5 Cell viability graphs showing the emergence of a resistant tumor. (**a**) The averaged levels of cell tolerance to damage (*black*) and cell accumulated damage (*grey*) shown over 1150 h of simulated time. The vertical lines indicate standard deviations. (**b–f**) Five snapshots showing individual cell viability levels defined as a difference between cell tolerance to damage and cell damage; top row: color-coded from blue (low viability) to green, to magenta, to cyan (high viability); bottom row: the corresponding images show cells which viability is color-coded from light green (more viable cells) to dark green (less viable cells). Circled cells represent hypoxic cells, and red cylinders and circles represent four vessels

amplified for each cell independently in response to the local drug concentration (i.e., damage tolerance is no longer constant). Therefore, for a tumor to be resistant, its tolerance to damage has to increase faster than the accumulated damage (on average). In this way, the average cell viability would eventually become a monotonically increasing function in time. Figure 8.5 graphically shows changes in tumor viability (Fig. 8.5a), and changes in the viability of each individual cell during the same simulation as shown in Fig. 8.4. Each vertical cylinder in Figs. 8.5b–f represents a cell and is placed spatially where the cell is found within the tissue (red cylinder represent the vessels). The height of the cylinder is equal to the cell viability; that is, the height represents difference between cell tolerance level and cell damage level. Thus the lower the cylinder is, the closer the cell is to death. The taller the cylinder, the larger the gap between cell damage and the tolerance level, and the more viable the cell is.

The cell viability plots in Fig. 8.5 help us to study the emergence of drug resistance. The initial 65 cells shown in Fig. 8.5b have no damage, and all cylinders have a height equal to the non-resistant damage tolerance level. However, upon exposure to the drug the damage level in most cells increases at a faster rate than their tolerance levels increase. Therefore most of cells die, leaving only a small population of viable cells (Fig. 8.5c), the same as in Fig. 8.4e. As it has been shown in Fig. 8.4f, the cells in hypoxic regions have negligible exposure to the drug, thus cell damage does not increase for these cells, and the cell damage repair is relatively more effective leading to increased viability. This is illustrated by the taller green cylinders in Fig. 8.5d, again with a taller height being indicative of an increase in cell viability. This is also illustrated by lighter green cells in the corresponding image in the bottom row of Fig. 8.5. Note that these more tolerant cells are located within the hypoxic region. Subsequently, it is visible in Fig. 8.5e that cells with increased viability are located along an off-diagonal region between the vessel in the right top corner and the three remaining vessels. Lower levels of drug and low but normoxic levels

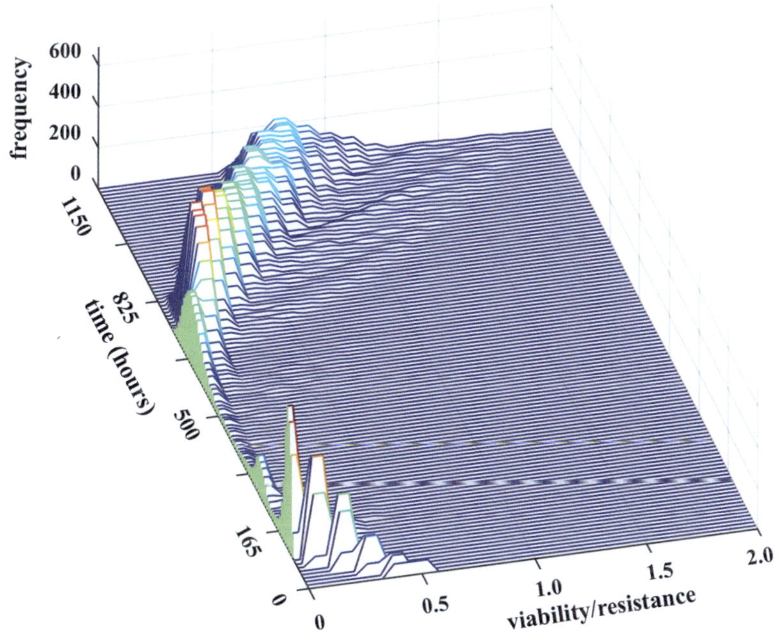

Fig. 8.6 Histogram surface of cell viability (with values between 0 and 2) recorded over the time of 1150 h (with frequency of 15 h). Initially, cell viability is diminished due to damage induced by the drug, leading to significant tumor reduction (between 165 and 500 h); however, at the later stages (after 825 h), there is a steady increase in cell viability leading to tumor resistance

of oxygen characterize this region, and represent common features of a tissue sanctuary, where drug penetration is limited.

The remaining cells, indicated by blue cylinders in Fig. 8.5f, are sensitive to the drug and die, leaving space for the more tolerant cells to divide. With time, this allows the population of more tolerant cells to expand. This is clearly indicated in the time course plot in Fig. 8.5a in which we observe a progressively increasing divergence between the averaged damage tolerance and the average accumulated damage (compare Figs. 8.5e, f). Since the disparity between two curves (and therefore, the cell viability) is continuously increasing, we interpret that as the emergence of acquired drug resistance. Moreover, in this particular case chemotherapy could not overcome acquired resistance. While individual cells can still be killed by the DNA damaging drug, the tumor as a whole survives and becomes nonresponsive to the drug.

Our observations are further confirmed by the histogram surface of cell viability shown in Fig. 8.6. Here, for the consecutive time points (from 0 to 1150 h) we constructed the histograms for cell viability with centers from 0.1 to 2.0 in intervals of 0.1. At the beginning of the simulation all cells had viability of 0.5. At early times, the viability of a large fraction of cells decreases (is in the range 0.1–0.5) due to increased drug-induced cell damage. However, with more time, cell viability increased in a majority of cells, and the fraction of cells with low viability drastically decreased (after 825 h, there are no cells with near zero viability). Again, the trend of increasing cell viability indicates that the tumor cells individually and as a whole population acquired resistance to the drug.

8.7 Tissue Niches and Sanctuaries and Their Relation to Resistance

Model simulations presented in this chapter showed that two tumor regions play an important role in the emergence of drug resistance – one with a low content of oxygen and drug (a hypoxic niche), and a second with limited drug penetration, but normoxic oxygenation (a tumor sanctuary). In particular, we have shown that when the tumor cells reach hypoxic niches they attain a dormant, non-proliferative state, which allowed them to overcome accumulated damage due to still active DNA repair mechanisms. Some such "persister" cells in our simulation were able to survive for more than 80 h. Our computational study also pointed to tissue sanctuaries in which tumor cells are protected from lethal effects of the drug. We showed that in these regions the tolerance to drug-induced damage is rising faster than the damage itself, which, in turn, leads the cells to acquire drug resistance.

While the concept of a tissue sanctuary has previously been reported in cancer biology, it was mostly used in reference to metastatic brain sanctuaries that arise due to the blood brain barrier [48, 56] and bone marrow sanctuaries that arise due to their protective structure [12, 41]. Recently, this concept of drug sanctuaries has been adopted in the context of acquired drug resistance [15]. In this work the authors argued that, in contrast to the belief that the development of resistance is caused by genetic heterogeneity, spatial heterogeneity in drug concentrations can be a driving force in the emergence of drug resistance. They showed that resistance is likely to arise in cells found in sanctuaries with poor drug penetration, and later these cells can invade non-sanctuary compartments. They also suggested that cell migration may be pivotal to this process, and that a certain combination of cell migration rate and spatial heterogeneity level can accelerate the development of resistance. Although our model does not include migration, the fact that resistance arises in low-drug sanctuaries and can later result in resistance in high-drug regions is consistent with our findings.

The concept of dormant persister tumor cell phenotype has been reported experimentally [23, 50, 57, 60], though not in hypoxic niches. In in vitro experiments, dormant persister tumor cells were able to survive treatment for a long time before gaining the ability to grow and proliferate in the presence of drug. Moreover, the surviving clones acquired various drug-induced resistance mechanisms. This has an important clinical application. The fact that they can survive in treatment for a prolonged time and develop various forms of resistance indicates that novel approaches should be developed towards exploiting vulnerabilities of such drug-tolerant persister cells. If such persister cells are not removed or suppressed, the typical anti-cancer chemotherapy will lead to the emergence of heterogeneously resistant persister-derived clones instead of leading to tumor eradication.

8.8 Summary and Outlook

The work presented in this chapter is a more detailed analysis of one particular example from the model developed by us in [16]. We considered a tumor with a parameter Δ_{death} (defining how cell tolerance to drug-induced damage is changing) equal to 9.5×10^{-5}, which we previously showed to be in a parameter regime that results in the emergence of a drug resistant tumor. As first discovered in [16] and detailed here, hypoxic niches and low drug sanctuaries play a very important role in the dynamics of acquired drug resistance.

However, in [16] we also showed that the microenvironment plays a less significant role in the tumor dynamics when resistance to the drug is pre-existing. In the pre-existing resistance case, a certain (small) population of tumor cells was not responsive to the drug even before the treatment was applied. We demonstrated that a simplified, spatially-invariant version of this model can predict long-term tumor response (tumor eradication, survival of only drug resistant clones, complete treatment failure) when resistance is pre-existing, indicating that the spatial (microenvironmental) considerations are not a key driving force in long-term dynamics. This occurs because the inherent fitness advantage that the cells with pre-existing resistance possess in the presence of drug can dominate (in the long-term) over any spatial considerations. In contrast for the acquired resistance case, the dynamics predicted by the spatially-invariant model are vastly different than those predicted by the full spatial model [16]. These differences point to the key role of niches and sanctuaries in treatment dynamics when drug resistance is acquired.

It is worth noting that in the recent years, various mathematical approaches have been developed to model drug resistance. Most of them deal with resistance on a whole tumor cell population level and they usually do not incorporate any components of tumor microenvironment. If the microenvironment is incorporated, it is typically treated as a homogeneous medium. Some excellent overviews of such methods can be found in [7, 13, 29, 30, 34, 54], and in our previous publications [16, 59]. There are only a few mathematical models that deal with microenvironmental heterogeneity in the context of chemotherapy and drug resistance, using either agent-based approaches [42, 52, 53, 61] or continuous equations [22, 26, 35, 37].

There is also an increasing interest in developing experimental and computational methods, to predict a tumor's potential for drug resistance before it emerges. We recently postulated a concept of Virtual Clinical Trials to assess tumor chemoresistance based on patient-specific biopsy data and computer simulations of therapeutic treatments [58]. While it is difficult and costly to perform extensive scheduling experiments in a laboratory, it is relatively easy and inexpensive to run large numbers of computational simulations to test various dosing and timing schedules of mono- and multi-drug therapies. Of particular interest is whether a specifically designed treatment schedules would be able to prevent or delay the development of acquired drug resistance. Moreover, following findings described in this chapter, it is of interest to investigate whether such protocols would limit the survival or proliferation of persister cells, and the development of tissue sanctuaries. For instance, response to various treatment protocols including maximum tolerated dose, metronomic and fractioned, has been studied in the model described herein. It was found that for micrometastases that harbor pre-existing resistant cells (up to a certain level of resistance) and for those that can acquire resistance in response to drug (up to a certain rate), a small number of fractioned dose protocols proved to be optimal in limiting (and in some cases, preventing) the emergence of drug resistant micrometastases [59].

While here we specifically consider drug resistance in tumors, there are parallels to therapy resistance in populations of bacterial cells [2, 32, 33]. In particular, it has been experimentally shown that the emergence of bacterial antibiotic resistance was accelerated in heterogeneous microenvironments [66]. Further, various strains of bacteria were shown to possess the capability to actively form niches to adapt to different environments, such as changes in nutrient availability [45]. It is also worth noting that the concept of persister cells was first described in the context of bacterial infections in which a small number of antibiotic-resistant mutants survived in a dormant non-dividing state [4, 36]. For bacteria, several factors such as growth stage [46], anaerobic adaptation [24], toxin–antitoxin modules as well as a number of genes and pathways [67], have been identified to be linked to persister formation.

However, in contrast to bacteria, it is not known how cancer persister cells arise. Nor is it known how the various mechanisms of resistance that have been observed experimentally can emerge in cell colonies arising from individual persisters. While it is likely that any adaptation strategies of drug-tolerant mother cells are passed along to daughter cells, why cancer persister cells remain dormant for months and what makes them finally leave the dormant state, has yet to be identified. More detailed investigations at the single cell level, such as tracking cell lineages to recreate different evolutionary dynamics: clonal, spatial, and pharmacological may shed light on these aspects (subject of ongoing work). Of particular interest is to correlate spatial dynamics of single cells and cell populations with the pre-existing and emerging microenvironmental niches and sanctuaries.

The importance of the microenvironment in tumor initiation, growth, invasion and metastasis is now well appreciated. The role that the microenvironment plays in tumors response to therapies and in the development of anti-treatment resistance is currently under intense investigation, both experimentally and mathematically. In this chapter we presented a scenario in which tumor cells can gain drug resistance in specific microenvironmental conditions, namely tissue niches and sanctuaries. However, it is still an open question whether the tumor cells that acquired drug resistance happened to reach these specific environments, were actively adapting to these new conditions, or were actively engaged in microenvironment remodeling. Understanding the ways in which tumor cell communities thrive in heterogeneous microenvironments and their strategies for surviving under extreme stress can be used for the development of new therapeutic treatments that target or control these interactions instead of directly targeting the cells themselves. This can pave new directions for research in cancer biology.

Acknowledgments This work was initiated during the Woman in Applied Mathematics (WhAM!) Research Collaboration Workshop at the Institute of Mathematics and Its Applications (IMA). KAR was supported in part by the U01 CA202229-01 grant from the National Institute of Health. The Bavarian State Ministry of Education and Culture, Science and Arts joint with the Technical University of Munich provided funding for JPV through the Laura Bassi Award. JPV also wants to thanks the German Research Foundation (Deutsche Forschungsgemeinschaft, DFG) for providing a travel grant (CE 243/1-1) to facilitate the initiation of this cooperation.

References

1. Baguley BC (2010) Multiple drug resistance mechanisms in cancer. Mol Biotechnol 46(3):308–316. doi:10.1007/s12033-010-9321-2
2. Baquero F, Coque TM, de la Cruz F (2011) Ecology and evolution as targets: the need for novel eco-evo drugs and strategies to fight antibiotic resistance. Antimicrob Agents Chemother 55(8):3649–3660. doi:10.1128/AAC.00013-11
3. Barcellos-Hoff MH, Lyden D, Wang TC (2013) The evolution of the cancer niche during multistage carcinogenesis. Nat Rev Cancer 13(7):511–518. doi:10.1038/nrc3536
4. Bigger JW (1946) Synergic action of penicillin and sulphathiazole on Bacterium typhosum. Lancet 1(6386):81–83
5. Borovski T, De Sousa EMF, Vermeulen L, Medema JP (2011) Cancer stem cell niche: the place to be. Cancer Res 71(3):634–639. doi:10.1158/0008-5472.CAN-10-3220
6. Branzei D, Foiani M (2008) Regulation of DNA repair throughout the cell cycle. Nat Rev Mol Cell Biol 9(4):297–308. doi:10.1038/nrm2351
7. Brocato T, Dogra P, Koay EJ, Day A, Chuang YL, Wang Z, Cristini V (2014) Understanding drug resistance in breast cancer with mathematical oncology. Curr Breast Cancer Rep 6(2):110–120. doi:10.1007/s12609-014-0143-2
8. Carmeliet P, Dor Y, Herbert JM, Fukumura D, Brusselmans K, Dewerchin M, Neeman M, Bono F, Abramovitch R, Maxwell P, Koch CJ, Ratcliffe P, Moons L, Jain RK, Collen D, Keshert E (1998) Role of HIF-1alpha in hypoxia-mediated apoptosis, cell proliferation and tumour angiogenesis. Nature 394(6692):485–490. doi:10.1038/28867
9. Chang DJ, Cimprich KA (2009) DNA damage tolerance: when it's OK to make mistakes. Nat Chem Biol 5(2):82–90. doi:10.1038/nchembio.139
10. Cheung-Ong K, Giaever G, Nislow C (2013) DNA-damaging agents in cancer chemotherapy: serendipity and chemical biology. Chem Biol 20(5):648–659. doi:10.1016/j.chembiol.2013.04.007
11. Cory TJ, Schacker TW, Stevenson M, Fletcher CV (2013) Overcoming pharmacologic sanctuaries. Curr Opin HIV AIDS 8(3):190–195. doi:10.1097/COH.0b013e32835fc68a
12. David E, Blanchard F, Heymann MF, De Pinieux G, Gouin F, Redini F, Heymann D (2011) The bone niche of chondrosarcoma: a sanctuary for drug resistance, tumour growth and also a source of new therapeutic targets. Sarcoma 2011:932451. doi:10.1155/2011/932451
13. Foo J, Michor F (2014) Evolution of acquired resistance to anti-cancer therapy. J Theor Biol 355:10–20. doi:10.1016/j.jtbi.2014.02.025
14. Friedberg EC (2005) Suffering in silence: the tolerance of DNA damage. Nat Rev Mol Cell Biol 6(12):943–953. doi:10.1038/nrm1781
15. Fu F, Nowak MA, Bonhoeffer S (2015) Spatial heterogeneity in drug concentrations can facilitate the emergence of resistance to cancer therapy. PLoS Comput Biol 11(3):e1004142. doi:10.1371/journal.pcbi.1004142
16. Gevertz JL, Aminzare Z, Norton KA, Perez-Velazquez J, Volkening A, Rejniak KA (2015) Emergence of anti-cancer drug resistance exploring the importance of the microenvironmental niche via a spatial model. In: Radunskaya A, Jackson T (eds) Applications of dynamial systems in biology and medicine, vol 158. Springer, Berlin, pp 1–34
17. Ghajar CM (2015) Metastasis prevention by targeting the dormant niche. Nat Rev Cancer 15(4):238–247. doi:10.1038/nrc3910
18. Ghajar CM, Peinado H, Mori H, Matei IR, Evason KJ, Brazier H, Almeida D, Koller A, Hajjar KA, Stainier DY, Chen EI, Lyden D, Bissell MJ (2013) The perivascular niche regulates breast tumour dormancy. Nat Cell Biol 15(7):807–817. doi:10.1038/ncb2767
19. Ghosal G, Chen J (2013) DNA damage tolerance: a double-edged sword guarding the genome. Transl Cancer Res 2(3):107–129. doi:10.3978/j.issn.2218-676X.2013.04.01
20. Gilkes DM, Semenza GL, Wirtz D (2014) Hypoxia and the extracellular matrix: drivers of tumour metastasis. Nat Rev Cancer 14(6):430–439. doi:10.1038/nrc3726
21. Gordan JD, Bertout JA, Hu CJ, Diehl JA, Simon MC (2007) HIF-2alpha promotes hypoxic cell proliferation by enhancing c-myc transcriptional activity. Cancer Cell 11(4):335–347. doi:10.1016/j.ccr.2007.02.006

22. Greene J, Lavi O, Gottesman MM, Levy D (2014) The impact of cell density and mutations in a model of multidrug resistance in solid tumors. Bull Math Biol 76(3):627–653. doi:10.1007/s11538-014-9936-8
23. Hata AN, Niederst MJ, Archibald HL, Gomez-Caraballo M, Siddiqui FM, Mulvey HE, Maruvka YE, Ji F, Bhang HE, Krishnamurthy Radhakrishna V, Siravegna G, Hu H, Raoof S, Lockerman E, Kalsy A, Lee D, Keating CL, Ruddy DA, Damon LJ, Crystal AS, Costa C, Piotrowska Z, Bardelli A, Iafrate AJ, Sadreyev RI, Stegmeier F, Getz G, Sequist LV, Faber AC, Engelman JA (2016) Tumor cells can follow distinct evolutionary paths to become resistant to epidermal growth factor receptor inhibition. Nat Med 22(3):262–269. doi:10.1038/nm.4040
24. Hemsley CM, Luo JX, Andreae CA, Butler CS, Soyer OS, Titball RW (2014) Bacterial drug tolerance under clinical conditions is governed by anaerobic adaptation but not anaerobic respiration. Antimicrob Agents Chemother 58(10):5775–5783. doi:10.1128/AAC.02793-14
25. Holohan C, Van Schaeybroeck S, Longley DB, Johnston PG (2013) Cancer drug resistance: an evolving paradigm. Nat Rev Cancer 13(10):714–726. doi:10.1038/nrc3599
26. Jackson TL, Byrne HM (2000) A mathematical model to study the effects of drug resistance and vasculature on the response of solid tumors to chemotherapy. Math Biosci 164(1):17–38
27. Kaplan RN, Psaila B, Lyden D (2007) Niche-to-niche migration of bone-marrow-derived cells. Trends Mol Med 13(2):72–81. doi:10.1016/j.molmed.2006.12.003
28. Karran P (2001) Mechanisms of tolerance to DNA damaging therapeutic drugs. Carcinogenesis 22(12):1931–1937
29. Komarova NL, Wodarz D (2005) Drug resistance in cancer: principles of emergence and prevention. Proc Natl Acad Sci U S A 102(27):9714–9719. doi:10.1073/pnas.0501870102
30. Komarova NL, Wodarz D (2007) Stochastic modeling of cellular colonies with quiescence: an application to drug resistance in cancer. Theor Popul Biol 72(4):523–538. doi:10.1016/j.tpb.2007.08.003
31. Koong AC, Mehta VK, Le QT, Fisher GA, Terris DJ, Brown JM, Bastidas AJ, Vierra M (2000) Pancreatic tumors show high levels of hypoxia. Int J Radiat Oncol Biol Phys 48(4):919–922
32. Korolev KS, Xavier JB, Gore J (2014) Turning ecology and evolution against cancer. Nat Rev Cancer 14(5):371–380. doi:10.1038/nrc3712
33. Lambert G, Estevez-Salmeron L, Oh S, Liao D, Emerson BM, Tlsty TD, Austin RH (2011) An analogy between the evolution of drug resistance in bacterial communities and malignant tissues. Nat Rev Cancer 11(5):375–382. doi:10.1038/nrc3039
34. Lavi O, Gottesman MM, Levy D (2012) The dynamics of drug resistance: a mathematical perspective. Drug Resist Updat 15(1–2):90–97. doi:10.1016/j.drup.2012.01.003
35. Lavi O, Greene JM, Levy D, Gottesman MM (2013) The role of cell density and intratumoral heterogeneity in multidrug resistance. Cancer Res 73(24):7168–7175. doi:10.1158/0008-5472.CAN-13-1768
36. Lewis K (2007) Persister cells, dormancy and infectious disease. Nat Rev Microbiol 5(1):48–56. doi:10.1038/nrmicro1557
37. Lorz A, Lorenzi T, Clairambault J, Escargueil A, Perthame B (2015) Modeling the effects of space structure and combination therapies on phenotypic heterogeneity and drug resistance in solid tumors. Bull Math Biol 77(1):1–22. doi:10.1007/s11538-014-0046-4
38. Lu H, Clauser KR, Tam WL, Frose J, Ye X, Eaton EN, Reinhardt F, Donnenberg VS, Bhargava R, Carr SA, Weinberg RA (2014) A breast cancer stem cell niche supported by juxtacrine signalling from monocytes and macrophages. Nat Cell Biol 16(11):1105–1117. doi:10.1038/ncb3041
39. Lu P, Weaver VM, Werb Z (2012) The extracellular matrix: a dynamic niche in cancer progression. J Cell Biol 196(4):395–406. doi:10.1083/jcb.201102147
40. Luoto KR, Kumareswaran R, Bristow RG (2013) Tumor hypoxia as a driving force in genetic instability. Genome Integr 4(1):5. doi:10.1186/2041-9414-4-5
41. Meads MB, Hazlehurst LA, Dalton WS (2008) The bone marrow microenvironment as a tumor sanctuary and contributor to drug resistance. Clin Cancer Res 14(9):2519–2526. doi:10.1158/1078-0432.CCR-07-2223
42. Menchon SA (2015) The effect of intrinsic and acquired resistances on chemotherapy effectiveness. Acta Biotheor 63(2):113–127. doi:10.1007/s10441-015-9248-x
43. Minchinton AI, Tannock IF (2006) Drug penetration in solid tumours. Nat Rev Cancer 6(8):583–592. doi:10.1038/nrc1893
44. Mumenthaler SM, Foo J, Choi NC, Heise N, Leder K, Agus DB, Pao W, Michor F, Mallick P (2015) The impact of microenvironmental heterogeneity on the evolution of drug resistance in cancer cells. Cancer Informat 14(Suppl 4):19–31. doi:10.4137/CIN.S19338
45. Ni L, Yang S, Zhang R, Jin Z, Chen H, Conrad JC, Jin F (2016) Bacteria differently deploy type-IV pili on surfaces to adapt to nutrient availability. Biofilms Microbiomes 2:15029
46. Nierman WC, Yu Y, Losada L (2015) The in vitro antibiotic tolerant persister population in Burkholderia pseudomallei is altered by environmental factors. Front Microbiol 6:1338. doi:10.3389/fmicb.2015.01338
47. Oxnard GR (2016) The cellular origins of drug resistance in cancer. Nat Med 22(3):232–234. doi:10.1038/nm.4058
48. Palmieri D, Chambers AF, Felding-Habermann B, Huang S, Steeg PS (2007) The biology of metastasis to a sanctuary site. Clin Cancer Res 13(6):1656–1662. doi:10.1158/1078-0432.CCR-06-2659

49. Peinado H, Lavotshkin S, Lyden D (2011) The secreted factors responsible for pre-metastatic niche formation: old sayings and new thoughts. Semin Cancer Biol 21(2):139–146. doi:10.1016/j.semcancer.2011.01.002
50. Pisco AO, Huang S (2015) Non-genetic cancer cell plasticity and therapy-induced stemness in tumour relapse: 'what does not kill me strengthens me'. Br J Cancer 112(11):1725–1732. doi:10.1038/bjc.2015.146
51. Plaks V, Kong N, Werb Z (2015) The cancer stem cell niche: how essential is the niche in regulating stemness of tumor cells? Cell Stem Cell 16(3):225–238. doi:10.1016/j.stem.2015.02.015
52. Powathil GG, Adamson DJ, Chaplain MA (2013) Towards predicting the response of a solid tumour to chemotherapy and radiotherapy treatments: clinical insights from a computational model. PLoS Comput Biol 9(7), e1003120. doi:10.1371/journal.pcbi.1003120
53. Powathil GG, Gordon KE, Hill LA, Chaplain MA (2012) Modelling the effects of cell-cycle heterogeneity on the response of a solid tumour to chemotherapy: biological insights from a hybrid multiscale cellular automaton model. J Theor Biol 308:1–19. doi:10.1016/j.jtbi.2012.05.015
54. Powathil GG, Swat M, Chaplain MA (2015) Systems oncology: towards patient-specific treatment regimes informed by multiscale mathematical modelling. Semin Cancer Biol 30:13–20. doi:10.1016/j.semcancer.2014.02.003
55. Primeau AJ, Rendon A, Hedley D, Lilge L, Tannock IF (2005) The distribution of the anticancer drug Doxorubicin in relation to blood vessels in solid tumors. Clin Cancer Res 11(24 Pt 1):8782–8788. doi:10.1158/1078-0432.CCR-05-1664
56. Puhalla S, Elmquist W, Freyer D, Kleinberg L, Adkins C, Lockman P, McGregor J, Muldoon L, Nesbit G, Peereboom D, Smith Q, Walker S, Neuwelt E (2015) Unsanctifying the sanctuary: challenges and opportunities with brain metastases. Neuro Oncol 17(5):639–651. doi:10.1093/neuonc/nov023
57. Ramirez M, Rajaram S, Steininger RJ, Osipchuk D, Roth MA, Morinishi LS, Evans L, Ji W, Hsu CH, Thurley K, Wei S, Zhou A, Koduru PR, Posner BA, Wu LF, Altschuler SJ (2016) Diverse drug-resistance mechanisms can emerge from drug-tolerant cancer persister cells. Nat Commun 7:10690. doi:10.1038/ncomms10690
58. Rejniak KA, Lloyd MC, Reed DR, Bui MM (2015) Diagnostic assessment of osteosarcoma chemoresistance based on Virtual Clinical Trials. Med Hypotheses 85(3):348–354. doi:10.1016/j.mehy.2015.06.015
59. Shah AB, Rejniak KA, Gevertz JL (Under review) Limiting the development of anti-cancer drug resistance in a spatial model of micrometastases. https://arxiv.org/abs/160103412
60. Sharma SV, Lee DY, Li B, Quinlan MP, Takahashi F, Maheswaran S, McDermott U, Azizian N, Zou L, Fischbach MA, Wong KK, Brandstetter K, Wittner B, Ramaswamy S, Classon M, Settleman J (2010) A chromatin-mediated reversible drug-tolerant state in cancer cell subpopulations. Cell 141(1):69–80. doi:10.1016/j.cell.2010.02.027
61. Silva AS, Gatenby RA (2010) A theoretical quantitative model for evolution of cancer chemotherapy resistance. Biol Direct 5:25. doi:10.1186/1745-6150-5-25
62. Sun JD, Liu Q, Wang J, Ahluwalia D, Ferraro D, Wang Y, Duan JX, Ammons WS, Curd JG, Matteucci MD, Hart CP (2012) Selective tumor hypoxia targeting by hypoxia-activated prodrug TH-302 inhibits tumor growth in preclinical models of cancer. Clin Cancer Res 18(3):758–770. doi:10.1158/1078-0432.CCR-11-1980
63. Thurber GM, Yang KS, Reiner T, Kohler RH, Sorger P, Mitchison T, Weissleder R (2013) Single-cell and subcellular pharmacokinetic imaging allows insight into drug action in vivo. Nat Commun 4:1504. doi:10.1038/ncomms2506
64. Vinegoni C, Dubach JM, Thurber GM, Miller MA, Mazitschek R, Weissleder R (2015) Advances in measuring single-cell pharmacology in vivo. Drug Discov Today 20(9):1087–1092. doi:10.1016/j.drudis.2015.05.011
65. Waters LS, Minesinger BK, Wiltrout ME, D'Souza S, Woodruff RV, Walker GC (2009) Eukaryotic translesion polymerases and their roles and regulation in DNA damage tolerance. Microbiol Mol Biol R 73(1):134 − +. doi:10.1128/MMBR.00034-08
66. Zhang Q, Lambert G, Liao D, Kim H, Robin K, Tung CK, Pourmand N, Austin RH (2011) Acceleration of emergence of bacterial antibiotic resistance in connected microenvironments. Science 333(6050):1764–1767. doi:10.1126/science.1208747
67. Zhang Y (2014) Persisters, persistent infections and the Yin-Yang model. Emerg Microbes Infect 3(1), e3. doi:10.1038/emi.2014.3

The Tumor Microenvironment as a Barrier to Cancer Nanotherapy

9

Louis T. Curtis and Hermann B. Frieboes

Abstract

Although extensive research effort and resources have been dedicated to the development of nanotherapeutics to treat cancer, few formulations have reached clinical application. A major reason is that the large number of parameters available to tune nanotherapy characteristics coupled with the variability in tumor tissue precludes evaluation of complex interactions through experimentation alone. In order to optimize the nanotechnology design and gain further insight into these phenomena, mathematical modeling and computational simulation have been applied to complement empirical work. In this chapter, we discuss modeling work related to nanotherapy and the tumor microenvironment. We first summarize the biology underlying the dysregulated tumor microenvironment, followed by a description of major nano-scale parameters. We then present an overview of the mathematical modeling of cancer nanotherapy, including evaluation of nanotherapy in multi-dimensional tumor tissue, coupling of nanotherapy with vascular flow, modeling of nanotherapy in combination with *in vivo* imaging, modeling of nanoparticle transport based on *in vitro* data, modeling of vasculature-bound nanoparticles, evaluation of nanotherapy using pharmacokinetic modeling, and modeling of nano-based hyperthermia. We conclude that an even tighter interdisciplinary effort between biological, material, and physical scientists is needed in order to eventually overcome the tumor microenvironment barrier to successful nanotherapy.

Keywords

Nanotherapy • Nanotechnology • Cancer • Tumor microenvironment • Mathematical modelling • Computational simulation

L.T. Curtis • H.B. Frieboes (✉)
Department of Bioengineering, University of Louisville,
Lutz Hall 419, Louisville, KY 40292, USA
e-mail: louis.curtis@louisville.edu;
hbfrie01@louisville.edu

K.A. Rejniak (eds.), *Systems Biology of Tumor Microenvironment*, Advances in Experimental Medicine and Biology 936, DOI 10.1007/978-3-319-42023-3_9

9.1 Dysregulation of the Tumor Microenvironment

The tumor microenvironment presents a formidable barrier that prevents adequate transport and uptake of systemically-administered nano-scale devices as well as conventional chemotherapeutics. Insufficient blood and lymphatic vascularization limit access to and from flow, and a denser-than-normal extracellular matrix (ECM) hinders interstitial transport. Consequently, solid tumors are typically characterized by hypoxic and necrotic tissue, metabolic waste buildup leading to an extracellular acidic pH, and high interstitial fluid pressure (IFP) [8, 100]. Irregular tumor vasculature as well as ECM arise from uncoordinated cellular regulation [90], mainly due to unbalanced stimuli by unscheduled tumor cell proliferation. Some tissue regions may be hyper-vascularized while others will remain hypo-vascularized. As tumor cells typically proliferate at higher rates than vascular endothelial cells, the vascular development is chronically insufficient to properly maintain the growing tissue [69]. The resulting oxygen and nutrient gradients promote heterogeneous cellular proliferation [38].

Cells distal from blood vessels beyond the diffusion limit of oxygen (100 μm), with impaired access to cell nutrients and oxygen, as well as insufficient drainage of surrounding extracellular fluid and metabolic waste, will become quiescent (non-proliferating) in order to promote survival. Quiescence increases cell resistance to chemotherapeutics as well as radiotherapy [20, 70], as it thwarts the action of cell-cycle dependent therapeutics. Post therapy, quiescent cells may be able to resume proliferation and thus resume tissue growth. Under sufficiently deteriorated conditions, these hypoxic cells will experience unregulated cell death (necrosis) [41], leading to spilling of toxic cellular contents into the surrounding microenvironment. Further, hypoxic or acidic microenvironments may render chemotherapeutics ineffective. Even if drugs were effective, the concentrations achieved distal to the vasculature may be insufficient for cytotoxicity [69].

An acidic microenvironment hinders chemotherapeutic delivery in multiple ways [98]. The low pH stems from the excessive build-up in the extracellular space of metabolites such as carbonic acid and lactic acid. The vascular and lymphatic vessels are unable to eliminate the metabolites due to overproduction in addition to insufficient drainage. In order to cross cellular membranes, drugs must be in an uncharged state. Weak bases such as doxorubicin will become ionized in an acidic environment, which prevents cellular uptake and diminishes overall efficacy [62].

The IFP optimizes the flow of fluid and nutrients between blood vessels and surrounding cells [100]. In normal tissue the IFP is slightly negative (−1 to −3 mmHg) [37], whereas in solid tumors it can be as high as 10–30 mmHg. An elevated IFP results from the synergistic combination of decreased blood and lymphatic vessel function, unbalanced osmosis, and stromal contraction [8]. The increased IFP affects therapeutic efficacy, with a positive correlation shown between poor prognosis and elevated IFP [16].

Solid tumors can have a significantly higher proportion of ECM proteins than non-cancerous tissue, resulting in decreased diffusivity of both medium and large molecular weight proteins through the interstitial space [69]. The ECM, composed of various proteins (e.g. collagen, hyaluronic acid, proteoglycans), accounts for cell-cell and cell-matrix adhesion, cell-to-cell communication, and tissue structural integrity. In particular, the slightly positive charge of collagen may attract anionic nanoparticles and drugs, decreasing their availability to tumor cells [87].

9.2 Nanotherapy as a Means to Address the Tumor Microenvironment

Nanoparticles of different materials and properties have been devised to target drugs more effectively to tumor tissue. In spite of numerous efforts, only a few formulations have reached clinical application, including liposomal Doxorubicin (Doxil) and albumin-bound paclitaxel (Abraxane).

Nanoparticles targeted for cancer therapeutic application typically include the following characteristics: (i) accumulation in tumor tissue via passive targeting, (ii) active targeting of cancer cells, (iii) controllable drug release kinetics tailored to treatment regimen or cancer type, (iv) bypassing of intrinsic cellular resistance mechanisms, (v) overcoming drug pharmacokinetic limitations, (vi) multiple drug loading, (vii) high tunability enabling modulation of properties such as shape, size, and surface charge, and (viii) potential theranostic capability, i.e., with both diagnostic and therapeutic functions.

(i) Passive targeting of nanoparticles is based upon the observation that accumulation is enhanced in tissue surrounding the irregular tumor vasculature due to fenestrations of the endothelial cell layer as well as limited lymphatic draining [36, 100]. This phenomenon has been termed the enhanced permeability and retention effect (EPR) [61, 63]. As the effect is time-dependent, simply increasing the NP circulation time leads to higher accumulation. However, the reticulo-endothelial system (RES), first reported in [4], filters most of circulating NPs out of the body. Thus, 40–50 % of injected nanoparticles will accumulate in the liver, 24–40 % in the spleen, and the remainder in the lungs, kidneys, and heart. Only 1–5 % of injected nanoparticles will extravasate into the tumor tissue, which can ultimately lead to therapeutic failure due to insufficient local concentration [45]. RES uptake can be decreased by coating nanoparticles with polymers such as poly(ethylene) glycol (PEG), which has been shown to potentially double the circulation time compared to non-coated particles [39, 54].

(ii) By specifically targeting cancer cells, systemic off target toxicity that is commonly associated with current chemotherapeutics can be minimized. This process is called "active targeting." In this case, nanoparticles are chemically modified with various surface modifications, such as proteins, peptides, antibodies, or other compounds that enhance uptake by the targeted cells. For example, nanoparticles have been targeted to cell surface proteins commonly up-regulated in cancer cells, such as epidermal growth factor receptor (EGFR) [14, 44, 68, 76]. As nanoparticles possess high surface area to volume ratio due to their small size and shape, attaching multiple targeting ligands onto a single nanoparticle may increase the probability of cellular interactions with the targeted tissue. Adding multiple targeting ligands, however, may not provide a synergistic effect on targeting capabilities and tumor accumulation [79].

(iii) NP drug release kinetics can be tailored to treat specific disease conditions. The kinetics depends upon several factors including interactions between the nanoparticle and the drug (e.g. covalent or non-covalent linkage), nanoparticle physiochemical properties, and the tissue microenvironment pH. Through surface modifications, slow and rapid release patterns can be achieved. A common method for analyzing drug release kinetics is with dialysis, along with analysis by high performance liquid chromatography (HPLC) or UV spectroscopy [104].

(iv) In addition to physiological factors, cellular mechanisms can alter sensitivity to drugs. This resistance may be intrinsic (pre-existent in cells pre-treatment) or induced by drugs (acquired post-treatment), and in either case can be a major stumbling block to successful chemotherapy. Cellular mechanisms include upregulated drug metabolism, downregulated apoptosis, altered molecular targets, upregulation of efflux pumps, and genetic alterations [75]. Efflux pumps such as P-glycoprotein are commonly associated with increased drug resistance to multiple classes of chemotherapeutics including doxorubicin, vinca alkaloids, and taxols [1, 33]. In particular, multidrug resistance (MDR) is a well-known mechanism through which cancer cells acquire resistant to multiple drugs [77]. Nanoparticles may help to overcome some mechanisms of cellular drug resistance. For example, it was shown that both sensitive and resistant cells would uptake the same amount of doxorubicin if delivered via nanoparticles [19]. Intracellular apoptotic mechanisms, such as Bcl-2, which is a protein inhibiting downstream apoptotic pathways and is up-regulated in most solid tumors, have also been targeted. In particular, this resistance can be decreased by encapsulating Bcl-2 siRNAs [6, 11, 12]. For example, liposomes containing two different siRNAs, Bcl2-siRNA and MRP1-siRNA, and doxorubicin have been developed for use in MDR lung cancer cells. Results showed effective co-delivery of doxorubicin and siRNAs with up to 95 % cell-death induction, with high suppression of cellular resistance [82].

(v) The biological effect of a drug depends highly upon its pharmacokinetic properties. Many chemotherapeutics have substandard pharmacokinetic profiles which limit their travel through systemic circulation, due to characteristics such as rapid metabolism or elimination. In particular, poorly water-soluble compounds suffer from uncontrollable precipitation after dosing, suboptimal dosing, lack of dose-response proportionality, use of harsh excipients, use of basic or acid conditions to enhance solubility, poor bioavailability, and patient non-compliance due to dosing strategies [66]. Nanoparticles can help overcome unfavorable drug harmacokinetic properties. Nanoencapsulation can help to address pharmacokinetic shortcomings of certain drugs such as paclitaxel, which has deficient water solubility and high plasma protein binding [57]. Paclitaxel is an effective FDA-approved chemotherapeutic with potent anti-cancer activity and relatively low EC_{50}, yet harsh solvents and minimal water solubility can hinder effective delivery and increase the risk of systemic toxicity. The albumin-bound formulation of paclitaxel has shown decreased systemic toxicity, enhanced effectiveness, and increased survival rates as well as time to disease progress [32].

(vi) Due to a high surface area to volume ratio, multiple targeting ligands and therapeutic agents can be loaded onto single nanoparticles. This can be advantageous since chemotherapeutic strategies commonly consist of two or more drugs in combination. For example, mesoporous silica nanoparticles were developed with two siRNA molecules targeted to MRP1 and BCL2 mRNA as well as two anti-cancer drugs (doxorubicin and cisplatin) to decrease cellular resistance [91]. Using the luteinizing hormone releasing peptide (LHRH) as an active targeting ligand, tumor accumulation was significantly enhanced, as the anti-cancer agents exhibited enhanced therapeutic effect for the treatment of non-small cell lung cancer (NSCLC) and the siRNA suppressed cellular resistance to the anticancer drugs. In another case, both hydrophilic and hydrophobic drugs were loaded onto the same nanoparticle by encapsulating doxorubicin with paclitaxel or doxorubicin with rapamycin into magnetic mesoporous silica nanoparticles [55]. Furthermore, by loading multiple drugs onto single nanoparticles, patient compliance may be improved by decreasing the number of medicines requiring intake.

(vii) The high tunability of nanoparticles enables their use in a variety of medical applications. Characteristics such as conformational shape, maximum absorbance, surface charge, and hydrodynamic size can be chemically modified to modulate efficacy. In particular, shape can dramatically influence cellular uptake, while maximum absorbance affects the efficacy of applications such as photoacoustic imaging and photothermal therapy. Surface charge impacts systemic travel and diffusion through tissue; cationic nanoparticles accumulate in tumor vasculature due to electrostatic interactions, but the cationic charge must be minimal for nanoparticles to extravasate from vasculature [46, 93]. Neutrally charged nanoparticles diffuse rapidly within tumor tissue as compared to charged nanoparticles, which may interact with extracellular proteins such as collagen or hyaluronan [53, 87].

Nanoparticle size affects tumor-specific accumulation and diffusion within tissue, but the relationship is not necessarily linearly proportional. In general, smaller nanoparticles (10–100 nm) have enhanced diffusivity compared to larger particles. Liposomes ranging in size from 100 to 200 nm have enhanced tumor accumulation compared to smaller liposomes (<100 nm) and larger nanoparticles (>200 nm) [95], which may be related to the EPR effect. Further studies have analyzed the tumor intra-tissue diffusion of very small nanoparticles, showing that 2 and 6 nm nanoparticles accumulate after intravenous injection at higher concentrations than slightly larger nanoparticles (15 nm) [40]. Both 2 and 6 nm particles were distributed throughout the cytoplasm and nucleus of cancer cells, while the larger 15 nm nanoparticles were only detected in the cytoplasm where they formed aggregates [40].

(viii) Theranostic applications require that nanoparticles function as imaging as well as therapeutic agents. Magnetic resonance imaging (MRI) has been extensively utilized with theranostic nanoparticles. New imaging methodologies, such as photoacoustic imaging with multispectral optoacoustic tomography (MSOT), have enabled monitoring nanoparticle circulation in real-time [58, 71, 92]. Recently, theranostic nanoparticles were developed possessing multiple modifications, including a cytotoxic compound (c-substituted diindolymethanes or DIM-C-pPhC6H5) for therapeutic activity, a peptide (Cys-Arg-Glu-Lys-Ala or CREKA) for active targeting to lung cancer, and PEG-coating to increase systemic circulation [74]. These particles showed a 40-fold increased transport in tumor vasculature compared to non-PEG and non-CREKA versions.

9.3 Mathematical Modeling of Cancer Nanotherapy

It has been proposed that similar to the 'rational drug design' concept applied in pharmaceutical community studies, optimization of nanoscale delivery systems should integrate predictive mathematical models, *in vitro* assays, and *in vivo* testing into a unified framework [22]. Extensive work modeling tumor growth and angiogenesis has been completed in recent years (see reviews [2, 5, 7, 9, 15, 23, 24, 26, 47, 56, 67, 72, 73, 80, 81, 89, 99, 101, 102] and associated references), with a subset of studies specifically focusing on nanotherapy [17, 18, 22, 27, 28, 30, 31, 43, 49–52, 96, 97, 103].

9.3.1 Multi-dimensional Simulation of Nanotherapy in Vascularized Tumors

Traditionally, tumor growth and therapy have been evaluated in one-dimensional systems which can afford significant insight into complex behavior, e.g., as in the classical work by [34, 35]. To further elucidate the role of multi-dimensional space in nanotherapy, evaluation of nanoparticle-based drug delivery and treatment response in a two-dimensional spatial model of vascularized tumor growth was pioneered in [84] based on work by [3, 105]. The potential efficacy of nanotherapeutics was shown to be heavily dependent on physical transport. The study considered very small (1–10 nm) particles that extravasate from the vasculature and diffuse through the interstitium as well as larger (>100 nm) particles that remain at their location of extravasation. The nanoparticle localization was assumed to be associated with fenestrations in the leaky tumor vasculature. The simulations showed that the nanoparticles evinced more uniform and larger concentrations of drug over longer periods of time compared to drug freely administered.

The local tumor mass growth rate is described as the divergence velocity field **u** of the tumor cells [105]:

$$\nabla \cdot \mathbf{u} = \lambda_M \frac{n}{n_v} - \lambda_D$$

where λ_M is the proliferation rate, n is the oxygen concentration, n_v is the oxygen concentration in the vasculature (representing a maximum), and λ_D is the cellular death rate also dependent on the local chemotherapeutic concentration. The velocity is set proportional to the tumor pressure p via Darcy-Stokes' law [94]:

$$\mathbf{u} = -\mu \nabla p$$

where μ is hydraulic conductivity obtained by describing the tumor and extracellular matrix as porous media.

The local concentration of drug S as a function of the local concentration s of 1–10 nm NPs is:

$$S = \nu_1 \left(\left(s_{\mathrm{v}} - s \right) / n_{\mathrm{v}} \right) \left(1 - p/p_{\mathrm{v}} \right) \, \delta - \eta s / n_{\mathrm{v}}.$$

where ν_1 is the transfer rate from the vasculature, S_V is the NP concentration in the vasculature, p_V is the pressure in the vasculature, δ is the indicator function for the vasculature (1 where it exists, and 0 otherwise), and η is the rate of oxygen uptake by the tumor cells.

The concentration s_v of NPs in the blood is assumed to be constant over the course of the therapy. The NPs are capable of convecting from the vasculature, diffusing through the tumor interstitium, and entering cancer cells. The transport of the rescaled variable s/n_v is governed by an equation analogous to the transport of oxygen [84]:

$$S = v_1 \frac{n_v - s}{n_v} \left(1 - \frac{p}{p_v} \right)^+ \delta - \eta \frac{s}{n_v}$$

Unlike the 1–10 nm NPs, the larger 100 nm NPs have limited diffusion through the interstitium. These NPs are modeled to remain near the neovasculature and to diffuse into the tissue on a short timescale.

The typical biphasic release of drug from the NPs (e.g., [86]) is modeled as $C_t = 1 - (Ae^{-\alpha t} + Be^{-\beta t})$ [13, 83], which can be approximated as $(1 - B) + B\beta t$ assuming that A and α are parameters corresponding to the rapidly released portion and B and β correspond to the slowly released portion of the fraction C_t of released drug at time t. For the larger particles, drug S is determined by $S = \nu_2 \delta - \eta s / n_{\mathrm{v}}$, where ν_2 is a constant flux of drug molecules into the tissue [84].

The effect of chemotherapy on cell proliferation is described by letting N be a cell population uniformly exposed (e.g., *in vitro*) to a NP concentration s and to an oxygen concentration n, letting N_0 be the initial cell population and t be the time of exposure:

$$\log N/N_0 = -\lambda_A \left(s/n_{\mathrm{v}} \right) \left(n/n_{\mathrm{v}} \right) t$$

In the study, a best case scenario was assumed with only one clone of tumor cells, all cells being drug-sensitive, with no host toxicity and no non-specific NP binding.

The simulations confirmed inadequate drug delivery *in vivo* compared to *in vitro* due to transport limitations induced by the heterogeneously vascularized tumor microenvironment. The results showed that drugs released from the nanoparticles would be affected by the same fundamental transport limitations as freely administered drug. Non-uniform drug delivery resulted from the interaction between vasculature density, supporting nanoparticle extravasation, and intratumoral pressure, hindering vascular extravasation. The inhomogeneous oxygen and nutrient supply to the tumor tissue coupled with the cellular proliferative response further exacerbated this non-uniformity. The simulation results additionally suggest that anti-angiogenic agents may improve transport but could also lead to tissue fragmentation and hence higher resistance due to decreased access to vasculature. Limitations of this study included that the simulation served as a best-case scenario, no distinction was made between interstitial fluid pressure and tumor pressure, complete disappearance of dead cells was assumed as opposed to a slower lysis with mass conservation, and simplified NP types were used with no specific information on drug loading.

9.3.2 Coupling of Nanotherapy Modeling with Vascular Flow

Interactions between the changing tumor vasculature and blood flow have been evaluated by coupling an updated version [59] of the multidimensional tumor model referenced above [105] with a more elaborate model of tumor-induced angiogenesis [3, 64, 65]. Similar to the study in [84], this multiscale model [60] incorporates oxygen/nutrients and angiogenic factors released by the neovasculature and tumor cells, respectively. Now, however, the flow of fluid in the vasculature, and the corresponding transport of oxygen/nutrients, is hindered by hydrostatic pressure exerted by the proliferating tumor, which in turn applies pressure to the neovasculature and alters the vascular flow pattern and associated shear stress. Further, the tumor growth and vascular network are coupled to the extracellular matrix (ECM) to simulate the role of ECM degradation during migration of tumor and endothelial cells.

Recently, this modeling framework [60] was extended in [103] to incorporate interstitial pressure and flow, a lymphatic system to model fluid drainage, and blood vessel leakage. Building upon this work, the microenvironment complexities that inhibit drug transport and internalization in tumors were analyzed in [103]. The vascular tumor growth model coupled with a discrete angiogenesis model [3, 64, 65] was supplemented with interstitial and lymphatic fluid dynamics to evaluate the role of hypertensive interstitial fluid pressure as a physical barrier hindering extravasation of agents into tumors. This system enables more biologically realistic simulation of the flow and distribution of NPs and drug molecules (collectively referred to as "agents") within tumor vasculature, as well as agent extravasation and diffusion through tumor tissue.

Blood flow is modeled using a generalized Poiseuille's law [29]:

$$Q_{qp} = \frac{\pi R_{qp}^4 \left(P_{v_q} - P_{v_p}\right)}{8\mu_{\text{apparent}} L}$$

where Q_{qp} is the blood flow from vessel node q to a neighboring node p, R_{qp} is the radius of the vessel segment between q and p, μ_{apparent} is the apparent plasma viscosity, P_{vq} and P_{vp} are the corresponding blood pressures. The transcapillary fluid flux is modeled by Starling's law [29]:

$$O_V = K_{V_f} S_V \left(P_V - P_f - \omega \left(\pi_V - \pi_i\right)\right)$$

where K_{Vf} is vascular hydraulic conductivity, S_V is the summation of all half-vessel segment surface areas connecting to neighboring vessel nodes, P_V is blood pressure, P_f is interstitial fluid pressure, ω is average osmotic reflection coefficient for plasma proteins, and π_V and π_i are osmotic pressures of plasma and interstitial fluid, respectively.

Following the interstitial fluid dynamics modeled in [103], Darcy's law is used to relate the interstitial fluid pressure and fluid flux:

$$\mathbf{v_f} = -K \nabla P_f$$

where K is the interstitial hydraulic conductivity. Conservation of mass is applied to obtain:

$$\nabla \cdot \left(\phi_f \mathbf{v_f}\right) = \phi_f \left(J_v - J_L\right) - \nabla \cdot \left(\phi_c \mathbf{v_c}\right)$$

where ϕ_f and ϕ_c are the water and cell volume fractions, respectively, v_c is the cell velocity and J_L is the lymphatic drainage, described as

$$J_L = \lambda_{fL} c\left(P_c, L\right)\left(P_f - P_L\right) \cdot \mathbf{1}_{P_f > P_L}$$

where $c\left(P_c, L\right)$ is the lymphatic drainage capacity, λ_{fL} is a constant rate and P_L is the critical IFP level below which there is no drainage. $\mathbf{1}_{P_f > P_L}$ is 1 for $P_f > P_L$ and 0 for $P_f \leq P_L$. The function $c\left(P_c, L\right)$ depends on the tumor hydrostatic pressure P_c and the lymphatic vessel density L. For small hydrostatic pressures P_c below a partial collapse threshold, $c\left(P_c, L\right)$ increases with P_c

and the drainage increases as hydrostatic pressure from the cells pushes fluid into the lymphatic vessels. When the pressure increases beyond this threshold, the lymphatic vessels responding to the pressure P_c begin to partially close, and $c\,(P_c, L)$ subsequently decreases, eventually reaching 0 at a maximum threshold at which the lymphatic vessels collapse.

This study first quantified agent availability in the tumor after systemic injection under different conditions of lymphatic resistance and drug permeability. It was shown that agent availability would increase for less permeable agents as lymphatic resistance increases, and oppositely will decrease for highly permeable agents. Additionally, as shown in Fig. 9.1, different conditions of interstitial and vascular hydraulic conductivity lead to varied interstitial fluid pressure, and consequently influencing the extravasation of therapeutics into the tumor. Tumors with elevated interstitial hydraulic conductivity are shown to require the strongest dosage to shrink. From these computational results, it was concluded that normalization of the intra-tumoral vascular and interstitial conditions could potentially lead to maximized effects of chemotherapy, in agreement with experimental and theoretical observations [10, 42]. Normalization of only one of these factors may be insufficient to overcome the physical transport barriers required to achieve optimal tumor shrinkage. Although this study focused mainly on free drug chemotherapy, the model framework incorporating interstitial pressure, vascular flow, and a lymphatic system make the results also applicable to the transport of nanotherapeutics.

9.3.3 Modeling of Nanotherapy in Combination with In Vivo Imaging

It is actually very difficult to visualize vasculature and flow of nanotherapeutics in tumors. In [97], nanoparticle-based drug delivery systems designed to circumvent biophysical barriers were evaluated in conjunction with intravital microscopy (IVM) data from *in vivo* mouse experiments. The study simulated tumor response to doxorubicin delivered via porous silicon particles in order to evaluate how much drug per particle and how many particles would need to be released from the tumor vasculature in order to achieve remission.

The model builds upon the vascularized tumor system in [103], which describes the transport of small molecules s such as oxygen ($s = \sigma$) and drug ($s = D$) at the point of release from the vasculature by quasi-steady reaction-diffusion equations. The model assumes that small molecules are supplied by the pre-existing vasculature as well as the neo-vasculature at rates $\lambda^s{}_{pre}$ and $\lambda^s{}_{neo}$, respectively, diffuse into the host and cancerous tissue with a diffusion coefficient D_s, are uptaken both by the normal cells (with a rate $\lambda^s{}_{tissue}$) and tumor cells ($\lambda^s{}_{tumor}$ in the proliferating region and q_s in the hypoxic region), and decay (with a rate $\lambda^s{}_N$) in the necrotic regions. The transport equations are:

$$0 = \nabla \cdot (D_s \nabla s) - \lambda^s(s)s + \lambda^s_{ev}\,(\boldsymbol{x},\ t,\ \mathbf{1}_{vessel}, p, s, h)$$

$$\lambda^s = \begin{cases} \lambda^s_{tissue} & \text{outside } \Omega \\ \lambda^S_{tumor} & \text{in } \Omega_P \\ q_s(s) & \text{in } \Omega_H \\ \lambda^s_N & \text{in } \Omega_N \end{cases}$$

where $\mathbf{x}$ is position in space, t is time, p is the tumor (solid) pressure, h is the hematocrit in the neo-vascular network related to oxygen extravasation, and q_s is a smooth interpolating function. The extravasation is modulated by the extra-vascular interstitial pressure p_i scaled by the effective pressure p_e, with k_{pi} representing the weight of the small molecule convective transport component:

$$\lambda^\sigma_{ev} = \overline{\lambda}^\sigma_{ev} \mathbf{1}_{vessel}\,(\boldsymbol{x}, t) \left(\frac{h}{\overline{H}_D} - \overline{h}_{min} \right)^+ \times \left(1 - k_{p_i} \frac{p_i}{p_e} \right) (1 - \sigma)$$

$$\lambda^D_{ev} = \overline{\lambda}^D_{ev} \mathbf{1}_{vessel}\,(\boldsymbol{x}, t) \left(1 - k_{p_i} \frac{p_i}{p_e} \right) \times \left(\frac{C^t}{\overline{C}_D} - \overline{C}_{min} - D \right)$$

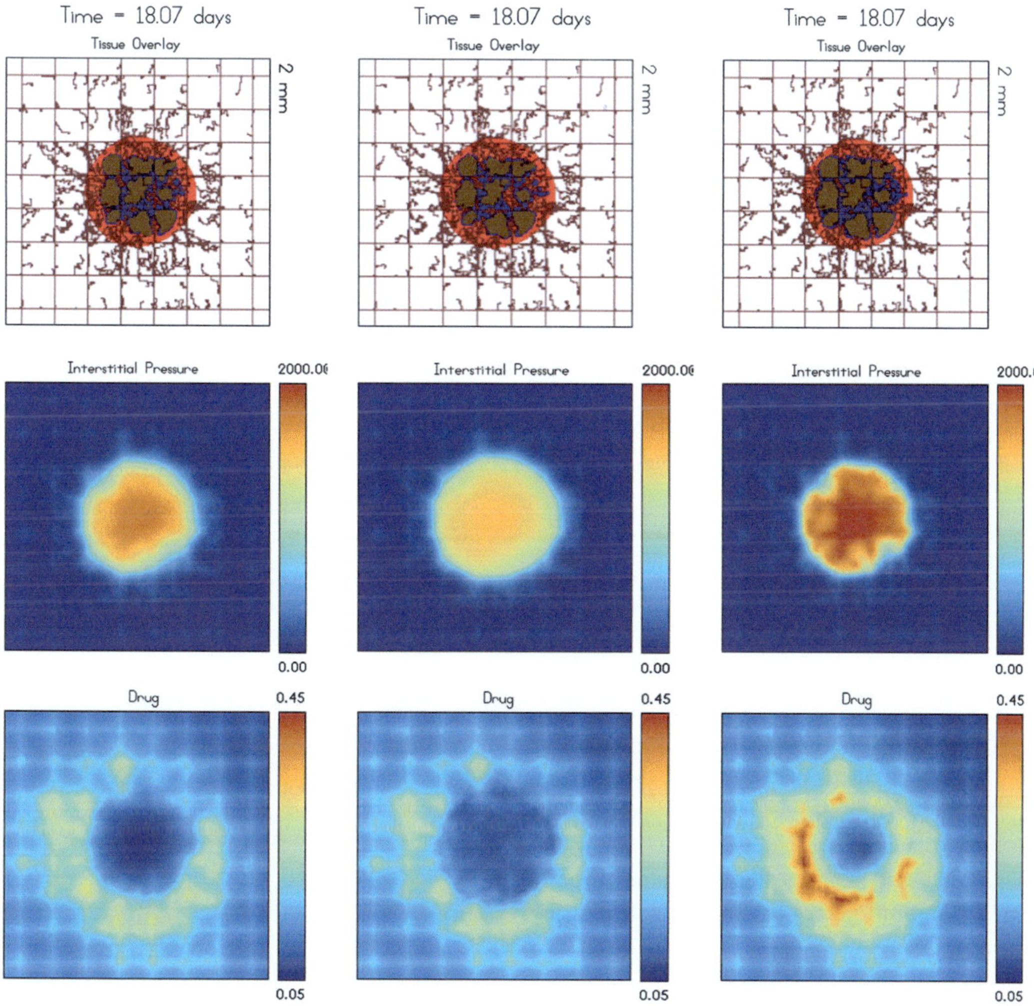

Fig. 9.1 Tumor vasculature, IFP and drug distribution immediately after injection on day 18 after tumor inception. *Red*: Viable (proliferating) tissue; *blue*: hypoxic (quiescent) tissue; *brown*: necrotic (dead) tissue. Existing capillary network is denoted by regularly spaced grid (*brown*), with vessels induced by angiogenesis shown as irregular lines growing towards the hypoxic tumor regions which act as a source of angiogenic stimuli. Normal tissue (not shown) surrounds the lesion. Each panel represents 2×2 mm^2. A pressure gradient is imposed on the domain boundary, 3750 Pa s at the lower left and 3000 Pas at the upper right, linearly decreasing between the two corners. Chemotherapy was simulated via constant injection of a generic macromolecule drug for 5 days. The tumor with elevated tumor interstitial hydraulic conductivity has a broad base plateau profile of IFP (column 2, row 2) whereas the IFP with elevated tumor vascular hydraulic conductivity (column 3) is more hypertensive compared to the control (column 1) due to excessive fluid extravasation. The broad base plateau profile contributes to a larger elevated IFP area and fluid flow away from the tumor. This decreases the drug concentration in and near the tumor (column 2) while the plateau profile itself makes the drug distribution more uniform inside the tumor compared to the control (column 2). Excessive fluid extravasation by an elevated tumor vascular hydraulic conductivity contributes to higher drug extravasation, thus increasing the concentration in the interstitium (column 3), but the distribution is heterogeneous and the concentration in the tumor remains low (though higher than the base case) (Reprinted from Journal of Theoretical Biology, Vol. 355, Wu et al. [103], The effect of interstitial pressure on therapeutic agent transport: Coupling with the tumor blood and lymphatic vascular systems, Page 203, Copyright (2014), with permission from Elsevier)

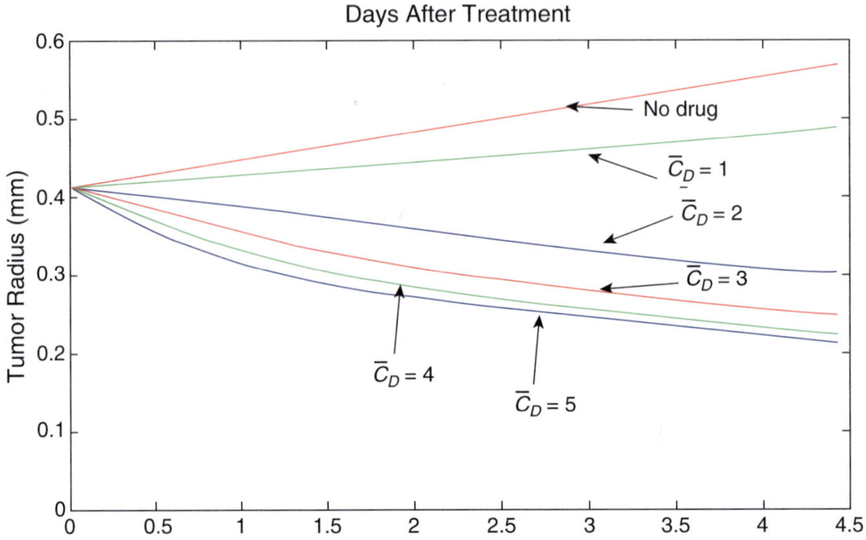

Fig. 9.2 Simulation results of tumor inhibition based on different values of initial local drug concentration $\overline{C}_D$ achieved by intra-vascularly administered NPs, illustrating a nonlinear relationship between drug concentration and tumor regression. Reprinted from van de Ven et al. [97], AIP Advances, Vol. 2, p. 11208, (2012); used in accordance with the Creative Commons Attribution 3.0 Unported License

where $\overline{\lambda}_{ev}^{s}$ is the constant transfer rate from tumor-induced and pre-existing vasculatre, and **1** is the vessel characteristic function. The constants $\overline{h}_{min}$ and $\overline{H}_D$ represent the minimum and normal blood hematocrit required for oxygen extravasation, respectively. Drug release is assumed to follow $C^t = \overline{C}_D\, e^{-\alpha t}$, where the decay α was estimated from the experiments. The drug is assumed to only affect proliferating cells when the level is above a threshold T_{drug}. The effect of the drug is included into the proliferation term λ_p, where λ_{effect} is a rescaling factor (units of effect per drug concentration), $\overline{C}_D$ rescales the drug concentration D, and A is the natural apoptosis rate:

$$\lambda_p = \sigma\left(1 - \lambda_{effect}\overline{C}_D\mathbf{1}_D > T_{drug}\right) - A,$$

which in turn determines the tumor growth velocity: $\nabla \cdot \mathbf{u} = \lambda_p$.

Simulations were run to determine an initial concentration of drug $\overline{C}_D$ to achieve an average 50 % tumor size remission (the "IC50"). This defines the number of NPs per volume as well as the drug per NP required to achieve this remission. Based on the experimentally measured IC50 for BL16-B6 murine melanoma cells *in vitro*, the simulations predicted that approximately 0.5–50 mg doxorubicin in ~10 billion silicon particles would be needed intratumorally. The modeling further showed that therapy-induced cell death is not necessarily linearly proportional to the concentration of drug delivered by the nanoparticles. A saturation phenomenon occurs in which increasing the amount of drug leads to a correspondingly smaller increase in the death effect (Fig. 9.2).

It is expected that drug concentration following an initial burst release from NPs would be insufficient to induce death for cells farther removed from the vasculature [48, 78]. However, tumor remission begins to level off after several days of treatment even when subjected to continuous drug release. Consequently, the relationship between tumor regression (as measured by the tumor radius) and the initial local drug concentration is nonlinear. This saturation effect has been observed in previous studies [88], where an increase in drug concentration is not necessarily linearly proportional to the death effect. One reason is that cell death and removal is a much slower process (typically 6–12 h) compared to drug penetration and decay modulated by transport barriers imposed by under-vascularization and tissue density [85]. Another reason may be attributed to the complex interaction between heterogeneities in the vascular and tumor structures, which constrain the transport of drug release from the nanoparticles. As tumor cells undergo apoptosis, additional regions of inadequate drug penetration may be created. Thus, simply delivering more nanoparticles or more drug per particle in a single injection may not necessarily provide improved response. This work suggests that optimal nanotherapy requires a strategy combining both experimental and mathematical modeling in order to overcome tumor microenvironment induced barriers.

In [31], optimal combinations of size, shape, and surface functionalization were studied to maximize nanovector localization within diseased microvasculature. The approach combines IVM, multiscale mathematical modeling, and *in vitro* assays. The IVM provides a quantitative assessment of microvascular phenomena in real-time, enabling the acquisition of data pertaining to the geometry, permeability, and hemodynamic conditions of the vascular network in animal models. An *in vitro* parallel plate flow chamber system enabled the quantitative adhesive properties of nanovectors to be measured. Using an external pumping system, NPs were flown through the apparatus, and their adhesive dynamics to the bottom cell layer were analyzed over time using fluorescence microscopy. The results of the mathematical modeling in terms of concentration of adhering and flowing NPs could be used to validate and improve upon the *in vitro* and *in vivo* experiments. The end goal of this approach is to identify efficacious nanovector properties for optimal tumor delivery as well as reduce the time and costs required for future optimization.

In [96], IVM was utilized to simultaneously acquire vascular perfusion information on breast cancer tumor xenografts in mice in conjunction with nanoparticle accumulation kinetics. In simplifying the analysis of heterogeneous perfusion measurements, a modeling framework was developed using a fuzzy c-means classifier to rank tumors. The ranking was based on two features identified through IVM observation of the vascular perfusion, namely the time to arterial peak and the venous delay. This ranking would not require underlying assumptions about the tissue structure, function, or internal heterogeneity. The rankings were then correlated inversely with experimental nanoparticle accumulation measurements, thus allowing for estimation of NP delivery into tumor tissue based on the perfusion rankings. The methodology was proposed to be applicable to study a variety of nanoparticles targeted in different tumor models.

9.3.4 Modeling of Nanoparticle Transport Based on In Vitro Data

The hypothesis that NP diffusive transport and spatial distribution in solid tumors can be predicted using mathematical models based on experimentally derived parameters was tested in [30] with pharynx FaDu cells using three different formulations: negatively charged polystyrene beads, near-neutral liposomes, and positively charged liposomes, with respective diameter of 20, 110, and 130 nm. The applied methodology consisted of five steps:

1. Develop a model for diffusive flux in 3-dimensional tumor interstitium, which is determined by several concentration and time-dependent processes including NP binding and internalization in cells.
2. Obtain model parameters for nanoparticle diffusivity, binding, and internalization from experiments with 2D monolayer cell culture.
3. Perform *in silico* studies to simulate NP diffusive transport in 3D tumor interstitium.
4. Experimentally determine the concentration-depth profiles of fluorescently-labeled NP in 3D tumor cell spheroids at three initial NP concentrations and three treatment times.
5. Evaluate model performance by comparison of *in silico* and *in vitro* diffusion profile studies.

The study of effect of treatment time used a fixed initial NP concentration (18.8 nM for polystyrene beads, 1.13 nM for HSPC liposomes, 0.55 nM for C20-5 liposomes). The evaluation of effect of initial NP concentration used a single treatment time (5 h for beads, 6 h for both liposomes). The results compared the model-simulated NP spatial distribution to experimentally observed data. The model was shown to be predictive of 3D transport in spheroids for the neutral and negative liposomal NPs with greater than 90 % accuracy, but not for the positive liposomal NPs. The study additionally showed the limited penetration of NPs in FaDu spheroids. This may be partly due to the limited diffusive transport of nanoparticles, the binding of nanoparticles to cells, and the presence of tight cell packing creating a barrier to diffusion. As complex and as accurate as this approach was shown to be, numerous improvements and potential limitations were noted. For example, the model did not consider convection, charge effects, efflux from cells, effect of NP shape, non-saturable binding of NPs, and heterogeneous tumor interstitium. The approach shows promise for further development and refinement of such predictive methods.

An interdisciplinary experimental/ computational modeling approach to evaluate diffusive transport of drug-loaded gold nanoparticles in heterogeneously-vascularized tumors was proposed in [17]. Gold nanoparticles with two or three layer formulations to enhance tumor tissue uptake and diffusivity, and loaded with either cisplatin (a hydrophilic drug) or paclitaxel (a hydrophobic drug), were evaluated for size, surface properties, release profiles of loaded drug, and cytotoxicity in non-small cell lung cancer (NSCLC) cell lines and in 3D spheroid cell culture. The cytotoxicity experiments demonstrated increased efficacy of the layered formulations over free drug, and confirmed the increased resistance conferred by the hypovascularized microenvironment in 3D cell culture. *In silico* analysis using computer simulations was applied to project the efficacy of these NP formulations in an *in vivo* vascularized environment. The simulations were calibrated to the *in vitro* data using the drug release profiles and cytotoxicity results. The mathematical equations for nanoparticle transport built upon the studies performed in [97], while the vascularized tumor component including vascular flow is based on [103]. The drug G is released at the location of extravasated nanoparticles and diffuses through the tissue with a coefficient D_G. The uptake by tumor and normal cells and the wash-out from the interstitial space are included as a combined effect in the rate $\overline{\lambda}^G_{decay}$, which reflects the drug half-life:

$$\frac{\partial G}{\partial t} = \nabla \cdot (D_G \nabla G) + \lambda^G_{release}(t, s) - \overline{\lambda}^G_{decay} G$$

where the drug release $\lambda^G_{release}$ from the nanoparticles is $\lambda^G_{release} = sC^G_t$. The release C^G_t in time is fitted to follow the profile experimentally observed for cisplatin or paclitaxel obtained in [25]. Locally, the drug release rate thus combines the effect of the nanoparticle concentration and the drug release profile.

The simulated injection of paclitaxel-loaded nanoparticles and the subsequent drug release are illustrated in Fig. 9.3. Over time, the drug concentration in the interstitium reaches a maximum around 12 h after treatment due to the steady release of drug from the NPs, followed by drug and NP washout. Figure 9.4 shows the response of the vascularized tumor to each potential treatment. The results highlight the importance of steady release and longer bioavailability of drug for effective response. Although the results were not verified *in vivo*, the application of computational modeling to evaluate drug-loaded gold nanoparticle performance based on dosage and tissue transport characteristics offers the possibility to guide the design of such systems with tumor-specific data.

9.3.5 Modeling of Vasculature-Bound Nanoparticles

NPs targeted to the tumor vasculature (instead of the tumor cells) provide a means for anti-tumor and anti-angiogenic therapy as well as

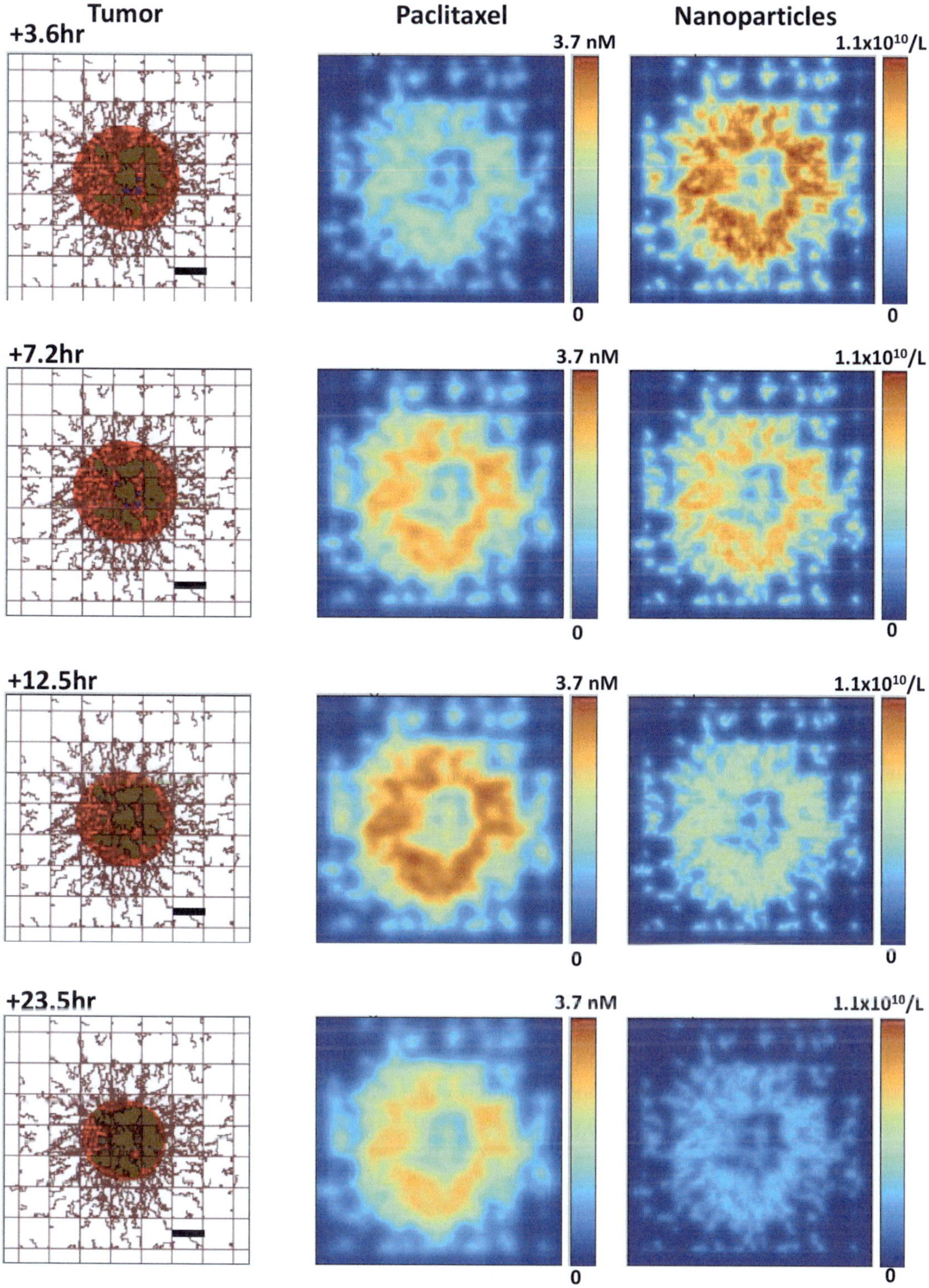

Fig. 9.3 Simulated treatment with paclitaxel-loaded three-layer gold nanoparticles (colors as in Fig. 9.1). Initially (at 3.6 h), the nanoparticle concentration is very high while the drug concentration is low. Over the course of 24 h, the nanoparticles wash out of the interstitium with the drug concentration actually peaking at ~12 h post injection. During this process, the tumor lesion begins to shrink as a result of the drug effect. Vascular flow is from lower left to upper right for each panel (Reprinted from Curtis et al. [17]: Curtis LT et al. (2016) An interdisciplinary computational/experimental approach to evaluate drug-loaded gold nanoparticle tumor cytotoxicity, with permission from Nanomedicine (Lond.) as agreed by Future Medicine Ltd.)

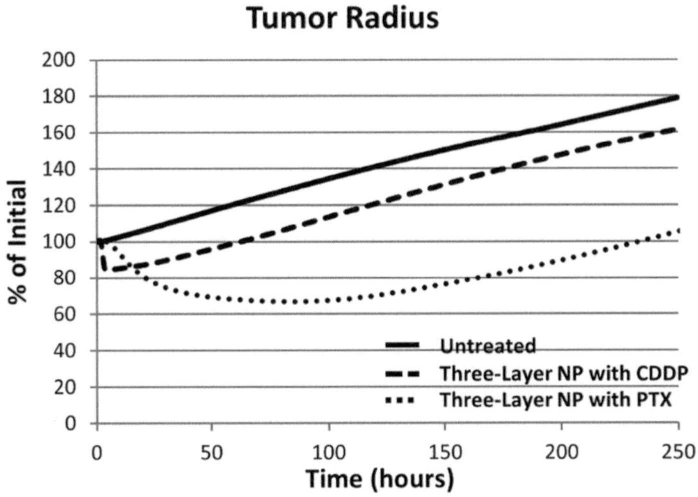

Fig. 9.4 Simulated change in vascularized tumor lesion radius after treatment with drug-loaded gold nanoparticles. Solid line: untreated (control) tumor; dashed: cisplatin-treated lesion; dotted: paclitaxel-treated lesion. The paclitaxel system achieved a 41.3 % decrease in lesion radius compared to the untreated control by 250 h post injection, while cisplatin attained a 9.5 % decrease (Reprinted from Curtis et al. [17]: Curtis LT et al. (2016) An interdisciplinary computational/experimental approach to evaluate drug-loaded gold nanoparticle tumor cytotoxicity, with permission from Nanomedicine (Lond.) as agreed by Future Medicine Ltd.)

cancer detection and imaging. The clinical application of this concept has been limited because of the complex interplay of tumor, drug, and NP parameters. The NP vascular affinity, a critical factor in this scheme, is a function of vascular receptor expression and NP size and surface-bound ligand density. Building upon the vascularized tumor growth model presented in [103], a computational model was developed in [28] to study the accumulation of vasculature-bound NPs in order to help elucidate the interplay between NP distribution and accumulation as a function of NP vascular affinity, size, and drug loading and release characteristics. The tumor response to drug released from such as system was evaluated in [18] by coupling with the nanotherapy model from [97]. The NPs are represented to be functionalized with surface-bound ligands targeting specific vascular endothelial receptors of the tumor lesions. The particle number N attached in each vessel segment is described by [21]:

$$N(d, R_u, S_u) = S_u \alpha d^{\delta_1} \exp\Big(-\beta \left(1 + \gamma d^{\delta_2}\right) Srt_u \Big)$$

where the shear rate is $Srt_u = 4Q_u/pR_u^3$ and Q_u is the flow rate. The parameters α, β, and γ are respectively proportional to [28]: (*i*) the surface density of receptors on endothelial cells and ligands on the NP, and the ligand-receptor affinity under zero external force; (*ii*) the characteristic length scale of the ligand-receptor bond and the viscosity of fluid; (*iii*) the inverse of the surface density of receptors. The values of α, β, γ are varied to modulate the NP affinity for the vascular endothelial cell receptors. Following this equation, the particle concentrations in the blood and on the vessel surface are quantified by mass conservation equations in the vessels and on the vessel surface between neighboring nodes [28].

In Fig. 9.5, tumor radius, blood area fraction, average vascular flow rate, and average wall shear rate are shown for both the neovasculature and pre-existing vasculature [28]. The tumor shows overall a 30-fold increase in radius over the course of 24 days. The blood area fraction is introduced as the ratio between the total area covered by the vessel network and the tumor area in a cross section. Interestingly, for the new vascular network originating with the tumor, the blood area fraction grows with time reaching a value close to 0.6 at 24 days, implying that more

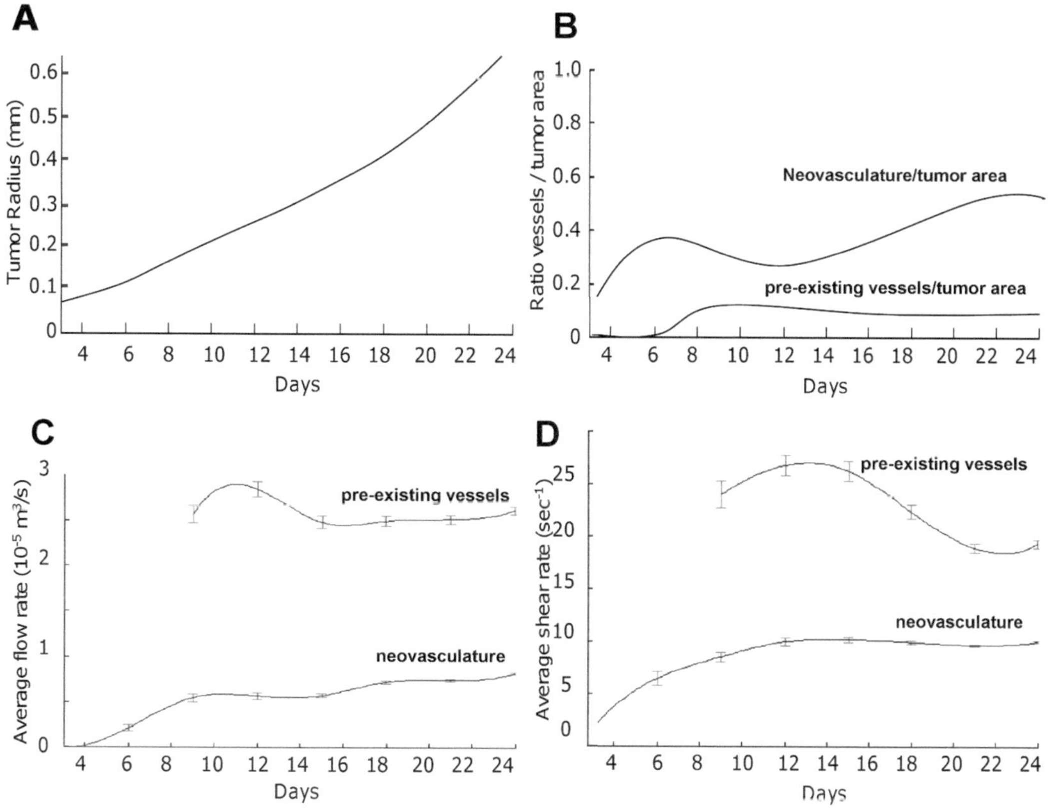

Fig. 9.5 Simulated tumor and vasculature growth presented as a function of the number of days post tumor inception. (**a**) Tumor radius; (**b**) blood area fraction, defined as ratio of the vasculature to the tumor area, (**c**) average vascular flow rate; and (**d**) average wall shear rate. In (**c**) and (**d**), bars denote SEM (standard error of the mean) values estimated over the number of vessel segments in the tumor area (Reprinted with permission from [28]: Frieboes HB et al. (2013) A computational model for predicting nanoparticle accumulation in tumor vasculature. PloS ONE 8 (2):e56876. doi:10.1371/journal.pone.0056876; used in accordance with Creative Commons Attribution (CC BY) license)

than 50 % of the tumor cross section is covered by blood vessels. On the other hand, the blood area fraction for the pre-existing vasculature holds steady over time, being always smaller than 10 %. Additionally, for the pre-existing vessels, little variation in flow rate and shear rate are observed over the 24 days. Differently, these two hydrodynamic parameters change dramatically for the neovasculature starting from the avascular phase of the tumor during the rapid growth over the first 10 days. These values are comparable with those observed for the pre-existing vasculature, implying that after 10 days the tumor neovasculature is fully functional.

Systemic injection of NPs was simulated and particle fraction adhering per tumor area was quantified (Fig. 9.6) over the period of tumor growth evaluated in Fig. 9.5. The total fraction of adhering NPs is shown to be not only dependent on the dynamic tumor vascular characteristics, but also on the NP parameters α, β, and NP diameter *d*. The variation in the fraction of adhering NPs presents a complex and interesting behavior (Fig. 9.6 – bottom row). For $\alpha = 10^{12}$ m^{-2}, the accumulation of the 100 nm particles grows rapidly over the first 10 days and then decreases slowly up to 24 days. Differently, for the 600 and 1000 nm particles, the fraction of adhering particles reaches a maximum at about 10 days and then decreases rapidly to almost zero up to 24 days. These results emphasize the importance of properly modulating the size of the NPs with

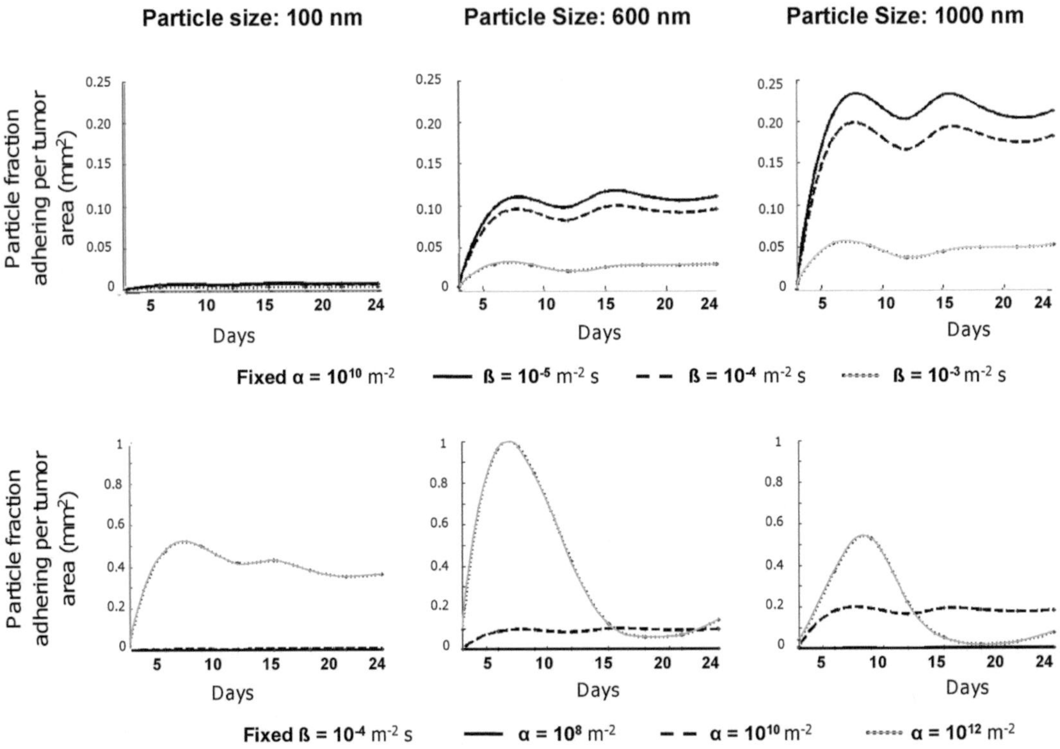

Fig. 9.6 Fraction of injected NPs adhering firmly at the blood vessel walls per tumor area (mm^2) as a function of the tumor development stage (days post tumor inception) for a range of magnitudes of the NP parameters α and β (Reprinted with permission from [28]: Frieboes HB et al. (2013) A computational model for predicting nanoparticle accumulation in tumor vasculature. PloS ONE 8 (2):e56876. doi:10.1371/journal.pone.0056876; used in accordance with Creative Commons Attribution (CC BY) license)

respect to the NP vascular affinity as well as the stage of the tumor growth in order to foster tumoritropic accumulation and a more uniform distribution of the NPs.

Building upon this study, scenarios of variable NP vascular affinity and diameter were further simulated in [18] in order to evaluate how these characteristics affect both the heterogeneities of drug distribution from drug-loaded NPs as well as the tumor response to the released drug. It was assumed that the initial NP drug loaded is linearly proportional to the diameter d, and that the release rate $\lambda^G_{release}$ in time is proportional to the square root of this diameter:

$$\lambda^G_{release} = kC_{pS}\sqrt{d}$$

where k is a proportionality constant describing the drug release and C_{pS} is the local NP concentration on the vessel surface [28]. Following [103], it is assumed that the net proliferation rate λ_P of the tumor is proportional to the amount of oxygen and nutrients present. This rate is modulated by the effect of the drug, $\overline{\lambda}_{effect}$:

$$\lambda_P = \begin{cases} 0 & \text{outside } \Omega \\ \lambda_M\sigma\left(1 - \overline{\lambda}_{effect}D\right) - \lambda_A & \text{in } \Omega_P \\ 0 & \text{in } \Omega_H \\ -\lambda_N & \text{in } \Omega_N \end{cases}$$

In order to simulate the typically cell-cycle dependent effects of chemotherapeutic drugs (e.g., as is the case with paclitaxel), the drug is assumed to only act upon proliferating cells. The term $\overline{\lambda}_{effect}$ is the rate of drug-induced cell death, D is the local drug concentration, λ_M is the mitosis rate, λ_A is the apoptosis rate, and λ_N is the rate of volume loss in the necrotic regions assuming

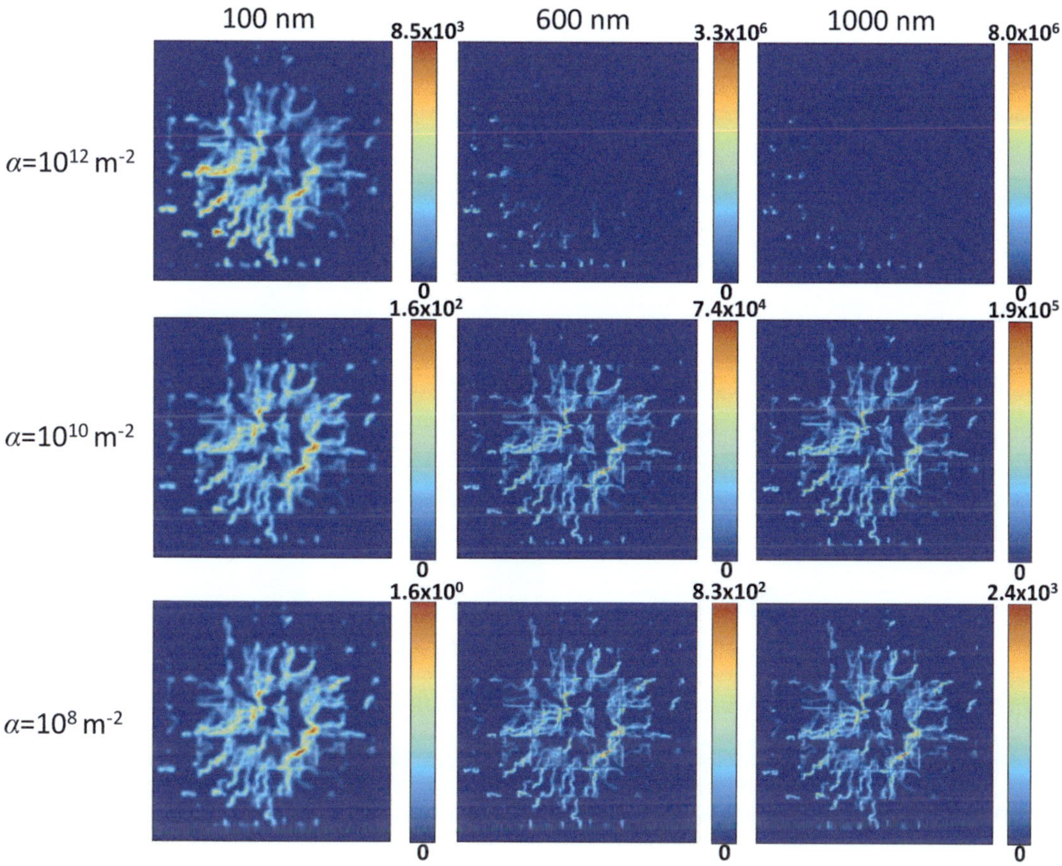

Fig. 9.7 Distribution of drug released by NPs 24 h after NP injection (colors as in Fig. 9.1). Vascular flow is from lower left to upper right in each panel. Drug concentration is in dimensionless units, with the maximum drug concentration decreasing about an order of magnitude for each 10^2 decrease in α (Reprinted with permission from [18]: Curtis LT et al. (2015) Computational Modeling of Tumor Response to Drug Release from Vasculature-Bound Nanoparticles. PloS ONE 10(12):e0144888; used in accordance with Creative Commons Attribution (CC BY) license)

that cellular debris is constantly degraded and the resulting fluid is removed. When $\overline{\lambda}_{effect}D \leq 1$, the net proliferation is reduced, and when $\overline{\lambda}_{effect}D > 1$, cell death is introduced and contributes to tumor regression.

As was shown in [28], immediately after injection the NPs distribute more uniformly throughout the vasculature when their affinity, primarily determined by the parameter α, is lower, but the maximum concentration is significantly decreased with lower α (~10^2 lower for each 10^2 decrease in α). For the same value of α, however, the maximum NP concentration increases with larger size. For the 600 and 1000 nm diameters, the NP vascular accumulation is highly heterogeneous. For $\alpha = 10^{12}$ m^{-2}, the NPs accumulate preferentially at the periphery of the tumor (at the vascular inflow side) as a result of high affinity for the neovasculature. Figure 9.7 illustrates the drug distribution as a function of NP diameter and the parameter α, proportional to the adhesive strength of NPs to the vasculature [18]. These results emphasize how excessively high vascular affinity can lead to NP distribution at the periphery of the tumor with poor homogeneous penetration throughout, leading to lower intra-tumor drug concentrations and poorly-effective treatments.

Additionally, the study in [18] quantified the tumor response to NP adherence, with the overall effectiveness of the drug represented as $1/\overline{\lambda}_{IC50}$,

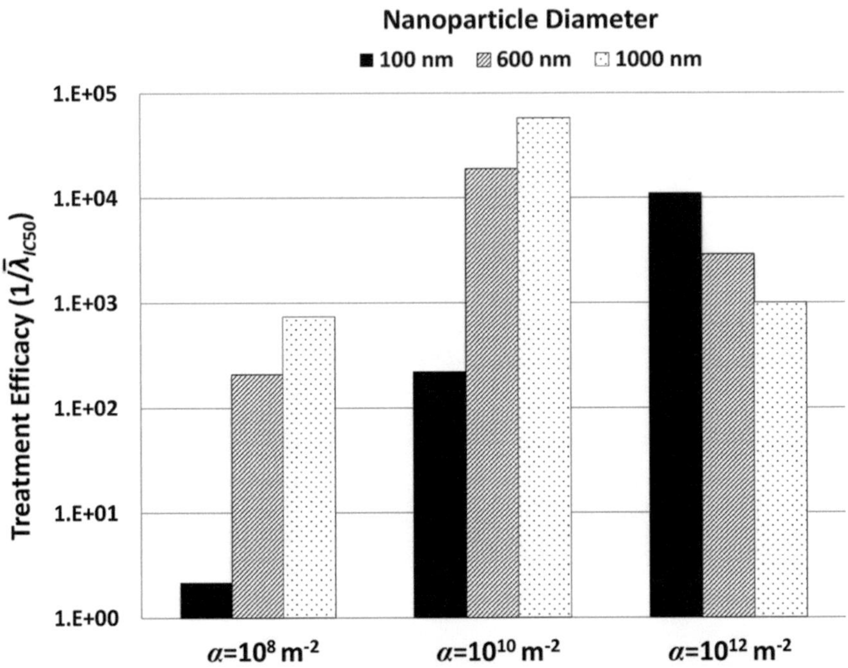

Fig. 9.8 Relative efficacy of all treatments types (three levels of parameter α, three levels of NP diameter d). Relative efficacy (non-dimensional value) is measured as $1/\bar{\lambda}_{IC50}$, with $\bar{\lambda}_{IC50}$ being the rate of drug-induced death necessary to reduce the tumor to 50 % of its original size. A larger value of $1/\bar{\lambda}_{IC50}$ thus denotes more effective treatment (Reprinted with permission from [18]: Curtis LT et al. (2015) Computational Modeling of Tumor Response to Drug Release from Vasculature-Bound Nanoparticles. PloS ONE 10(12):e0144888; used in accordance with Creative Commons Attribution (CC BY) license)

where $\bar{\lambda}_{IC50} = \bar{\lambda}_{effect}$ for 50 % tumor remission. The results in Fig. 9.8 show that although larger NP have greater drug loading, drug release rates, and higher maximal adherence in tissue, they do not necessarily correspond to highest treatment efficacy. For the lowest and medium affinities ($\alpha = 10^8$ and $\alpha = 10^{10}$, respectively), larger NP diameters yielded increased efficacy. The smallest diameter (100 nm) performed best at the highest affinity, while the two larger diameters (600 and 1000 nm) were most efficacious with the medium affinity. Regardless of NP size, efficacy was superior for medium than for lowest affinity. This trend was reversed for highest affinity, for which an increased NP diameter decreased the treatment efficacy due to lower uptake of larger NPs. All NPs were less effective than at higher affinities at the lowest affinity ($\alpha = 10^8$). These results indicate that effective nanotherapy requires a delicate balance between vascular affinity and NP diameter dependent upon the stage of tumor development and vascularity.

9.3.6 Evaluation of Nanotherapy Using Pharmacokinetic Modeling

Physiologically based pharmacokinetic modeling (PBPK) has been employed to seek improvement of efficacy of pharmacological agents [49], including recently for nanotherapeutics. These models are designed to study the absorption, distribution, metabolism, and excretion (ADME) of small agents such as drugs or nanoparticles, and enable interspecies extrapolation from animal data to humans. PBPK models organize individual organs or tissues of an anatomical structure as separate compartments, where each compartment is interconnected through mass

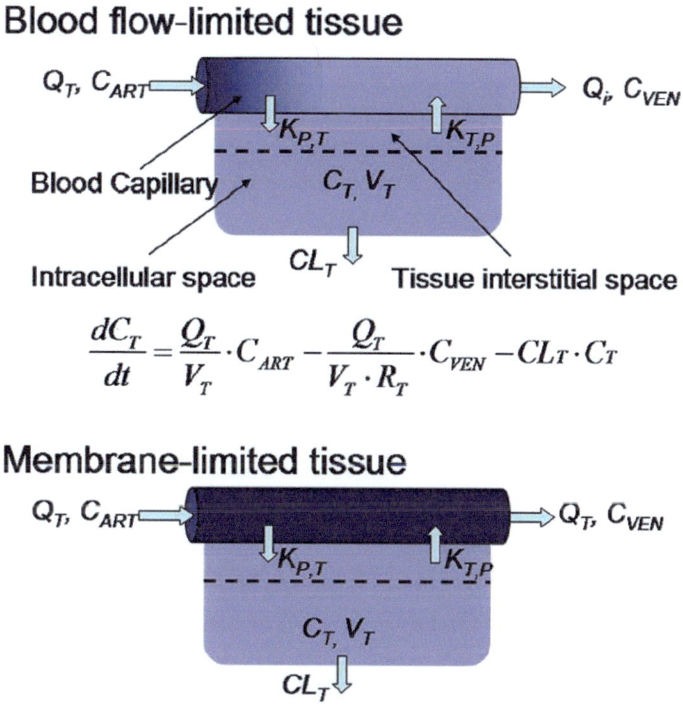

$$\frac{dC_T}{dt} = \frac{Q_T}{V_T} \cdot C_{ART} - \frac{Q_T}{V_T \cdot R_T} \cdot C_{VEN} - CL_T \cdot C_T$$

$$\frac{dC_T}{dt} = K_{P,T} \cdot C_{ART} - \frac{K_{T,P}}{R_T} \cdot C_T - CL_T \cdot C_T$$

Fig. 9.9 Diagrams and equations for a blood flow-limited tissue (upper panel) and a membrane-limited tissue with the vascular membrane as the limiting membrane (bottom panel). C = concentration; CL = clearance; R = tissue-to-plasma partition coefficient; Q = blood flow; V = volume, K = transportation coefficient; subscripts *ART*, *VEN*, *P*, and *T* indicate arterial, venous, plasma, and tissue, respectively. For tissues without elimination, the *CL* will be zero (Reprinted with permission from [49]: Li et al., Physiologically based pharmacokinetic modeling of nanoparticles. ACS Nano 4 (11):6303–6317. Copyright (2010) American Chemical Society)

transport. Time-dependent concentrations of drugs and metabolites in these compartments are described by mathematical equations. Traditional pharmacokinetic models cannot provide mass-time profiles of individual tissues and organs. An example of PBPK modeling of transport mechanisms is shown in Fig. 9.9 [49]. The blood flow-limited model assumes that every organ is a well-stirred compartment and that nanoparticle distribution between blood and tissue instantly reaches equilibrium. This means that the only limiting factor in the transportation between blood and tissue is the blood flow rate. For membrane-limited models, it is assumed that transportation between blood and tissue is limited by permeability of the blood capillary membrane or tissue cell membrane [51].

PBPK modeling has been successfully applied to therapeutic agents for over 30 years, and more recently to the modeling of nanotherapy. Nanoparticles differ from drugs in their absorption, distribution, metabolism, and excretion in physiological systems. Larger nanoparticles are involved in potentially more complex processes such as opsonization in blood, cellular recognition and internalization, enzymatic degradation, and property changes, which may not be relevant to freely administered drug molecules.

Pharmacokinetic modeling was employed in [50] to quantitatively represent and understand *in vivo* behavior of nano-sized drug carriers. Using paclitaxel incorporated in poly (lactic-co-glycolic acid) nanoparticles chemically labeled with a fluorescence probe, the NPs were delivered to human breast cancer MCF7 cells and the PK profiles were quantified. A mathematical model was used to describe the intracellular PK of paclitaxel delivered by the nanoparticles. The model accurately represented the effects of relative factors, such as drug dose, drug-release kinetics, and nanoparticle PK, on the PK of paclitaxel delivered by the nanoparticles. In [52] a PBPK model was developed to describe the process of absorption, distribution, metabolism, and elimination of nanoparticles after intravenous, oral, and pulmonary exposure. As with typical PBPK models, the human body was divided into compartments and mass-time profiles were simulated using mathematical equations describing the inter-compartmental transport of nanoparticles, assuming first order kinetics.

In [51], a physiologically based pharmacokinetic model was utilized to interpret the effects of nanoparticle properties on previously published biodistribution data for five poly(lactic-co-glycolic) acid (PLGA) nanoparticle formulations prepared with varied content of monomethoxy-poly (ethyleneglycol) (mPEG) (PLGA, PLGA-mPEG256, PLGA-mPEG153, PLGA-mPEG51, PLGA-mPEG34) after intravenous injection in mice. Multivariate regression analysis was performed on the data to build the relationship between nanoparticle properties (size, zeta potential, and number of PEG molecules per unit surface area) and biodistribution parameters. Based on the experimental data collected, the body was divided into compartments as outlined in Fig. 9.10.

Rate equations for each compartment were based on the principle of mass balance conservation: the net amount of nanoparticles in an organ equals the sum of all nanoparticles that enter the organ minus all nanoparticles that exit the organ. This can be generalized for each separate organ as follows [51]:

$$V_i\frac{dC_i}{dt} = \frac{dM_i}{dt} = k_i\left(C_p - \frac{C_i}{R_i}\right) - CL_iC_i$$
$$= k_i\left(\frac{M_P}{V_P} - \frac{M_i}{R_iV_i}\right) - CL_i\frac{M_i}{V_i}$$

where C_p and M_p refer to the nanoparticle concentration and amount in the blood, V_i, C_i, M_i, k_i, R_i, and CL_i refer to the tissue volume, nanoparticle concentration, nanoparticle amount, diffusion coefficient, tissue–blood partition coefficient, and excretion coefficient in tissue *i*, respectively. After the parameters were determined, the equations describing the mass-time profiles of the five nanoparticle formulations were solved numerically. Simulations were then performed to analyze mass-time curves for comparison with experimental data. This model system simulated well the experimental results of tissue concentration-time curves, providing insight into the mechanisms of nanoparticle biodistribution and the relationship between nanoparticle properties and biodistribution.

To verify the predictive ability of the model, characterized physicochemical properties of PLGA-mPEG495 nanoparticles (a sixth formulation) were used to predict biodistribution profiles. It was found that the predicted values for PLGA-mPEG495 were close to experimental data for all tissues. Further applications of this work could include nanoparticle toxicity

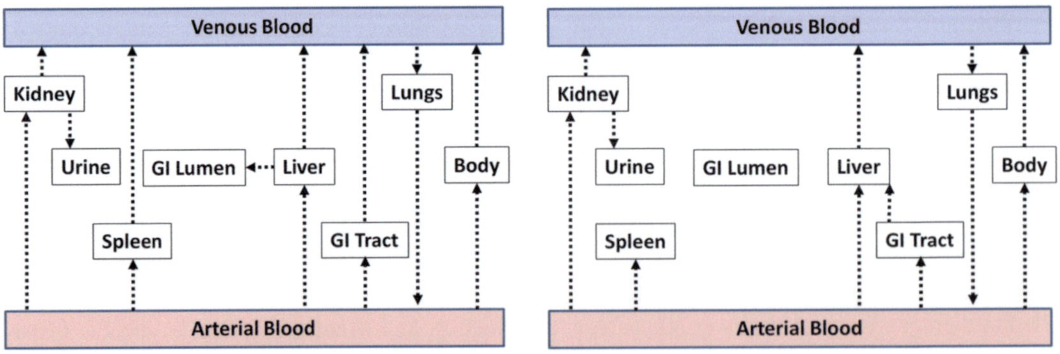

Fig. 9.10 Schematic diagrams of (**a**) membrane- limited model and (**b**) blood flow-limited model. Arrows indicate the transport of nanoparticles, with the set of transport equations being different between the two cases (Adapted from Li et al. [51])

assessment, rational carrier design for drug delivery and imaging, and data interpretation from *in vivo* bioimaging/biomarker systems.

9.3.7 Modeling of Nanotherapy Induced Hyperthermia

In contrast to the drug-related NP studies discussed so far, magnetic and gold nanoparticle-mediated hyperthermia has also been computationally modeled, as reviewed in [43]. In this approach, the nanoparticle itself, rather than a nanoparticle-encapsulated drug, is the basis for the therapy. Hyperthermia, or thermal therapy, is a form of treatment in which the tissue is heated, typically between 40 and 45 °C, causing thermally-induced tissue damage. Hyperthermia may be applied locally or systemically to the whole body. There are multiple possible mechanisms for hyperthermia, such as ultrasound, thermal chambers, laser light, or nanoparticles. As a result of heating, a number of molecular effects take place, including protein denaturation and the induction of cellular apoptosis. Cancer cells are not intrinsically more vulnerable to hyperthermia than normal cells; however, *in vivo*, the typically low pH and hypoxic tumor microenvironment increases cell sensitivity to heat. Nanoparticle-based hyperthermia allows concentration of the local heat source (the nanoparticles) within the tumor, so heat-induced damage to healthy cells can be minimized. In addition, NP-based hyperthermia could potentially be combined with chemotherapeutic delivery of drugs to take advantage of chemosensitization effects.

Computational models of tumor hyperthermia treatment have been developed by multiple groups (see [43] and associated references), some more focused on the physics of the process while others are more focused on the therapeutic aspect of hyperthermia on the biological tissues. There are several important variables and parameters that characterize typical models, including nanoparticle physical properties such as material and shape, the excitation source, tissue properties such as bioheat transference and tissue heterogeneity, and nanoparticle distribution in tissue. Current limitations include lack of consideration of tumor heterogeneities, oversimplification of nanoparticle accumulation and transport within tumors, lack of experimental verification of *in silico* studies integrated with *in vitro and in vivo* experiments, and insufficient understanding of the effect of nanoparticle-based hyperthermia on chemotherapeutics. Despite these limitations, the prospective value of computational modeling is high, especially to help elucidate the complex and non-intuitive interactions between thermal, drug, NP, and tumor tissue parameters. In particular, models could be developed for optimization of nanoparticle properties and excitation, as well as the assessment of therapeutic strategies tailored to patient-specific tumors.

9.4 Conclusion

The work reviewed in this chapter shows that the complexity of the interactions between the typically dysregulated tumor microenvironment and the plethora of available nanotherapy parameters can be more comprehensively addressed via complementary experimental and computational modeling efforts. The presented results illustrate examples of recent modeling approaches. However, much more is required in order to facilitate the bridging of nanotechnology to cancer clinical treatment, including a tighter *a priori* coordinated integration between experiments and computational model parameters, further development of mathematical models capable of capturing the multi-scale biophysics from the nano- to the patient tissue-scales, and the establishment of more uniform methods for nanotherapy design and validation. In order to succeed, this effort will require even more intensive interdisciplinary collaboration and exchange between biological, material, and physical scientists, as well as with researchers in related disciplines. Such considerations provide a realistic basis for hope that the tumor microenvironment barrier to nanotherapy will eventually be overcome and thus lead to effective cancer treatments.

References

1. Allen JD, Brinkhuis RF, van Deemter L, Wijnholds J, Schinkel AH (2000) Extensive contribution of the multidrug transporters P-glycoprotein and Mrp1 to basal drug resistance. Cancer Res 60(20):5761–5766
2. Andasari V, Gerisch A, Lolas G, South AP, Chaplain MA (2011) Mathematical modeling of cancer cell invasion of tissue: biological insight from mathematical analysis and computational simulation. J Math Biol 63(1):141–171. doi:10.1007/s00285-010-0369-1
3. Anderson AR, Chaplain MA (1998) Continuous and discrete mathematical models of tumor-induced angiogenesis. Bull Math Biol 60(5):857–899. doi:10.1006/bulm.1998.0042
4. Aschoff L, Kiyono K (1913) Zur frage der grossen Mononuclearen. Folia Haematol 15:383–390
5. Bachmann J, Raue A, Schilling M, Becker V, Timmer J, Klingmuller U (2012) Predictive mathematical models of cancer signalling pathways. J Intern Med 271(2):155–165. doi:10.1111/j.1365-2796.2011.02492.x
6. Beh CW, Seow WY, Wang Y, Zhang Y, Ong ZY, Ee PLR, Yang YY (2009) Efficient delivery of Bcl-2-targeted siRNA using cationic polymer nanoparticles: downregulating mRNA expression level and sensitizing cancer cells to anticancer drug. Biomacromolecules 10(1):41–48. doi:10.1021/Bm801109g
7. Byrne HM (2010) Dissecting cancer through mathematics: from the cell to the animal model. Nat Rev Cancer 10(3):221–230. doi:10.1038/Nrc2808
8. Cairns R, Papandreou I, Denko N (2006) Overcoming physiologic barriers to cancer treatment by molecularly targeting the tumor microenvironment. Mol Cancer Res 4(2):61–70. doi:10.1158/1541-7786.mcr-06-0002
9. Chaplain MAJ (2011) Multiscale mathematical modelling in biology and medicine. IMA J Appl Math 76(3):371–388. doi:10.1093/imamat/hxr025
10. Chauhan VP, Jain RK (2013) Strategies for advancing cancer nanomedicine. Nat Mater 12(11):958–962. doi:10.1038/nmat3792
11. Chen AM, Zhang M, Wei DG, Stueber D, Taratula O, Minko T, He HX (2009) Co-delivery of doxorubicin and Bcl-2 siRNA by mesoporous silica nanoparticles enhances the efficacy of chemotherapy in multidrug-resistant cancer cells. Small 5(23):2673–2677. doi:10.1002/smll.200900621
12. Cheng D, Cao N, Chen JF, Yu XS, Shuai XT (2012) Multifunctional nanocarrier mediated co-delivery of doxorubicin and siRNA for synergistic enhancement of glioma apoptosis in rat. Biomaterials 33(4):1170–1179. doi:10.1016/j.biomaterials.2011.10.057
13. Systems CDD (1994) Drugs and the pharmaceutical sciences, 1st edn. CRC Press/Marcel-Dekker, Inc., New York/Basel/Hong Kong
14. Creixell M, Bohorquez AC, Torres-Lugo M, Rinaldi C (2011) EGFR-targeted magnetic nanoparticle heaters kill cancer cells without a perceptible temperature rise. ACS Nano 5(9):7124–7129. doi:10.1021/Nn201822b
15. Cristini V, Lowengrub J (2010) Multiscale modeling of cancer: an integrated experimental and mathematical modeling approach. Cambridge University Press, Cambridge
16. Curti BD, Urba WJ, Alvord WG, Janik JE, Smith JW 2nd, Madara K, Longo DL (1993) Interstitial pressure of subcutaneous nodules in melanoma and lymphoma patients: changes during treatment. Cancer Res 53(10 Suppl):2204–2207
17. Curtis LT, England CG, Wu M, Lowengrub J, Frieboes HB (2016) An interdisciplinary computational/experimental approach to evaluate drug-loaded gold nanoparticle tumor cytotoxicity. Nanomedicine (Lond) 11(3):197–216. doi:10.2217/nnm.15.195
18. Curtis LT, Wu M, Lowengrub J, Decuzzi P, Frieboes HB (2015) Computational modeling of tumor response to drug release from vasculature-bound nanoparticles. PLoS One 10(12):e0144888. doi:10.1371/journal.pone.0144888, eCollection 2015
19. Cuvier C, Roblot-Treupel L, Millot JM, Lizard G, Chevillard S, Manfait M, Couvreur P, Poupon MF (1992) Doxorubicin-loaded nanospheres bypass tumor cell multidrug resistance. Biochem Pharmacol 44(3):509–517. doi:10.1016/0006-2952(92)90443-M
20. Dean M, Fojo T, Bates S (2005) Tumour stem cells and drug resistance. Nat Rev Cancer 5(4):275–284. doi:10.1038/nrc1590
21. Decuzzi P, Ferrari M (2006) The adhesive strength of non-spherical particles mediated by specific interactions. Biomaterials 27(30):5307–5314. doi:10.1016/j.biomaterials.2006.05.024
22. Decuzzi P, Pasqualini R, Arap W, Ferrari M (2009) Intravascular delivery of particulate systems: does geometry really matter? Pharm Res 26(1):235–243. doi:10.1007/s11095-008-9697-x
23. Deisboeck TS, Wang Z, Macklin P, Cristini V (2011) Multiscale cancer modeling. Annu Rev Biomed Eng 13:127–155. doi:10.1146/annurev-bioeng-071910-124729
24. Edelman LB, Eddy JA, Price ND (2010) In silico models of cancer. Wiley Interdiscip Rev Syst Biol Med 2(4):438–459. doi:10.1002/Wsbm.75
25. England CG, Miller MC, Kuttan A, Trent JO, Frieboes HB (2015) Release kinetics of paclitaxel and cisplatin from two and three layered gold nanoparticles. Eur J Pharm Biopharm 92:120–129
26. Frieboes HB, Chaplain MA, Thompson AM, Bearer EL, Lowengrub JS, Cristini V (2011) Physical oncology: a bench-to-bedside quantitative and predictive approach. Cancer Res 71(2):298–302. doi:10.1158/0008-5472.CAN-10-2676
27. Frieboes HB, Sinek JP, Nalcioglu O, Fruehauf JP, Cristini V (2006) Nanotechnology in cancer drug

therapy: a biocomputational approach. In: Lee AP, Lee LJ, Ferrari M (eds) BioMEMS and biomedical nanotechnology. Springer, New York, pp 435–460. doi:10.1007/978-0-387-25842-3_15
28. Frieboes HB, Wu M, Lowengrub J, Decuzzi P, Cristini V (2013) A computational model for predicting nanoparticle accumulation in tumor vasculature. PLoS One 8(2):e56876. doi:10.1371/journal.pone.0056876
29. Fung YC (1997) Biomechanics: Circulation. Springer, University of California, San Diego
30. Gao Y, Li M, Chen B, Shen Z, Guo P, Wientjes MG, Au JL (2013) Predictive models of diffusive nanoparticle transport in 3-dimensional tumor cell spheroids. AAPS J 15(3):816–831. doi:10.1208/s12248-013-9478-2
31. Godin B, Driessen WH, Proneth B, Lee SY, Srinivasan S, Rumbaut R, Arap W, Pasqualini R, Ferrari M, Decuzzi P (2010) An integrated approach for the rational design of nanovectors for biomedical imaging and therapy. Adv Genet 69:31–64. doi:10.1016/S0065-2660(10)69009-8
32. Gradishar WJ (2006) Albumin-bound paclitaxel: a next-generation taxane. Expert Opin Pharmacother 7(8):1041–1053. doi:10.1517/14656566.7.8.1041
33. Grant CE, Valdimarsson G, Hipfner DR, Almquist KC, Cole SPC, Deeley RG (1994) Overexpression of multidrug resistance-associated protein (Mrp) increases resistance to natural product drugs. Cancer Res 54(2):357–361
34. Greenspan HP (1972) Models for the growth of a solid tumor by diffusion. Stud Appl Math 51:317–340
35. Greenspan HP (1976) On the growth and stability of cell cultures and solid tumors. J Theor Biol 56:229–242
36. Greish K (2007) Enhanced permeability and retention of macromolecular drugs in solid tumors: a royal gate for targeted anticancer nanomedicines. J Drug Target 15(7-8):457–464. doi:10.1080/10611860701539584
37. Heldin CH, Rubin K, Pietras K, Ostman A (2004) High interstitial fluid pressure – an obstacle in cancer therapy. Nat Rev Cancer 4(10):806–813. doi:10.1038/nrc1456
38. Hirst DG, Denekamp J (1979) Tumour cell proliferation in relation to the vasculature. Cell Tissue Kinet 12(1):31–42
39. Hong MH, Zhu SJ, Jiang YY, Tang GT, Pei YY (2009) Efficient tumor targeting of hydroxycamptothecin loaded PEGylated niosomes modified with transferrin. J Control Release 133(2):96–102. doi:10.1016/j.jconrel.2008.09.005
40. Huang K, Ma H, Liu J, Huo S, Kumar A, Wei T, Zhang X, Jin S, Gan Y, Wang PC, He S, Zhang X, Liang XJ (2012) Size-dependent localization and penetration of ultrasmall gold nanoparticles in cancer cells, multicellular spheroids, and tumors in vivo. ACS Nano 6(5):4483–4493. doi:10.1021/nn301282m
41. Izuishi K, Kato K, Ogura T, Kinoshita T, Esumi H (2000) Remarkable tolerance of tumor cells to nutrient deprivation: possible new biochemical target for cancer therapy. Cancer Res 60(21):6201–6207
42. Jain RK (2013) Normalizing tumor microenvironment to treat cancer: bench to bedside to biomarkers. J Clin Oncol 31(17):2205–2218. doi:10.1200/JCO.2012.46.3653
43. Kaddi CD, Phan JH, Wang MD (2013) Computational nanomedicine: modeling of nanoparticle-mediated hyperthermal cancer therapy. Nanomedicine (Lond) 8(8):1323–1333. doi:10.2217/nnm.13.117
44. Kao HW, Lin YY, Chen CC, Chi KH, Tien DC, Hsia CC, Lin MH, Wang HE (2013) Evaluation of EGFR-targeted radioimmuno-gold-nanoparticles as a theranostic agent in a tumor animal model. Bioorg Med Chem Lett 23(11):3180–3185. doi:10.1016/j.bmcl.2013.04.002
45. Konno T, Maeda H, Iwai K, Tashiro S, Maki S, Morinaga T, Mochinaga M, Hiraoka T, Yokoyama I (1983) Effect of arterial administration of high-molecular-weight anti-cancer agent SMANCS with lipid lymphographic agent on hepatoma – a preliminary-report. Eur J Cancer Clin Oncol 19(8):1053–1065. doi:10.1016/0277-5379(83)90028-7
46. Krasnici S, Werner A, Eichhorn ME, Schmitt-Sody M, Pahernik SA, Sauer B, Schulze B, Teifel M, Michaelis U, Naujoks K, Dellian M (2003) Effect of the surface charge of liposomes on their uptake by angiogenic tumor vessels. Int J Cancer 105(4):561–567. doi:10.1002/Ijc.11108
47. Kreeger PK, Lauffenburger DA (2010) Cancer systems biology: a network modeling perspective. Carcinogenesis 31(1):2–8. doi:10.1093/carcin/bgp261
48. Lankelma J, Dekker H, Luque FR, Luykx S, Hoekman K, van der Valk P, van Diest PJ, Pinedo HM (1999) Doxorubicin gradients in human breast cancer. Clin Cancer Res 5(7):1703–1707
49. Li M, Al-Jamal KT, Kostarelos K, Reineke J (2010) Physiologically based pharmacokinetic modeling of nanoparticles. ACS Nano 4(11):6303–6317. doi:10.1021/nn1018818
50. Li M, Czyszczon EA, Reineke JJ (2013) Delineating intracellular pharmacokinetics of paclitaxel delivered by PLGA nanoparticles. Drug Deliv Transl Res 3(6):551–561. doi:10.1007/s13346-013-0162-y
51. Li M, Panagi Z, Avgoustakis K, Reineke J (2012) Physiologically based pharmacokinetic modeling of PLGA nanoparticles with varied mPEG content. Int J Nanomedicine 7:1345–1356
52. Li M, Reineke J (2011) Mathematical modelling of nanoparticle biodistribution: extrapolation among intravenous, oral and pulmonary administration routes. Int J Nano Biomater 3(3):222–238

53. Lieleg O, Baumgartel RM, Bausch AR (2009) Selective filtering of particles by the extracellular matrix: an electrostatic bandpass. Biophys J 97(6):1569–1577. doi:10.1016/j.bpj.2009.07.009
54. Litzinger DC, Buiting AMJ, Vanrooijen N, Huang L (1994) Effect of liposome size on the circulation time and intraorgan distribution of amphipathic poly(ethylene glycol)-containing liposomes. Biochim Et Biophys Acta-Biomembranes 1190(1):99–107. doi:10.1016/0005-2736(94)90038-8
55. Liu Q, Zhang J, Sun W, Xie QR, Xia W, Gu H (2012) Delivering hydrophilic and hydrophobic chemotherapeutics simultaneously by magnetic mesoporous silica nanoparticles to inhibit cancer cells. Int J Nanomedicine 7:999–1013. doi:10.2147/ijn.s28088
56. Lowengrub JS, Frieboes HB, Jin F, Chuang YL, Li X, Macklin P, Wise SM, Cristini V (2010) Nonlinear modelling of cancer: bridging the gap between cells and tumours. Nonlinearity 23(1):R1–R9
57. Ma P, Mumper RJ (2013) Paclitaxel nano-delivery systems: a comprehensive review. J Nanomed Nanotechnol 4(2):1000164. doi:10.4172/2157-7439.1000164
58. Ma R, Taruttis A, Ntziachristos V, Razansky D (2009) Multispectral optoacoustic tomography (MSOT) scanner for whole-body small animal imaging. Opt Express 17(24):21414–21426
59. Macklin P, Lowengrub J (2007) Nonlinear simulation of the effect of microenvironment on tumor growth. J Theor Biol 245(4):677–704. doi:10.1016/j.jtbi.2006.12.004
60. Macklin P, McDougall S, Anderson AR, Chaplain MA, Cristini V, Lowengrub J (2009) Multiscale modelling and nonlinear simulation of vascular tumour growth. J Math Biol 58(4–5):765–798. doi:10.1007/s00285-008-0216-9
61. Maeda H, Ueda M, Morinaga T, Matsumoto T (1985) Conjugation of poly(styrene-co-maleic acid) derivatives to the antitumor protein neocarzinostatin - pronounced improvements in pharmacological properties. J Med Chem 28(4):455–461. doi:10.1021/Jm00382a012
62. Mahoney BP, Raghunand N, Baggett B, Gillies RJ (2003) Tumor acidity, ion trapping and chemotherapeutics. I. Acid pH affects the distribution of chemotherapeutic agents in vitro. Biochem Pharmacol 66(7):1207–1218
63. Matsumura Y, Maeda H (1986) A new concept for macromolecular therapeutics in cancer-chemotherapy - mechanism of tumoritropic accumulation of proteins and the antitumor agent smancs. Cancer Res 46(12):6387–6392
64. McDougall SR, Anderson AR, Chaplain MA (2006) Mathematical modelling of dynamic adaptive tumour-induced angiogenesis: clinical implications and therapeutic targeting strategies. J Theor Biol 241(3):564–589. doi:10.1016/j.jtbi.2005.12.022
65. McDougall SR, Anderson AR, Chaplain MA, Sherratt JA (2002) Mathematical modelling of flow through vascular networks: implications for tumour-induced angiogenesis and chemotherapy strategies. Bull Math Biol 64(4):673–702. doi:10.1006/bulm.2002.0293
66. Merisko-Liversidge EM, Liversidge GG (2008) Drug nanoparticles: formulating poorly water-soluble compounds. Toxicol Pathol 36(1):43–48. doi:10.1177/0192623307310946
67. Michor F, Liphardt J, Ferrari M, Widom J (2011) What does physics have to do with cancer? Nat Rev Cancer 11(9):657–670. doi:10.1038/Nrc3092
68. Milane L, Duan ZF, Amiji M (2011) Pharmacokinetics and biodistribution of lonidamine/paclitaxel loaded, EGFR-targeted nanoparticles in an orthotopic animal model of multi-drug resistant breast cancer. Nanomedicine 7(4):435–444. doi:10.1016/j.nano.2010.12.009
69. Minchinton AI, Tannock IF (2006) Drug penetration in solid tumours. Nat Rev Cancer 6(8):583–592. doi:10.1038/nrc1893
70. Moulder JE, Rockwell S (1987) Tumor hypoxia: its impact on cancer therapy. Cancer Metastasis Rev 5(4):313–341
71. Ntziachristos V, Razansky D (2010) Molecular imaging by means of multispectral optoacoustic tomography (MSOT). Chem Rev 110(5):2783–2794. doi:10.1021/cr9002566
72. Osborne JM, Walter A, Kershaw SK, Mirams GR, Fletcher AG, Pathmanathan P, Gavaghan D, Jensen OE, Maini PK, Byrne HM (2010) A hybrid approach to multi-scale modelling of cancer. Philos Trans R Soc A-Math Phys Eng Sci 368(1930):5013–5028. doi:10.1098/rsta.2010.0173
73. Palladini A, Nicoletti G, Pappalardo F, Murgo A, Grosso V, Stivani V, Ianzano ML, Antognoli A, Croci S, Landuzzi L, De Giovanni C, Nanni P, Motta S, Lollini PL (2010) In silico modeling and in vivo efficacy of cancer-preventive vaccinations. Cancer Res 70(20):7755–7763. doi:10.1158/0008-5472.Can-10-0701
74. Patel AR, Chougule MB, Lim E, Francis KP, Safe S, Sachdeva M (2013) Theranostic tumor homing nanocarriers for the treatment of lung cancer. Nanomedicine. doi:10.1016/j.nano.2013.12.002
75. Patel NR, Pattni BS, Abouzeid AH, Torchilin VP (2013) Nanopreparations to overcome multidrug resistance in cancer. Adv Drug Deliv Rev 65(13–14):1748–1762. doi:10.1016/j.addr.2013.08.004
76. Peng XH, Wang YQ, Huang DH, Wang YX, Shin HJ, Chen ZJ, Spewak MB, Mao H, Wang X, Wang Y, Chen Z, Nie SM, Shin DM (2011) Targeted delivery of cisplatin to lung cancer using ScFvEGFR-heparin-cisplatin nanoparticles. ACS Nano 5(12):9480–9493. doi:10.1021/Nn202410f
77. Persidis A (1999) Cancer multidrug resistance. Nat Biotechnol 17(1):94–95. doi:10.1038/5289
78. Primeau AJ, Rendon A, Hedley D, Lilge L, Tannock IF (2005) The distribution of the anticancer drug

Doxorubicin in relation to blood vessels in solid tumors. Clin Cancer Res 11(24 Pt 1):8782–8788. doi:10.1158/1078-0432.CCR-05-1664
79. Rangger C, Helbok A, Sosabowski J, Kremser C, Koehler G, Prassl R, Andreae F, Virgolini IJ, von Guggenberg E, Decristoforo C (2013) Tumor targeting and imaging with dual-peptide conjugated multifunctional liposomal nanoparticles. Int J Nanomedicine 8:4659–4670. doi:10.2147/Ijn.S51927
80. Rejniak KA, Anderson ARA (2011) Hybrid models of tumor growth. Wiley Interdisciplinary Reviews-Systems Biology and Medicine 3(1):115–125. doi:10.1002/Wsbm.102
81. Rejniak KA, McCawley LJ (2010) Current trends in mathematical modeling of tumor-microenvironment interactions: a survey of tools and applications. Exp Biol Med 235(4):411–423. doi:10.1258/cbm.2009.009230
82. Saad M, Garbuzenko OB, Minko T (2008) Co-delivery of siRNA and an anticancer drug for treatment of multidrug-resistant cancer. Nanomedicine (Lond) 3(6):761–776. doi:10.2217/17435889.3.6.761
83. Siepmann J, Göpferich A (2001) Mathematical modeling of bioerodible, polymeric drug delivery systems. Adv Drug Deliv Rev 48:229–247
84. Sinek J, Frieboes H, Zheng X, Cristini V (2004) Two-dimensional chemotherapy simulations demonstrate fundamental transport and tumor response limitations involving nanoparticles. Biomed Microdevices 6(4):297–309. doi:10.1023/B:BMMD.0000048562.29657.64
85. Sinek JP, Sanga S, Zheng X, Frieboes HB, Ferrari M, Cristini V (2009) Predicting drug pharmacokinetics and effect in vascularized tumors using computer simulation. J Math Biol 58(4–5):485–510. doi:10.1007/s00285-008-0214-y
86. Soppimath KS, Aminabhavi TM, Kulkarni AR, Rudzinski WE (2001) Biodegradable polymeric nanoparticles as drug delivery devices. J Control Release 70(1–2):1–20
87. Stylianopoulos T, Poh MZ, Insin N, Bawendi MG, Fukumura D, Munn LL, Jain RK (2010) Diffusion of particles in the extracellular matrix: the effect of repulsive electrostatic interactions. Biophys J 99(5):1342–1349. doi:10.1016/j.bpj.2010.06.016
88. Sumner DD, Stevens JT (1994) Pharmacokinetic factors influencing risk assessment: saturation of biochemical processes and cofactor depletion. Environ Health Perspect 102(Suppl 11):13–22
89. Swanson KR, Rockne RC, Claridge J, Chaplain MA, Alvord EC, Anderson ARA (2011) Quantifying the role of angiogenesis in malignant progression of gliomas: in silico modeling integrates imaging and histology. Cancer Res 71(24):7366–7375. doi:10.1158/0008-5472.Can-11-1399
90. Szala S, Jarosz M (2011) Tumor blood vessels. Postepy Hig Med Dosw (Online) 65: 437–446
91. Taratula O, Garbuzenko OB, Chen AM, Minko T (2011) Innovative strategy for treatment of lung cancer: targeted nanotechnology-based inhalation co-delivery of anticancer drugs and siRNA. J Drug Target 19(10):900–914. doi:10.3109/1061186x.2011.622404
92. Taruttis A, Herzog E, Razansky D, Ntziachristos V (2010) Real-time imaging of cardiovascular dynamics and circulating gold nanorods with multispectral optoacoustic tomography. Opt Express 18(19):19592–19602. doi:10.1364/oe.18.019592
93. Thurston G, McLean JW, Rizen M, Baluk P, Haskell A, Murphy TJ, Hanahan D, McDonald DM (1998) Cationic liposomes target angiogenic endothelial cells in tumors and chronic inflammation in mice. J Clin Invest 101(7):1401–1413. doi:10.1172/Jci965
94. Truskey G, Yuan F, Katz D (2004) Transport phenomena in biological systems. Pearson Prentice Hall, Upper Saddle River
95. Unezaki S, Maruyama K, Hosoda J, Nagae I, Koyanagi Y, Nakata M, Ishida O, Iwatsuru M, Tsuchiya S (1996) Direct measurement of the extravasation of polyethyleneglycol-coated liposomes into solid tumor tissue by in vivo fluorescence microscopy. Int J Pharm 144(1):11–17. doi:10.1016/S0378-5173(96)04674-1
96. van de Ven AL, Abdollahi B, Martinez CJ, Burey LA, Landis MD, Chang JC, Ferrari M, Frieboes HB (2013) Modeling of nanotherapeutics delivery based on tumor perfusion. New J Phys 15:55004. doi:10.1088/1367-2630/15/5/055004
97. van de Ven AL, Wu M, Lowengrub J, McDougall SR, Chaplain MA, Cristini V, Ferrari M, Frieboes HB (2012) Integrated intravital microscopy and mathematical modeling to optimize nanotherapeutics delivery to tumors. AIP Adv 2(1):11208. doi:10.1063/1.3699060
98. Vandenberg AP, Wikehooley JL, Vandenbergblok AE, Vanderzee J, Reinhold HS (1982) Tumor pH in human mammary-carcinoma. Eur J Cancer Clin Oncol 18(5):457–462. doi:10.1016/0277-5379(82)90114-6
99. Vineis P, Schatzkin A, Potter JD (2010) Models of carcinogenesis: an overview. Carcinogenesis 31(10):1703–1709. doi:10.1093/carcin/bgq087
100. Waite CL, Roth CM (2012) Nanoscale drug delivery systems for enhanced drug penetration into solid tumors: current progress and opportunities. Crit Rev Biomed Eng 40(1):21–41
101. Wang Z, Butner JD, Cristini V, Deisboeck TS (2015) Integrated PK-PD and agent-based modeling in oncology. J Pharmacokinet Pharmacodyn 42(2):179–189. doi:10.1007/s10928-015-9403-7
102. Wang Z, Deisboeck TS (2014) Mathematical modeling in cancer drug discovery. Drug Discov Today 19(2):145–150. doi:10.1016/j.drudis.2013.06.015
103. Wu M, Frieboes HB, Chaplain MA, McDougall SR, Cristini V, Lowengrub J (2014) The effect of interstitial pressure on therapeutic agent transport:

coupling with the tumor blood and lymphatic vascular systems. J Theor Biol 355:194–207. doi:10.1016/j.jtbi.2014.04.012
104. Zhang Z, Feng SS (2006) Nanoparticles of poly(lactide)/vitamin E TPGS copolymer for cancer chemotherapy: synthesis, formulation, characterization and in vitro drug release. Biomaterials 27(2):262–270. doi:10.1016/j.biomaterials.2005.05.104
105. Zheng X, Wise SM, Cristini V (2005) Nonlinear simulation of tumor necrosis, neo-vascularization and tissue invasion via an adaptive finite-element/level-set method. Bull Math Biol 67(2):211–259. doi:10.1016/j.bulm.2004.08.001

Microenvironment-Mediated Modeling of Tumor Response to Vascular-Targeting Drugs

10

Jana L. Gevertz

Abstract

The tumor-associated microvasculature is one of the key elements of the microenvironment that helps shape, and is shaped by, tumor progression. Given the important role of the vasculature in tumor progression, and the fact that tumor and normal vasculature are physiologically and molecularly distinct, much effort has gone into the development of vascular-targeting drugs that in theory should target tumors without significant risk to normal tissue. In this chapter, a multiscale hybrid mathematical model of tumor-vascular interactions is presented to provide a theoretical basis for assessing tumor response to vascular-targeting drugs. Model performance is calibrated to quantitative clinical data on tumor response to angiogenesis inhibitors (AIs), preclinical data on response to a cytotoxic chemotherapy, and qualitative preclinical data on response to vascular disrupting agents (VDAs). The calibrated model is then used to explore two questions of clinical interest. First, the hypothesis that AIs and VDAs are complementary treatments, rather than redundant, is explored. The model predicts a minimal increase in antitumor activity as a result of adding a VDA to an AI treatment regimen, and in fact at times the combination can exert less antitumor activity than stand-alone AI treatment. Second, the question of identifying an optimal dosing strategy for treating with an AI and a cytotoxic agent is addressed. Using a stochastic optimization scheme, an intermittent schedule for both chemotherapy and AI administration is identified that can eradicate the simulated tumors. We propose that this schedule may have increased clinical antitumor activity compared to currently used treatment protocols.

J.L. Gevertz (✉)
Department of Mathematics and Statistics, The College of New Jersey, Ewing, NJ 08628, USA
e-mail: gevertz@tcnj.edu

K.A. Rejniak (eds.), *Systems Biology of Tumor Microenvironment*, Advances in Experimental Medicine and Biology 936, DOI 10.1007/978-3-319-42023-3_10

Keywords
Tumor-vasculature interactions • Hybrid cellular automaton model • Angiogenesis inhibitors • Vascular disrupting agents • Cytotoxic chemotherapy

10.1 Introduction

Solid tumors maintain their vascular needs through a combination of co-option of existing host blood vessels and the creation of new blood vessels from existing ones in a process termed angiogenesis. While a tumor can partially fill its vascular needs through vessel co-option, growth beyond a few millimeters in diameter and invasion beyond the primary tumor site requires angiogenesis [4]. Therefore, the tumor-associated vasculature is a key component of the microenvironment that helps shape, and is shaped by, tumor progression.

The angiogenic process is influenced by a broad range of pro- and anti-angiogenic molecules, and can also be triggered by microenvironmental cues such as hypoxia and mechanical stress [4]. While angiogenesis is a speedy process for directing oxygen and nutrients to a tumor mass, the rapid nature of the process results in significant differences between angiogenic and normal vasculature. While the architecture of normal tissue vasculature is generally mature with an organized structure, the angiogenic vasculature is typically immature, leaky and tortuous [4]. These physiological differences, along with the different repertoire of proteins expressed by angiogenic vessels, make the angiogenic vasculature and process an exciting target for drug developers [24]. In particular, drugs that target the angiogenic vessels or process hold the promise of interfering with tumor growth and spread while having limited impact on normal tissue and its associated vasculature.

Drugs developed for this purpose are termed vascular-targeting compounds, and they can generally be classified into one of two categories. First, there are the angiogenesis inhibitors (AIs) which are designed to inhibit the formation of new blood vessels via the angiogenic process. These drugs are expected to limit the oxygen and nutrient supply of the tumor, and as a result indirectly inhibit tumor growth and metastasis. Many different aspects of the angiogenic process have been targeted through AIs, including inhibiting pro-angiogenic molecules, delivering anti-angiogenic molecules, blocking angiogenic receptors, and inhibiting the proliferation of endothelial cells (ECs) that comprise blood vessels [4, 24]. Hundreds of AIs have been or are currently undergoing clinical trial testing, and in the United States five AIs have been approved by the Food and Drug Administration: bevacizumab, sorafenib, sunitinib, pazopanib, and everolimus.[1] Bevacizumab (Avastin®), the first approved AI, has resulted in increased progression free survival times for several cancer types. However, the impacts of the drug are transient, with regrowth typically occurring several months after treatment [1].

The second class of vascular-targeting drugs are the vascular disrupting agents (VDAs). These drugs are designed to selectively target and collapse the vascular structure inside of an existing solid tumor [14, 17]. Therefore, they differ in their mode of action from AIs, which attempt to prevent new blood vessel formation without targeting pre-existing vasculature. No VDAs have been clinically approved for use in the United States, although a number of VDAs can be found at various stages in the clinical trial process. A non-exhaustive list includes the microtubule destabilizing agents derived from combretastatin (fosbretabulin, OXI4503, ombrabulin), plinabulin and vadimezan [19]. In vitro experiments with tubulin binding VDAs show that they induce

[1] http://www.cancer.gov/about-cancer/treatment/types/immunotherapy/angiogenesis-inhibitors-fact-sheet

rapid changes in EC morphology that cause these cells to retract, round and detach [17]. Preclinical studies in animal models have shown significant reductions in blood flow through the tumor; as one example, vadimezan has been shown to selectively decrease tumor blood flow and cause necrosis [19]. As another, studies using fosbretabulin triggered death in as much as 95 % of the tumor mass [27]. Although these preclinical results are promising, independent of the VDA tested a thin rim of cancer cells survive at the tumor periphery [27], limiting the antitumor activity of this class of drugs.

Although AIs have not achieved the promising clinical outcomes that were initially anticipated, and although no VDAs have received clinical approval, there is still significant research being conducted to determine the optimal schedules, doses and even the combination of these vascular-targeting drugs [20, 22]. The purpose of this work is to demonstrate how a validated mathematical model can be calibrated to predict the efficacy of a hypothetical AI and VDA. With such a calibrated model, we then ask a number of questions related to drug dosing, scheduling and combinations. Here, we particularly explore what additional clinical benefits, if any, are achieved through treating with a combination of an AI and VDA. Using a stochastic optimization scheme, we also identify an optimal dosing protocol for treating with an AI and a standard cytotoxic chemotherapeutic agent.

Before explaining the model we have developed in any detail, it is essential to understand the co-option-regression-growth experimental framework for tumor evolution in well-vascularized microenvironments [12]. This framework hypothesizes that a growing tumor mass co-opts the normal blood vessels in the associated healthy tissue. These vessels express a range of proteins that contribute to their maturation and stabilization, including the constitutive expression of angiopoietin-1 (Ang-1) [16]. The process of tumor expansion then leads to the upregulation of the antagonist of Ang-1, angiopoietin-2 (Ang-2), in the tumor and the surrounding microenvironment. Upregulation of Ang-2 creates competition with Ang-1 for binding to their target receptor, Tie-2 [12]. This competition can result in destabilization of the vasculature. A third protein, the hypoxia-inducible VEGF, determines vessel response to destabilization. Not only is VEGF an anti-apoptotic signal for blood vessel ECs, it also promotes EC proliferation and chemotaxis [3]. Vessels destabilized by the balance of Ang-2 to Ang-1 can therefore survive in the presence of VEGF, yet are marked for regression without sufficient VEGF levels [12]. Vessel regression in the absence of vessel growth leads to the formation of hypoxic tumor regions that subsequently upregulate VEGF production. In turn, the upregulation of VEGF promotes the formation of new blood vessels via angiogenesis.

10.2 Mathematical Model of Tumor Progression and Treatment

The vascular microenvironment of a solid tumor mass is highly heterogeneous. The properties of individual tumor cells are partially dictated by their proximity to this heterogeneous vasculature. For these reasons we used an explicit spatial representation of the blood vessels and a discrete model of cancer cells to study the complex interactions that occur when a solid tumor grows in a heterogeneous vascular microenvironment [10].

10.2.1 Spatial Model of Tumor-Vascular Interactions

The mathematical framework we have developed to study tumor-vascular interactions is a spatial hybrid cellular automaton model. It is initialized by defining two-dimensional tissue space using a discrete grid of automaton cells, with each automaton cell representing a cluster of approximately seven biological cells. Overlaid on top of the cells in the tissue space is a discrete network of microvessels. Tumor growth is initialized by designating a single automaton cell at the center of the space to be a proliferative cancer cell. As time progresses in the algorithm, this automaton cell and its progeny can divide according to their cell doubling time, provided the cell remains in normoxic conditions. As the cells respond to the oxygen conditions determined by the microvessels, the vascular network is also modified by the tumor cells. In particular, the spatial configuration of the cancer cells feeds into a system of partial differential equations (PDEs) that describe the dynamics of the key ligands involved in vessel evolution: Ang-1, Ang-2, VEGF, and their receptors. The numerically-determined concentration profile of these proteins at the next day is input into a biologically-inspired set of rules to determine vessel response (growth or regression) to its protein levels. In turn, the change in vessel configuration alters the oxygen levels at each cell site, and therefore determines whether each cancerous automaton cell is normoxic (and capable of proliferating), hypoxic (low oxygen state without any proliferation), or necrotic (dead) [10]. The algorithm is summarized in Fig. 10.1.

To provide more details, the algorithm can be divided into the following steps:

1. *Generate discrete grid of cells.* We use a Voronoi tessellation of points generated via random sequential addition of hard circular disks to determine the underlying two-dimensional lattice for the model. Length scales are determined by defining tissue space to be 24 × 24 mm. Under this assumption, the disks that determine the size of the automaton cells have an average diameter of 120 μm. Assuming an average cancer cell diameter of 40 μm, each of the automaton cells contains approximately seven cancer cells [10].
2. *Overlay normal vascular network.* A modified Krogh cylinder model is used to place randomized parallel line segments representing microvessels on a triangular grid until all cells in the space are well-vascularized (within a specified distance from a vessel). To determine when a cell is well-vascularized, a charac-

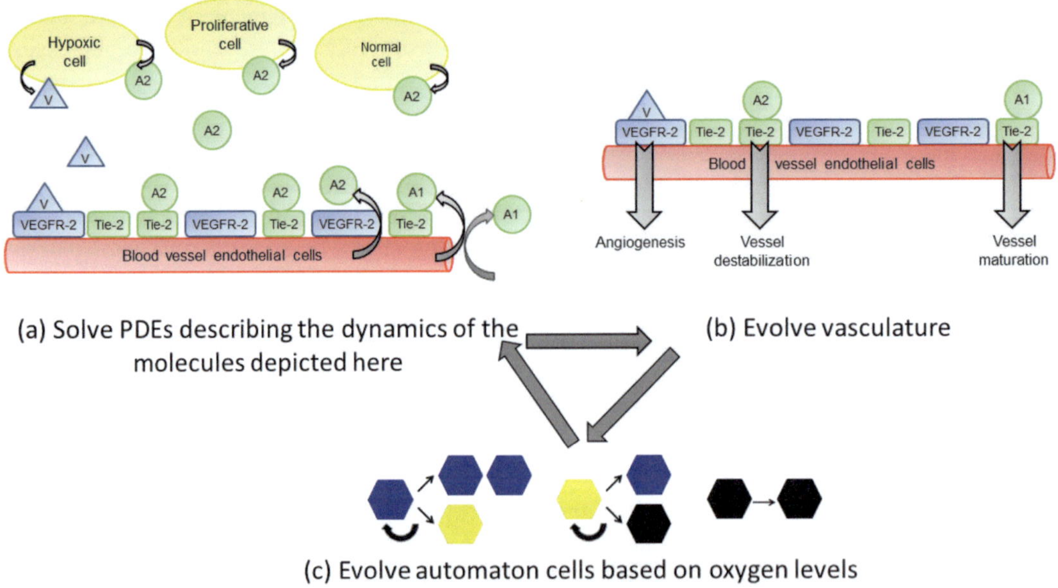

Fig. 10.1 Summary of the tumor-vasculature interaction model, partially adapted from [8]. (**a**) depicts the interactions modeled in the system of PDEs given in Eqs. (10.1), (10.2), (10.3), (10.4), (10.5), (10.6), (10.7), and (10.8), where *V* represents VEGF, *A*1 is Ang-1, *A*2 is Ang-2. *Curved arrows* indicate the cell type that produced the referenced protein. (**b**) depicts the physiological change in the vasculature in response to ligand-receptor binding. (**c**) illustrates the potential transition between cell types, where normoxic cells are *blue*, hypoxic cells are *yellow*, and necrotic cells are *black*. The processes in (**a**) through (**c**) are repeated for as many days of tumor growth as desired

teristic diffusion length of oxygen/nutrients in tissue of 250 μm is used [30]. When a cell is within this distance from a vessel, it is considered normoxic. Otherwise, it is considered hypoxic or necrotic (necrosis occurs when the distance to the nearest vessel exceeds 1500 μm). The line segments representing vasculature are constrained so that no two parallel vessels can be too close together, and that three vessels cannot intersect at one lattice site. More details can be found in [10].

3. *Introduce tumor.* The center automaton cell is designated as a proliferative cancer cell, with all remaining cells representing healthy tissue.
4. *Tumor and vessel evolution algorithm.* Time is discretized into one-day units. At each time point, the following processes are repeated, as illustrated in Fig. 10.1.

   a. *Numerically solve PDEs.* A stable forward difference scheme on a triangular grid is employed to solve the system of PDEs given in Eqs. (10.1), (10.2), (10.3), (10.4), (10.5), (10.6), (10.7), and (10.8) that describes the evolution of the key proteins involved in vessel evolution. Variables are defined as follows: VEGF concentration v, unbound VEGFR-2 (receptor for VEGF) r_{v0}, VEGF/VEGFR-2 complex r_v, Ang-1 a_1, Ang-2 a_2, unbound Tie-2 r_{a0}, Ang-1/Tie-2 complex r_{a1}, and Ang-2/Tie-2 complex r_{a2}.

$$\frac{\partial v}{\partial t} = \underbrace{D_v \Delta v}_{\textit{diffusion}} + \underbrace{b_v h_i \left(h - v^2/K_v\right)}_{\textit{production}} - \underbrace{k_0 v r_{v0}}_{\textit{complex formation}} + \underbrace{k_{-0} r_v}_{\textit{breakdown}} - \underbrace{\mu_v v}_{\textit{decay}} \tag{10.1}$$

$$\frac{\partial a_1}{\partial t} = \underbrace{b_{a1} e_i (p_i + h_i + n_i)\left(e_0 - a_1^2/K_a\right)}_{\textit{production}} - \underbrace{k_1 a_1 r_{a0}}_{\textit{complex formation}} + \underbrace{k_{-1} r_{a1}}_{\textit{breakdown}} - \underbrace{\mu_{a1} a_1}_{\textit{decay}} \tag{10.2}$$

$$\begin{aligned}\frac{\partial a_2}{\partial t} = {} & \underbrace{D_{a2} \Delta a_2}_{\textit{diffusion}} + \underbrace{b_{a2} e_i (p_i + h_i + n_i)\left(e_0 - a_2^2/K_a\right)}_{\textit{production}} + \underbrace{\bar{b}_{a2} h_i \left(h - a_2^2/K_a\right)}_{\textit{production}} \\ & - \underbrace{k_2 a_2 r_{a0}}_{\textit{complex formation}} + \underbrace{k_{-2} r_{a2}}_{\textit{breakdown}} - \underbrace{\mu_{a2} a_2}_{\textit{decay}}\end{aligned} \tag{10.3}$$

$$\frac{\partial r_{v0}}{\partial t} = \underbrace{-k_0 v r_{v0}}_{\textit{complex formation}} + \underbrace{k_{-0} r_v}_{\textit{breakdown}} \tag{10.4}$$

$$\frac{\partial r_{a0}}{\partial t} = \underbrace{-k_1 a_1 r_{a0}}_{\textit{complex formation}} + \underbrace{k_{-1} r_{a1}}_{\textit{breakdown}} - \underbrace{k_2 a_2 r_{a0}}_{\textit{complex formation}} + \underbrace{k_{-2} r_{a2}}_{\textit{breakdown}} \tag{10.5}$$

$$\frac{\partial r_v}{\partial t} = \underbrace{k_0 v r_{v0}}_{\textit{complex formation}} - \underbrace{k_{-0} r_v}_{\textit{breakdown}} \tag{10.6}$$

$$\frac{\partial r_{a1}}{\partial t} = \underbrace{k_1 a_1 r_{a0}}_{\textit{complex formation}} - \underbrace{k_{-1} r_{a1}}_{\textit{breakdown}} \tag{10.7}$$

$$\frac{\partial r_{a2}}{\partial t} = \underbrace{k_2 a_2 r_{a0}}_{\textit{complex formation}} - \underbrace{k_{-2} r_{a2}}_{\textit{breakdown}} \tag{10.8}$$

The equation for the three ligands VEGF, Ang-1, and Ang-2 all include the production by the appropriate cell types (either proliferative/normoxic cells p_i, hypoxic cells h_i or necrotic cells n_i, where the subscript i denotes an indicator function) with a carrying capacity term, as well as a linear decay term. Since Ang-1 is thought to act in a paracrine manner upon ECs, diffusion is only considered for VEGF and Ang-2. For the receptors VEGFR-2 and Tie-2, the equations represent the association and dissociation of the ligand-receptor complex. All variable concentrations are initially set to zero, with the exception of unbound Tie-2 and the Ang-1/Tie-2 complex at a blood vessel site, as Ang-1 is constitutively expressed by healthy vessels. Identical (Dirichlet) conditions are imposed at the boundary of the tissue space [10]. The schematic in Fig. 10.1a graphically illustrates the processes being modeled in Eqs. (10.1), (10.2), (10.3), (10.4), (10.5), (10.6), (10.7), and (10.8). A complete list of parameter definitions and values is found elsewhere [10].

b. *Evolve vasculature.* Vessels with a concentration of Ang-2/Tie-2 six times greater than that of Ang-1/Tie-2 regress [16], provided the concentration of VEGF/VEGFR-2 is below its critical value. Vessel tips with a sufficient amount of VEGF/VEGFR-2 sprout along the VEGF gradient [10].

c. *Evolve automaton cancer cells.* A nonmalignant cell undergoes apoptosis if vessel regression causes the oxygen concentration of the cell to drop below a critical threshold (distance to the nearest vessel is greater than 250 μm) [10]. Other than this, nonmalignant cells are inert. Similarly, necrotic cells are inert.

Hypoxic cells can transition to either a necrotic or normoxic state. They turn necrotic if they are further than 1500 μm from a vessel, and they turn normoxic (with the potential to proliferate) if the growth of a vessel results in the cell being within 250 μm from an oxygen/nutrient source [10].

Normoxic cells can transition to a hypoxic state if oxygen levels fall too low (cell is further than 250 μm from a vessel). If the cell remains normoxic, it may attempt to divide into the space of a nonmalignant cell. The probability of division, p_{div} depends on the location of the cell from the tumor center r, reflecting the effects of confinement pressure in the tissue space. Assuming a maximum tumor extent of R_{max}, the following equation is used to describe the probability of division:

$$p_{div} = p_0 \left(1 - \frac{r}{R_{max}}\right), \qquad (10.9)$$

where the base probability of division p_0 depends on the distance to the nearest vessel, and is calibrated so that the average cell doubling time corresponds to approximately four days [11].

10.2.2 Treatment Protocols

The model has previously been calibrated to data on the growth of glioblastoma multiforme (GBM), the most aggressive of the tumors arising from the glial cells in the central nervous system [10]. Herein, we will discuss how the model has been expanded to include treatment with vascular-targeting drugs and a cytotoxic chemotherapy.

10.2.2.1 Vascular-Targeting Drug 1: Angiogenesis Inhibitors

The simulated angiogenesis inhibitor will be based on the first clinically-approved AI in the United States, bevacizumab. Bevacizumab is a monoclonal antibody that binds to and inhibits VEGF [1]. To incorporate such an AI into our model, a drug will be considered that inhibits the production rate of VEGF (the b_v parameter in Eq. (10.1)) by a factor of T_{AI} [8]. The impact of varying the inhibition of VEGF over three orders of magnitude (from 10-fold to 1000-fold inhibition) will be explored.

Unless otherwise specified, the simulated AI is administered every two weeks [15]. This time interval has been chosen because the half-life of bevacizumab is approximately 20 days, with effective concentrations being found in the brain two to three weeks after administration [15]. For the sake of simplicity, drug is available immediately upon injection and remains at therapeutic levels (decreasing b_v by the factor T_{AI}) for the entire two week period of time.

10.2.2.2 Vascular-Targeting Drug 2: Vascular Disrupting Agents

The simulated VDA will be based on one of the most well-tested of these agents, fosbretabulin. Fosbretabulin is a compound that interacts with the colchicine binding site of β-tubulin. The resulting microtubule destabilization disrupts the cellular cytoskeleton and mitotic machinery of tumor-associated vessels [19]. To incorporate such a VDA into our model, it assumed that during each period of drug administration, the VDA destroys angiogenic vessels with probability p_{VDA} $(0 \le p_{VDA} \le 1)$ [8].

Unless otherwise specified, the simulated VDA will be administered once every three weeks, as this was the protocol followed in a Phase I pharmacokinetic study of fosbretabulin given as single-dose [7]. The VDA will only exert its destructive effects on the vasculature the day the drug is administered, as preclinical studies have demonstrated that in vivo effects peak 4–6 h after drug exposure, and are sustained for up to one day [5]. Just as with the AI, drug is available immediately upon injection, and remains at therapeutic levels (destroys angiogenic vessels with probability p_{VDA}) for the entire day.

10.2.2.3 Cytotoxic Chemotherapy

The simulated cytotoxic chemotherapy will be based on temozolomide, as this drug is standard of care for treating patients with GBM. Temozolomide is an alkylating agent that preferentially targets rapidly dividing cells, like most cytotoxic chemotherapies [26]. To incorporate such a chemotherapy in our model, a drug will be introduced that kills a certain proportion p_{chemo} $(0 \le p_{chemo} \le 1)$ of normoxic/proliferative cancer cells each day the drug is administered [8].

To calibrate this parameter value, we consider a preclinical setup in which mice genetically engineered to develop gliomas are treated with temozolomide for five days [18]. In this system, the log cell kill was calculated to be 0.4, where log cell kill satisfies the following equation:

$$L = \log_{10}\left(\frac{V_0}{V_5}\right) \approx \log_{10}\left(\frac{N_0}{N_5}\right). \qquad (10.10)$$

In this equation, V represents tumor volume either pre- ($t = 0$) or post-treatment ($t = 5$), and N represents the number of tumor cells. Under the assumption that each cell has a fixed volume, the expression for log cell kill can be expressed in terms of cell number, instead of cell volume.

To extract a value for p_{chemo} from L, we further assume that per day during the five day treatment window, each glioma cell divides at a rate r, and is killed by drug with probability p_{chemo}:

$$N_{t+1} = (r - p_{chemo})N_t. \qquad (10.11)$$

In Eq. (10.11), we use the fact that the doubling time for glioma cells is approximately four days [13] to compute a growth rate (without treatment, so $p_{chemo} = 0$) of $r = 2^{1/4}$. Using this information and an initial population size N_0, the solution to the difference equation after five days is:

$$N_5 = N_0\left(2^{1/4} - p_{chemo}\right)^5. \qquad (10.12)$$

Solving (10.10) for N_5 with the measured value of $L = 0.4$ and equating it to N_5 in Eq. (10.12) allows us to approximate the chemotherapy-induced death rate as:

$$p_{chemo} = 2^{1/4} - 10^{-2/25} \approx 0.36. \qquad (10.13)$$

For this reason, simulation experiments will be conducted with p_{chemo} close to this value.

From numerous years of use in the clinic, it has been established that a continuous administration schedule of temozolomide can be sustained for six to seven weeks [26]. For this reason, any simulations we conduct with chemotherapy will be constrained by the fact that the chemotherapy can be administered for, at maximum, six consecutive weeks in a row. Given that the half-life of temozolomide is 1.8 h [23], it can be assumed that therapeutic concentrations of the drug are maintained each day the cytotoxic agent is administered (meaning p_{chemo} stays fixed in a one day window of time). As done previously, we assume drug is available immediately upon injection and we further assume no drug is available the next day, unless drug is re-administered.

10.3 Benchmarking Model Performance Against Preclinical and Clinical Data

Now that we have introduced treatment with vascular-targeting drugs and a cytotoxic chemotherapy, here we will explain how the model has been calibrated to some available clinical and preclinical data on the treatment of GBM or other solid tumors. Note that for any treatment results presented, ten simulations have been run and the average tumor response is reported. Treatment is only applied once a tumor achieves a 4 mm radius (assumed size at detection).

10.3.1 Angiogenesis Inhibitor in Isolation

To benchmark model performance when simulating treatment with AIs, in silico predictions will be compared to a clinical dataset treating 12 GBM patients with bevacizumab. For each patient, an MRI was taken before the start of treatment, and a follow-up MRI was taken 40–83 days later (with the average being 53 days). By performing image analysis on each pair of MRIs, the relative changes in the volume of Gd-enhancement ($\Delta_{Gd} = V_{start} - V_{end}$), edema and necrosis were determined [21].

Here, we will focus on comparing the relative change in the Gd-enhancement volume to model predictions on tumor response to a simulated AI. In order to compare the clinical data to in silico data, we have assumed that the Gd-enhanced region corresponds to areas of high vascularity and therefore corresponds to the normoxic region of our simulated tumors. Among the 12 patients, the average relative change in the Gd-enhanced volume (calculated as Δ_{Gd}/V_{start}) was 59.5 % with a standard deviation of 19.2 % [21].

To compare these numbers to model predictions, for each AI model parameter ($T_{AI} = 10, 1000, 1000$), ten simulations were conducted. For each simulation, the start of AI treatment time was recorded (this time is determined by when the unique realization of the tumor achieves a radius of 4 mm), as is the area of the proliferative/normoxic tumor region (assumed to correlate with Gd-enhanced volume) at that time. AI is administered for 53 days (the average length of time in the patient study [21]), and the area of the proliferative tumor region is again recorded. From there, the relative change in proliferative tumor area is calculated for that unique tumor. This process is repeated for all ten tumor realizations per parameter value and the average and standard deviation in the relative change of the proliferative area is calculated. This is compared to the patient data in Table 10.1.

By comparing in silico results to clinical data, we find that $T_{AI} = 100$ gives a similar reduction in the proliferative tumor area (average of 58.2 %) as is observed in the clinical data (average of

Table 10.1 Relative change in proliferative tumor area after 53 days of AI treatment. Patient data is averaged over 12 patients. For a fixed model parameter, the percent change is averaged over 10 simulations from the precise start of treatment for each of those simulations

	Relative change in proliferative area 53 days post-AI
Patient data (from [21])	59.5 % ± 19.2 % (Gd-enhanced tumor volume)
$T_{AI} = 10$	−9.6 % ± 34.3 %
$T_{AI} = 100$	58.2 % ± 7.7 %
$T_{AI} = 1000$	72.7 % ± 7.2 %

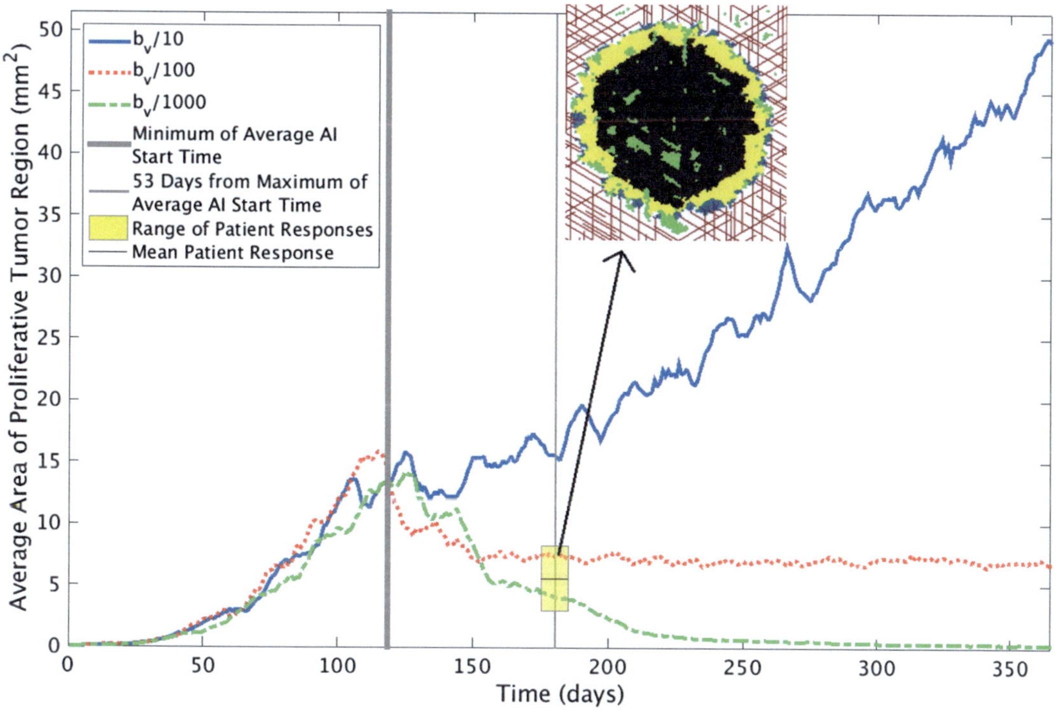

Fig. 10.2 Average area of the proliferative tumor region (averaged over 10 simulations) per AI treatment parameter is shown as a function of time. The *thick grey line* indicates the minimal average start time of treatment, and the *thin grey line* indicates 53-days post the maximal average start of treatment. It is on this *thin grey line* that the model predictions can be compared to clinical data. The *yellow box* contains the average size of the proliferative tumor region that would be predicted by the clinical dataset (within one standard deviation from the mean), with the *solid black line* running horizontally through the *yellow box* indicating the mean response predicted by the clinical data. Also included is a snapshot of one tumor treated with $T_{AI} = 100$ that happens to be 44 days into AI treatment at day 180. In this visualization, viable nonmalignant cells are labeled *white*, nonmalignant cells that have undergone apoptosis are *green*, necrotic tumor cells are *black*, hypoxic tumor cells are *yellow*, and normoxic (potentially proliferating) tumor cells are *blue*. Further, vessels that are originally part of the healthy tissue vascular network are *red*, and vessels that grew via angiogenesis are *purple*

59.5 %) [21]. On the other hand, reducing VEGF production by a factor of $T_{AI} = 10$ actually results in an *increase* in the relative change in the proliferative area. In other words, the average proliferative area is continuing to expand at this level of drug efficacy (Fig. 10.2), although the rate of expansion is very slow compared to the case of no treatment (data not shown). Finally, reducing VEGF production by a factor of $T_{AI} = 1000$ gives an average reduction in the proliferative tumor area that falls within one standard deviation of the mean of the patient data; that is, it falls within the range [40.3 %, 78.7 %] [21].

Figure 10.2 also allows us to compare model performance for different values of VEGF inhibition to the clinical dataset in Table 10.1. However, the numbers in the table cannot be directly extrapolated from the numbers in the figure. In the figure, the average proliferative tumor area is being compared from the minimum average start of treatment (that is, the average start time of treatment per 10 simulations is determined for each parameter value, and the minimum of those three values is plotted in the thick grey line.) The thin grey line then indicates the latest end of the treatment window (that is, the maximum of the average start times plus 53 days). The average proliferative tumor area among all 30 tumors at the start of treatment (thick grey line in Fig. 10.2) is determined; it works out to $13.76\,\text{mm}^2$.

Assuming the clinical prediction that 53 days of treatment should result in a 59.5±19.2 % reduction in average proliferative tumor area [21], the model predictions should be in the range 2.93–8.21 mm^2 (yellow box in Fig. 10.2), with an average value of 5.57 mm^2 illustrated as the center line in the box. So, the data in Table 10.1 is more precise, as each individual simulation is compared from its unique start time to 53-days post-treatment initiation. However, Fig. 10.2 still allows us to see that $T_{AI} = 10$ does not give results comparable to the clinical data, whereas $T_{AI} = 100$ and 1000 give predictions comparable to that observed clinically.

The time course plots in Fig. 10.2 also allows us to understand the sensitivity of the model to the AI treatment parameter. We observe that for $T_{AI} \geq 100$, the normoxic tumor region appears to stabilize during the course of treatment. The stabilized area of this region does depends on the strength of the drug, with the tumor stabilizing at a smaller size for larger values of VEGF-inhibition. However in neither case is the proliferative tumor region fully eliminated by the AI. The reason for tumor survival in spite of the use of a perfectly-penetrating VEGF inhibitor (which is already a theoretical idealization) is as follows: AI treatment leaves a number of proliferative cells surviving at the tumor periphery that obtain their oxygen and nutrients from the co-opted vasculature [8]. This can be seen in the tumor insert in Fig. 10.2, which clearly still contains normoxic cells with proliferative potential (labeled blue) at the tumor periphery. The fact that tumor survival is largely a consequence of the well-vascularized environment in which the mass grows highlights the important role the microenvironment plays in treatment response.

10.3.2 Vascular Disrupting Agents

To benchmark model performance when simulating treatment with a VDA, model simulations will be compared to some qualitative and quantitative preclinical data. Generally speaking, VDAs qualitatively result in characteristic patterns of widespread central necrosis which can extend to as much as 95 % of the tumor. However, a thin rim of normoxic cells typically survive at the tumor periphery, which ultimately results in tumor regrowth [27].

Quantitative data has been extracted from a preclinical study using ZD6126 (one of many tubulin-binding VDAs) to treat a range of solid tumors grown as xenografts in nude mice [2]. Tumors were grown to approximately 200–700 mm^3 in volume, treated with ZD6126, and after one day, tumors were excised, processed and stained. Level of necrosis was then scored on 10-point grading system, with 1 representing 0–10 % necrosis, and each additional point representing the next 10 % range, up to 10 representing >90–100 % necrosis. The median necrosis score was calculated for four or more tumors over a range of ZD6126 doses grown from the following human cell lines: Calu-6 lung cancer, HT-29 colorectal cancer, and LoVo colorectal cancer. The median necrosis score ranged from 4 to 9.5 depending on the drug dose and tumor type, while the median necrosis score without treatment is in the range of 1–4 [2].

To compare this preclinical data our in silico output, we first observe that the model predicts a median necrosis score of 4 in the case of no treatment. This is on the high end, but still is consistent, with preclinical measurements. Contrarily, the model predicts a median necrosis score of 5 for $p_{VDA} \geq 0.3$, and a median necrosis score of 4 when $p_{VDA} \leq 0.2$. In other words, the model does not exhibit the wide range of median necrosis scores seen in the experimental tumor models over a range of VDA doses. The narrow range of median necrosis scores predicted by the model compared to the preclinical data may indeed be indicative of the strong role of the microenvironment in determining response to VDAs: the cell lines were implanted into nude mice, making the microenvironment quite distinct from malignancies that develop de novo within the tumor microenvironment.

Although this preclinical data doesn't assess the impact of multiple rounds of treatment, clinical trials with the VDA fosbretabulin are looking to assess the longer-term antitumor activity of VDAs. In the first trial done with fosbretabulin, the drug is administered every three weeks for 8–40 cycles [7]. Here, we explore the antitumor activity of different doses of a VDA (controlled by the p_{VDA} parameter, $0 \leq p_{VDA} \leq 1$). An average of ten cycles of treatment will be used, but the precise number will depend on how much time passes between the initiation of treatment (when the tumor attains the "detectable" radius of 4 mm) and one year from the initiation of simulated tumor growth.

The antitumor activity of different "doses" of the VDA over approximately ten cycles is found in Fig. 10.3. We find that the VDA does limit tumor growth relative to the untreated case, but tumor growth does persist in spite of treatment. Consistent with qualitative preclinical observations [27], the sustained growth of the in silico tumors in spite of VDA treatment can be explained by a thin rim of normoxic cells at the tumor periphery (see insert in Fig. 10.3). Somewhat surprisingly, we also found that increasing the VDA dose (modeled as increasing fraction of angiogenic vessels the VDA can collapse from 10 % to 100 %) does not significantly increase the antitumor activity of the VDA (Fig. 10.3). The lack of a dose-dependent effect is attributed to the fact that destroying angiogenic vessels does not interfere with the survival of the proliferative rim. This rim is overwhelmingly dependent on the co-opted vasculature, and not the physiologically distinct angiogenic vasculature. Elsewhere, we have also shown that reducing the spacing between doses to less than three weeks does not significantly improve VDA efficacy [9], as the proliferative rim is still maintained in spite of more frequent dosing.

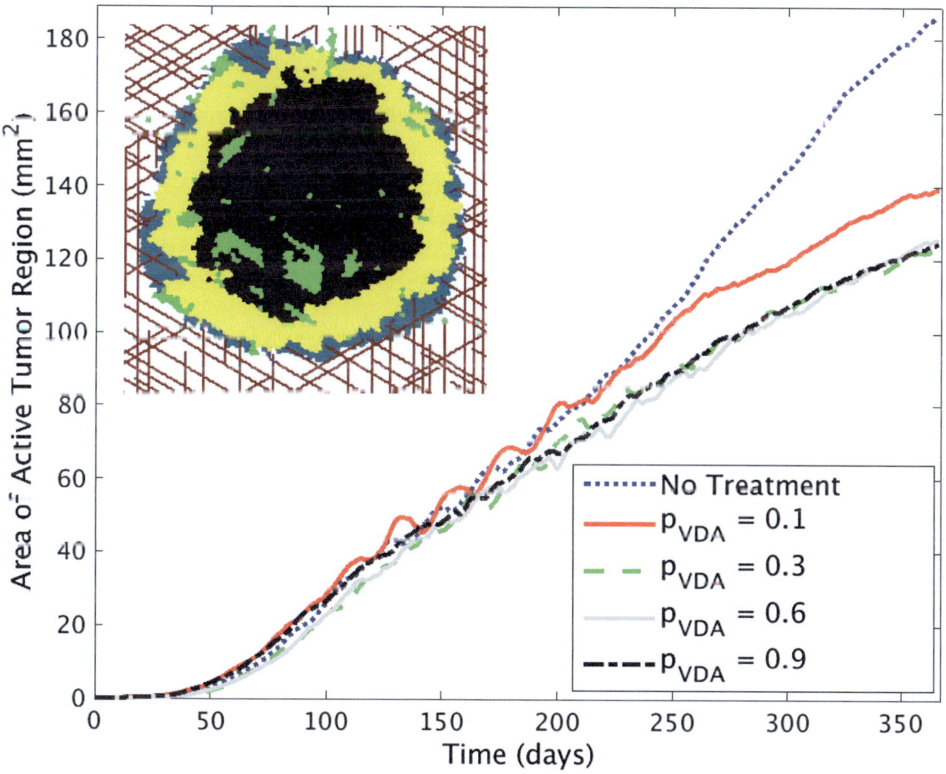

Fig. 10.3 Dose-response curves to administering a VDA once every three weeks for approximately 10 cycles. The average area of the active tumor region (normoxic + hypoxic) predicted by the algorithm is shown for each parameter value [8]. The *insert* shows a snapshot of a tumor one day after VDA administration when $p_{VDA} = 0.6$. Note the lack of angiogenic (*purple*) vessels and the presence of the normoxic (*blue*) rim at the tumor periphery

10.4 Antitumor Activity of Combination Therapies

Rarely is a new cancer drug given as a stand-alone treatment. For instance, in the treatment of GBM, bevacizumab is typically administered in combination with a cytotoxic chemotherapeutic agent that targets actively dividing cells, such as temozolomide [6]. And although no VDAs are yet approved for use in the clinic, numerous clinical studies have been conducted combining AIs and VDAs. For instance, a phase I study showed that combining bevacizumab with fosbretabulin produced a 39 % persistent reduction in vascularization [22]. Given that the treatment was also generally well tolerated, a randomized phase II trial is being conducted to see if there are survival benefits for adding fosbretabulin to a bevacizumab regimen.

In this section, the benchmarked model will first be utilized to quantify the benefits of adding a VDA to an AI treatment protocol. Second, we will focus on optimizing a combination therapy involving an AI and cytotoxic chemotherapy, subject to toxicity constraints.

10.4.1 Angiogenesis Inhibitors with Vascular Disrupting Agents

Preclinical studies have shown that the addition of an AI to a VDA regimen reduces the revascularization at the rim of the treated tumor, and thus significantly increases antitumor activity [25]. Further, a Phase I trial of fosbretabulin in combination with bevacizumab in patients with advanced solid tumors demonstrated this combination to be safe and found that the vascular changes induced by fosbretabulin alone were maintained when bevacizumab is added to the treatment protocol [22]. Our mathematical model can be used to explore the hypothesis that targeting the tumor vasculature with AIs and VDAs is complementary, rather than redundant [24].

With this goal in mind, the impact of adding an AI to a VDA treatment protocol (Fig. 10.4a) and the impact of adding a VDA to an AI treatment protocol (Fig. 10.4b) have been explored. The effect of both drugs will be tested over a range of "doses", with values of the VDA parameter p_{VDA} ranging from 0.1 to 1 and values of the AI parameter ranging from $T_{AI} = 100.9$ to 1000. Note that a normalized value for T_{AI} is presented on the x-axes in Fig. 10.4, where the normalized value is computed as $\bar{T}_{AI} = (T_{AI} - 1)/999$. As the VDA parameter is defined to be the proportion of angiogenic vasculature shut down each day of drug administration, by definition it is already normalized to values in the range [0, 1].

Ten simulations are run per normalized treatment value in the following instances:

- Treat with VDA only for $p_{VDA} = 0.1j, \quad \forall j \in \{i \in \mathbb{Z} : 1 \le i \le 10\}$
- Treat with AI only for $\bar{T}_{AI} = 0.1j, \quad \forall j \in \{i \in \mathbb{Z} : 1 \le i \le 10\}$
- Treat with AI and VDA at the same normalized treatment value. That is, if $p_{VDA} = x$, then $\bar{T}_{AI} = x$ as well, where as previously $x = 0.1j, \quad \forall j \in \{i \in \mathbb{Z} : 1 \le i \le 10\}$.

As done in Sect. 10.3, treatment is initiated when a tumor reaches a size of 4 mm in radius. AI is administered once every two weeks, and therapeutic levels are maintained throughout the treatment window. VDA is administered once every three weeks [7], and therapeutic levels are only maintained the day the drug is administered. Treatment is continued for four months, and the average tumor size as a function of time at each normalized drug level is determined.

First we will consider the impact of adding an AI to a VDA dosing protocol, as was explored in [22]. To this end, we have computed the average percent reduction in tumor size gained by adding an AI to a VDA treatment protocol:

$$\Delta_{perc}^{+AI}(t) = \frac{A_{VDA}(t) - A_{VDA+AI}(t)}{A_{VDA}(t)} \times 100, \tag{10.14}$$

where $A_{VDA}(t)$ is the average active tumor area (normoxic + hypoxic) t days from the initiation of VDA treatment, and $A_{VDA+AI}(t)$ is the average active tumor area t days from the initiation of treatment with a VDA and AI. Subsequently,

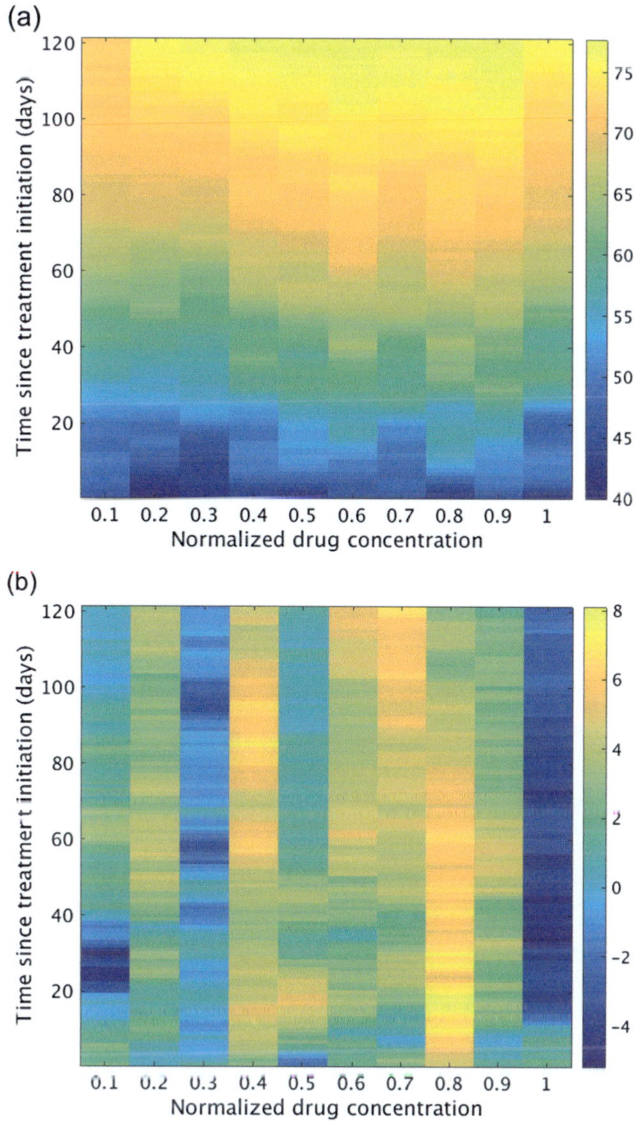

Fig. 10.4 The x axis shows the normalized drug concentration of both the AI and the VDA. (**a**) Percent reduction in the active tumor area as a function of time as a result of adding an AI to a VDA treatment regimen ($\Delta_{perc}^{+AI}(t)$). (**b**) Percent reduction in the active tumor area as a function of time as a result of adding a VDA to an AI treatment regimen ($\Delta_{perc}^{+VDA}(t)$)

we will consider the impact of adding a VDA to an AI dosing protocol. The metric used is Δ_{perc}^{+VDA}, the average percent reduction in tumor size gained by adding a VDA to an AI treatment protocol:

$$\Delta_{perc}^{+VDA}(t) = \frac{A_{AI}(t) - A_{VDA+AI}(t)}{A_{AI}(t)} \times 100, \tag{10.15}$$

where $A_{AI}(t)$ is the average active tumor area t days from the initiation of AI treatment. The larger the value of Δ_{perc}, the greater the impact that adding the second drug has on the antitumor effects of the first drug.

The model predicts a sustained benefit that increases with time when an AI is added to a VDA treatment protocol over the full range of normalized drug concentrations. Near the start of treatment, tumor burden is reduced by approximately 40 % due to the addition of the AI, and after four months a reduction of approximately 75 % is observed (Fig. 10.4a). However, the same benefit is not observed when adding a VDA to an AI treatment protocol (Fig. 10.4b). In that case, the maximum reduction in tumor burden gained by adding a VDA to an AI protocol is approximately 8 %. Further, there are multiple drug concentrations and time points for which the addition of the VDA actually negatively impacts tumor response (that is, measured active tumor area is actually larger when treating with an AI and VDA than just treating with an AI).

These observations allow us to conclude that, in the model, tumor response to the combination protocol of an AI and VDA is largely driven by the AI. There is little to no benefit gained by adding the VDA to the AI regimen (Fig. 10.4b), and antitumor activity is significantly improved when the AI is added to the VDA protocol (Fig. 10.4a). This strongly suggests that AIs and VDAs may not be as complementary as has been proposed (see [24]); the efficacy of a bevacizumab-like AI in the model is not significantly enhanced by the inclusion of a VDA [9].

10.4.2 Optimal Dosing Schedule for Angiogenesis Inhibitor and Chemotherapy

Given the limited benefit of adding a VDA to an AI treatment protocol, in this section we will only consider treatment with an AI and a cytotoxic chemotherapy. Our goal here is to identify an 8-week treatment protocol that minimizes the average number active (normoxic + hypoxic) tumor cells at the end of two treatment cycles (a 16-week window). To accomplish this goal, simulated annealing, a stochastic optimization scheme, will be employed. This approach allows us to begin with an initial treatment parameter set (specifying the number of days that the AI and chemotherapy will be given in isolation or in combination) and to evolve the treatment parameter set until an optimal protocol is identified [9].

More particularly, the goal is to minimize the average number (averaged over ten realizations) of active tumor cells remaining after two cycles of an 8-week treatment protocol, $\Lambda(t)$. The parameters that determine the value of this objective function are:

- λ_{AI}: the number of days in the 8-week treatment protocol that therapeutic levels of AI are maintained in the tissue without having therapeutic levels of chemotherapy. We will use $T_{AI} = 1000$ (a stronger AI gives quicker convergence to the optimal solution), and initially we assume $\lambda_{AI} = 28$ days.
- λ_{AIC}: the number of days in the 8-week treatment protocol that both the AI and the chemotherapeutic are simultaneously maintained at therapeutic levels in the tissue. We will use $p_{chemo} = 0.44$ (again, a strong chemotherapeutic gives quicker convergence to the optimal solution), and initially we assume $\lambda_{AIC} = 28$ days.
- λ_C: the number of days in the 8-week treatment protocol that only chemotherapy is maintained at therapeutic levels in the tissue. Initially, we assume $\lambda_C = 0$ to ensure the first treatment tested is an 8-week protocol.

Since at best the treatment can remove all active cancer cells, the aim is to solve

$$\begin{aligned} \text{Minimize} \quad & z = (\Lambda(t) - 0)^2 \\ \text{subject to} \quad & \lambda_{AIC} + \lambda_C \leq 36 \\ & \lambda_{AI}, \lambda_{AIC}, \lambda_C \geq 0. \end{aligned} \tag{10.16}$$

Notice that minimizing z is equivalent to $\min[\Lambda(t)]$, and that the second constraint simply indicates that no treatment can be given for a negative number of days. The first constraint is to limit the toxic side effects of chemotherapy. In particular, temozolomide can only be safely tolerated approximately six week [26], so 36 days was chosen to remain below this threshold, as the simulated treatment protocol is being administered twice [9].

Simulated annealing proceeds by perturbing λ_{AI} and λ_{AIC} by random integers in the range $[-4, 4]$, and λ_C is then calculated to preserve the 8-week treatment window. Provided the new protocol satisfies the constraints in Eq. (10.16), the 8-week protocol is twice applied to ten different growing tumors, and the average number of active cells surviving treatment is recorded as $\hat{\Lambda}(t)$

[9]. The probability that the new treatment parameter set is accepted depends on $\Delta\Lambda = \hat{\Lambda}(t) - \Lambda(t)$ and the Metropolis acceptance rule [28]:

$$p(\Delta\Lambda) = \begin{cases} 1, & \Delta\Lambda \le 0 \\ \exp(-\Delta\Lambda/(51.5(0.94)^k)), & \Delta\Lambda > 0, \end{cases} \quad (10.17)$$

where k is equal to the number of parameter sets tested. According to this rule, any treatment protocol which decreases the number of actively surviving cancer cells ($\Delta\Lambda < 0$) is accepted, and those protocols that increase the number of actively surviving cells ($\Delta\Lambda > 0$) are accepted with some nonzero probability in order to avoid converging to a local minimum.

New protocols are created and accepted with probability $p(\Delta\Lambda)$ (Eq. (10.17)) until the objective function is within a specified tolerance of the optimal value of zero (we use $z = \Lambda(112) \le 2$) [9]. Figure 10.5a shows the change in the three treatment parameters as a function of the accepted annealing step $\bar{k}$, along with the corresponding average number of active automaton cancer cells that remained at the end of the treatment window for each accepted parameter set.

The optimal treatment discovered is a *pulsed* or *intermittent* protocol in that neither the AI or the chemotherapy are maintained at therapeutic levels during the entire treatment window [9]:

- Days 1–20: Maintain therapeutic levels of AI only ($\lambda_{AI} = 20$)
- Days 21–32: Maintain therapeutic levels of AI and chemotherapy ($\lambda_{AIC} = 12$)
- Days 33–56: Maintain therapeutic levels of chemotherapy only ($\lambda_C = 24$).

On average, this left only 1.4 active automaton cells (approximately 10 active cancer cells) remaining after two cycles of the 8-week treatment protocol (Fig. 10.5a). Next, we considered what happens when we apply this 8-week protocol over four cycles at the treatment parameter values calibrated from data ($T_1 = 100$ and $p_{chemo} = 0.34$, with the latter being very close to the average value of 0.36 calculated from preclinical data). Tumor response for this optimal protocol is compared to a more "naive" 8-week protocol that maintains AI at therapeutic levels for the entire eight week period, and for the final four weeks, chemotherapy is also maintained at therapeutic levels. Figure 10.5b shows that unlike the more naive protocol that results in tumor regrowth at the cessation of treatment, the optimal protocol can eradicate the entire active tumor mass and therefore prevent regrowth [9].

As both the naive and optimal treatment protocol give chemotherapy intermittently to avoid toxicity, it must be the pulsing of the AI that explains the drastic difference in tumor response between the two dosing schedules. What removing the AI for certain periods of time does is allow angiogenesis to occur during the course of treatment. This has two positive impacts on treatment response. First, it increases blood flow through the tumor, and therefore allows chemotherapy access to parts of the tumor that would otherwise receive little to no drug due to poor vascularization. Second, angiogenesis also increases the flow of oxygen through the tumor. This permits a portion of the hypoxic cells (that cannot be targeted by the chemotherapy) to enter a normoxic state, allowing them to be targeted by the chemotherapy [9].

The success of the optimal protocol in eradicating the simulated tumors certainly does not suggest that the protocol would have equal success in the clinic. A number of simplifying assumptions are inevitably built into any mathematical model, and these limit the direct quantitative translation of in silico predictions to the clinic. However, the model provides strong evidence that a treatment schedule that pulses AI administration may do better to prolong progression-free survival than currently utilized protocols.

10.5 Conclusions

A previously validated hybrid spatial model has been extended to account for the activity of three cancer drugs. In particular, the model is calibrated to accurately predict antitumor activity of an angiogenesis inhibitor such as bevacizumab. Within the range of calibrated parameter values, the model correctly predicts that treatment with an AI

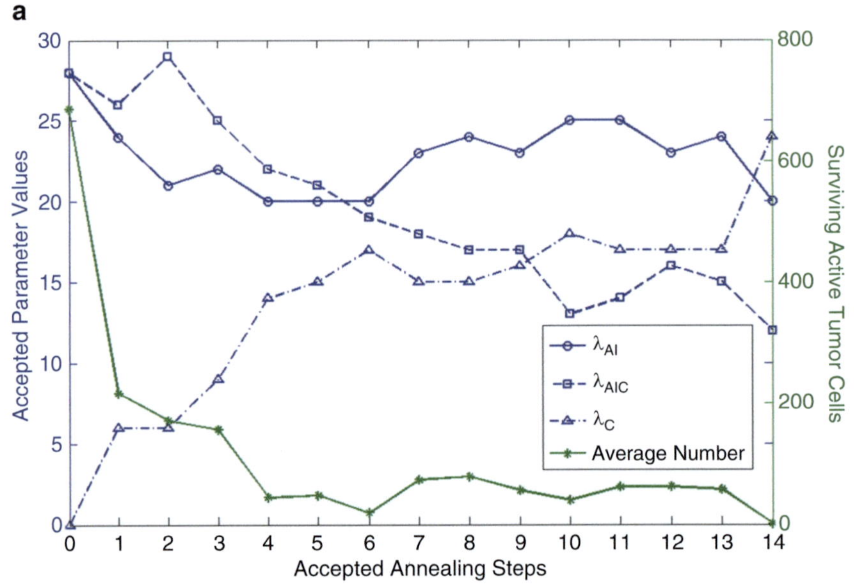

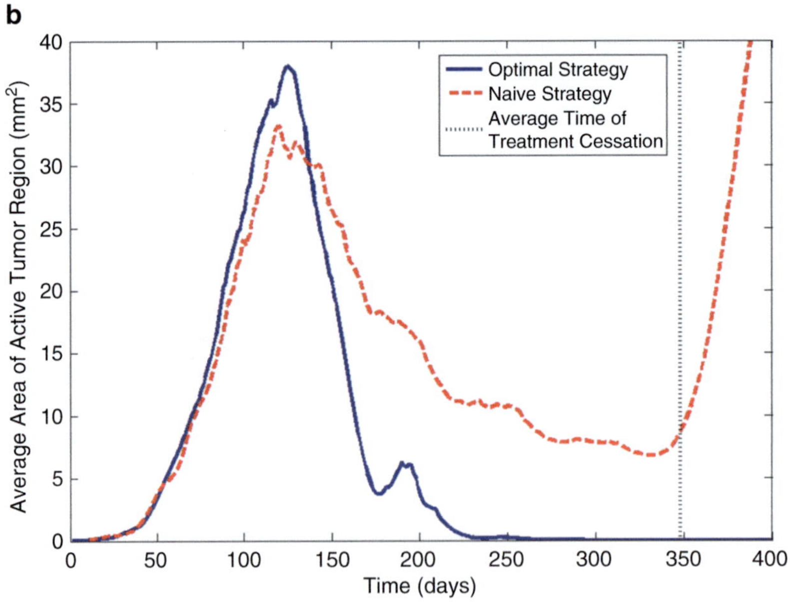

Fig. 10.5 Simulated annealing results. (**a**) Left axis (*blue*) shows the change in the treatment parameters λ_{AI}, λ_{AIC}, and λ_C as a function of accepted annealing step $\bar{k}$. Right axis (*green stars*) shows the change in the average number of active automaton cells remaining after 16 weeks of treatment, also as a function of accepted annealing step. Optimization was conducted using treatment parameters $T_1 = 1000$ and $p_{chemo} = 0.44$. (**b**) Comparison of tumor response to four cycles of the initial "naive" 8-week treatment protocol (4 weeks of AI followed by 4 weeks of AI + chemotherapy) and the optimal protocol. Simulations were conducted using treatment parameters $T_{AI} = 100$ and $p_{chemo} = 0.34$

results in a period of growth inhibition, followed by eventual regrowth. The model also made predictions regarding the efficacy of a VDA such as fosbretabulin that agreed with preclinical and clinical observations that this treatment allows a rim of proliferative cells at the tumor periphery to survive and maintain tumor growth. Finally, the implementation of a cytotoxic chemotherapy in the model was calibrated to the log cell kill value for glioma cells treated with temozolomide. Model predicted responses to drug administration are highly dependent on the vascular nature of the tumor microenvironment, in which vessel co-option occurs alongside angiogenesis.

As the model predicted limited antitumor activity for VDAs, we sought to explore if this was a result of the drug "dose". Surprisingly, we did not find a strong dose-dependent change in response to treatment with VDAs. The relative inefficacy of VDAs was further uncovered when used in combination with an AI. In particular, while the addition of an AI to a VDA protocol resulted in great reductions in the active tumor area (between 40 % and 75 % reduction depending on the time passed since treatment initiation), the addition of a VDA to an AI protocol did not result in significant reductions in the active tumor area. Taken together, this suggests that these two approaches to targeting the vasculature may not be complementary.

Finally, our mathematical model was also extended to search for an optimal delivery schedule for treating a solid tumor with an AI and cytotoxic chemotherapy. Using a simulated annealing algorithm to minimize the active tumor size subject to toxicity constraints, an optimal intermittent dosing strategy was uncovered. The protocol is intermittent in that it specifies that therapeutic levels of AI are, counterintuitively, not maintained at all times during treatment. This results in increased tumor vascularization with the consequence that some hypoxic cells are able to re-enter a normoxic state, making them more vulnerable to killing by the chemotherapy (which is also now more widely distributed throughout the tumor tissue due to angiogenesis that can occur when the AI is not at therapeutic levels). Three to four rounds of the optimal intermittent protocol resulted in permanent growth inhibition of the simulated tumors, strongly suggesting that this intermittent schedule may prolong progression free survival compared to the currently utilized dosing schedule for an AI and chemotherapeutic. This result warrants testing in a preclinical setting, and such experiments are now feasible since noninvasive methods have been developed to visualize and assess angiogenesis and vascular changes in live tumor-bearing animals [29].

References

1. Bergers G, Hanahan D (2008) Modes of resistance to antiangiogenic therapy. Nat Rev Cancer 8:592–603
2. Blakey DC, Russell Westwood F, Walker M, Hughes GD, Davis PD, Aston SE, Ryan AJ (2002) Antitumor activity of the novel vascular targeting agent ZD6126 in a panel of tumor models. Clin Cancer Res 8:1974–1983
3. Brekken RA, Thorpe PE (2001) VEGF-VEGF receptor complexes as markers of tumor vascular endothelium. J Control Release 74:173–181
4. Carmeliet P, Jain RK (2000) Angiogenesis in cancer and other diseases. Nature 407:249–257
5. Cooney MM, Ortiz J, Bukowski RM, Remick SC (2005) Novel vascular targeting/disrupting agents: combretastatin A4 phosphate and related compounds. Curr Oncol Rep 7:90–95
6. Desjardins A, Reardon DA, Coan A, Marcello J, Herndon JE, Bailey L, Peters KB, Friedman HS, Vredenburgh JJ (2012) Bevacizumab and daily temozolomide for recurrent glioblastoma. Cancer 118:1302–1312
7. Dowlati A, Robertson K, Cooney M, Petros WP, Stratford M, Jesberger J, Rafie N, Overmoyer B, Makkar V, Stambler B, Taylor A, Waas J, Lewin JS, McCrae KR, Remick SC (2002) A phase I pharmacokinetic and translational study of the novel vascular targeting agent combretastatin a 4 phosphate on a single-dose intravenous schedule in patients with advanced cancer. Cancer Res 62:3408–3416
8. Gevertz JL (2011) Computational modeling of tumor response to vascular-targeting therapies – part I: validation. Comput Math Methods Med 2011:830515
9. Gevertz JL (2012) Optimization of vascular-targeting drugs in a computational model of tumor growth. Phys Rev E 85:041914
10. Gevertz JL, Torquato S (2006) Modeling the effects of vasculature evolution on early brain tumor growth. J Theor Biol 243:517–531
11. Gevertz JL, Torquato S (2009) Growing heterogeneous tumors in silico. Phys Rev E 80:051910
12. Holash P, Maisonpierre PC, Compton D, Boland P, Alexander CR, Zagzag D,Yancoupoulos GD, Wie-

gand SJ (1997) Vessel cooption, regression, and growth in tumors mediated by angiopoietins and VEGF. Science 284:1994–1998
13. Hoshino T, Wilson CB (1979) Cell kinetic analyses of human malignant brain tumors (gliomas). Cancer 44:956–962
14. LoRusso PM, Boerner SA (2011) Clinical development of vascular disrupting agents: what lessons can we learn from ASA404? J Clin Oncol 29:2952–2955
15. Lu JF, Bruno R, Eppler S, Novotny W, Lum B, Gaudreault J (2008) Clinical pharmacokinetics of bevacizumab in patients with solid tumors. Cancer Chemother Pharmacol 62:779–786
16. Maisonpierre PC, Suri C, Jones PF, Bartunkova S, Wiegand SJ, Radziejewski C, Compton D, McClain J, Aldrich TH, Papadopoulos N, Daly TJ, Davis S, Sato TN, Yancopoulos GD (1997) Angiopoietin-2, a natural antagonist for Tie2 that disrupts in vivo angiogenesis. Science 277:55–60
17. Mason RP, Zhao D, Liu L, Trawick ML, Pinney KG (2011) A perspective on vascular disrupting agents that interact with tubulin: preclinical tumor imaging and biological assessment. Integr Biol 3:375–387
18. McConville P, Hambardzumyan D, Moody JB, Leopold WR, Kreger AR, Woolliscroft MJ, Rehemtulla A, Ross BD, Holland EC (2007) Magnetic resonance imaging determination of tumor grade and early response to temozolomide in a genetically engineered mouse model of glioma. Clin Cancer Res 13:2897–2904
19. Mita MM, Sargsyan L, Mita AC, Spear M (2013) Vascular-disrupting agents in oncology. Expert Opin Investig Drugs 22:317–328
20. Motzer RJ, Hutson TE, Olsen MR, Hudes GR, Burke JM, Edenfield WJ, Wilding G, Agarwal N, Thompson JA, Cella D, Bello A, Korytowsky B, Yuan J, Valota O, Martell B, Hariharan S, Figlin RA (2012) Randomized phase II trial of sunitinib on an intermittent versus continuous dosing schedule as first-line therapy for advanced renal cell carcinoma. J Clin Oncol 30:1371–1377
21. Najafi M, Soltanian-Zadeh H, Jafari-Khouzani K, Scarpace L, Mikkelson T (2012) Prediction of glioblastoma multiforme response to bevacizumab treatment using multi-parametric MRI. PLoS ONE 7:e29945
22. Nathan P, Zweifel M, Padhani AR, Koh D-M, Ng M, Collins DJ, Harris A, Carden C, Smythe J, Fisher N, Taylor NJ, Stirling JJ, Lu S-P, Leach MO, Rustin GJS, Judson I (2012) Phase I trial of combretastatin A4 phosphate (CA4P) in combination with bevacizumab in patients with advanced cancer. Clin Cancer Res 18:3428–3439
23. Rudek MA, Donehower RC, Statkevich R, Batra VK, Cutler DL, Baker SD (2004) Temozolomide in patients with advanced cancer: phase I and pharmacokinetic study. Pharmacotherapy 24:16–25
24. Siemann DW (2006) Tumor vasculature: a target for anticancer therapies. In: Seimann DW (ed) Vascular-targeted therapies in oncology. John Wiley, Chichester, pp 1–8
25. Siemann DW, Shi W (2008) Dual targeting of tumor vasculature: combining Avastin and vascular disrupting agents (CA4P or OXi4503). Anticancer Res 28:2027–2031
26. Stuff R, Dietrich PY, Kraljevic SO, Pica A, Maillard I, Maeder P, Meuli R, Janzer R, Pizzolato G, Mirabell R, Porchet F, Regli L, de Tribolet N, Mirimanoff RO, Leyvraz S (2002) Promising survival for patients with newly diagnosed glioblastoma multiforme treated with concomitant radiation plus temozolomide followed by adjuvant temozolomide. J Clin Oncol 20:1375–1382
27. Thorpe PE (2004) Vascular targeting agents as cancer therapeutics. Clin Cancer Res 10:415–427
28. Torquato S (2002) Random heterogeneous materials: microstructure and microscopic properties. Springer, New York
29. Xing HR, Zhang Q (2012) Real-time visualization and characterization of tumor angiogenesis and vascular response to anticancer therapies. In: Hoffman RM (ed) In vivo cellular imaging using fluorescent proteins: methods and protocols, methods in molecular biology, vol 872. Springer, New York, pp 115–127
30. Zheng X, Wise SM, Cristini V (2005) Nonlinear simulation of tumor necrosis, neo-vascularization and tissue invasion via an adaptive finite-element/level-set method. Bull Math Biol 67:211–259

Optimizing Chemotherapeutic Anti-cancer Treatment and the Tumor Microenvironment: An Analysis of Mathematical Models

11

Urszula Ledzewicz and Heinz Schaettler

Abstract

We review results about the structure of administration of chemotherapeutic anti-cancer treatment that we have obtained from an analysis of minimally parameterized mathematical models using methods of optimal control. This is a branch of continuous-time optimization that studies the minimization of a performance criterion imposed on an underlying dynamical system subject to constraints. The scheduling of anti-cancer treatments has all the features of such a problem: treatments are administered in time and the interactions of the drugs with the tumor and its microenvironment determine the efficacy of therapy. At the same time, constraints on the toxicity of the treatments need to be taken into account. The models we consider are low-dimensional and do not include more refined details, but they capture the essence of the underlying biology and our results give robust and rather conclusive *qualitative information about the administration of optimal treatment protocols* that strongly correlate with approaches taken in medical practice. We describe the changes that arise in optimal administration schedules as the mathematical models are increasingly refined to progress from models that only consider the cancerous cells to models that include the major components of the tumor microenvironment, namely the tumor vasculature and tumor-immune system interactions.

U. Ledzewicz (✉)
Department of Mathematics and Statistics,
Southern Illinois University Edwardsville,
Edwardsville, IL, USA 62026

Institute of Mathematics, Lodz University of Technology,
Lodz, Poland
e-mail: uledzew@siue.edu

H. Schaettler
Department of Electrical and Systems Engineering,
Washington University, St. Louis, MO, USA 63130
e-mail: hms@wustl.edu

K.A. Rejniak (eds.), *Systems Biology of Tumor Microenvironment*, Advances in Experimental Medicine and Biology 936, DOI 10.1007/978-3-319-42023-3_11

Keywords
Mathematical modeling • Optimal control • Optimization of cancer treatments • Tumor microenvironment • Anti-angiogenic therapy • Tumor immune system interactions • Metronomic chemotherapy

11.1 Introduction: Anti-cancer Treatment as an Optimal Control Problem

The question how anti-cancer therapies should be administered in order to be at the same time safe for the patient and to maximize their potential effects is a fundamental one, but also one very difficult to answer conclusively [1, 6, 7, 14, 36, 45]. The reasons for this lie in the complexities of the interactions of the various constituents that form a tumor and its microenvironment [4, 5, 11–14, 35]. If a specific aspect of the overall system is considered in isolation, sometimes clear and simple answers about the structure of optimal therapy protocols can be given. For example, if one only considers the tumor itself and assumes that it consists of a homogeneous population of chemotherapeutically sensitive cells, it seems to be intuitively clear that it would be best to give as much drugs as possible as soon as possible, the so-called MTD (maximum tolerated dose) paradigm in medicine. Simple mathematical models confirm such strategies as optimal [23, 28, 42–44]. But as increasingly more complex features of the overall system are taken into account such as tumor heterogeneity, tumor immune system interactions and tumor angiogenesis, clear answers still remain elusive [14, 36, 45]. It is here that mathematical modeling and analysis (*in silico*) can be useful by providing a framework for cancer progression and its response to various treatment options. These approaches can both be pursued within a general framework [2, 3, 25, 27] or tailored to specific diseases, like, for example, chronic myeloid leukemia in [31, 32].

The administration of anti-cancer therapies has all the characteristic features of what are called *optimal control* problems in the mathematical and engineering literature. Such problems deal with the minimization of a performance criterion imposed on an underlying dynamical system subject to constraints. Figure 11.1 illustrates the main components involved in such a problem: *controls* (input) are functions in time that describe allowable outside influences on the system which, when applied, induce a *system response* (output). Based on this response, a mathematical objective is evaluated which is taken as the *performance measure* for the behavior of the system. It generally includes terms related to the running cost of the system as well as penalty terms on the system response that are designed to induce a desired performance. Optimal control theory addresses the question of *optimizing* this objective. Naturally, the true response of the real system will also be shaped by other aspects that have not been included in the modeling and possibly random disturbances. Hence the true system response will be different from the one computed within the model and thus questions about robustness of the solutions and other stability properties become of great importance. The scheduling of cancer treatments has all the features of such a problem: *treatments are administered in time and the interactions of the therapeutic agents both with the tumor, its microenvironment and the healthy cells determine the efficacy of therapy*. At the same time, *constraints on the toxicity of the treatments must be taken into account to guarantee its safety*.

In this chapter, we review results that we have obtained about the administrations of anti-cancer therapies from an analysis of minimally parameterized mathematical models using optimal control [38, 39]. These models are low-dimensional and thus do not include more refined biological details, but they capture the essence of the underlying biology and our results give rather conclusive *qualitative information about the administration of optimal treatment proto-*

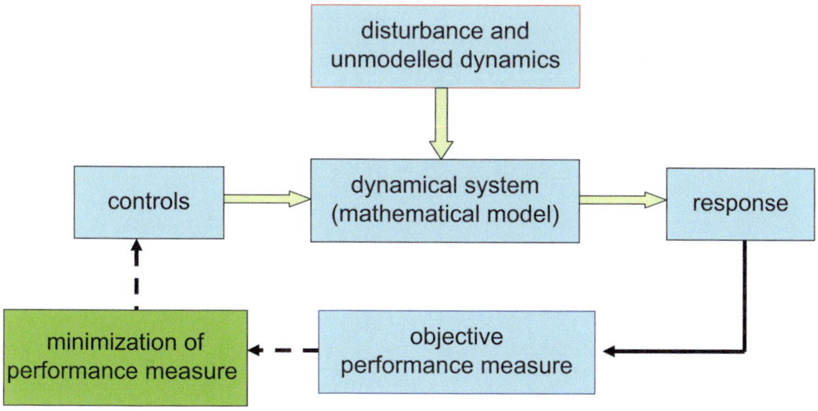

Fig. 11.1 Schematic representation of the structural elements of an optimal control problem

cols that strongly correlate with many approaches taken in medical practice. The chapter is organized as follows: In Sect. 11.2, as a basic reference, we briefly summarize mathematical results about the structure of optimal protocols for cell-cycle specific models for cancer chemotherapy if only the cancerous cells are considered. Already here it makes a significant difference if the tumor population is homogeneous or heterogeneous. Then we include specific aspects of a tumor's microenvironment separately, namely angiogenic signaling and anti-angiogenic treatments in Sect. 11.3 and tumor immune system interactions in Sect. 11.4. We then combine these effects in Sect. 11.5 into an overall mathematical model that considers these two principal components of a tumor's microenvironment in connection with metronomic chemotherapy.

11.2 Optimal Administration of Cancer Chemotherapy for Homogeneous and Heterogeneous Tumor Populations

Cancer is a widely symptomless disease that often is only detected in an advanced stage. This makes it imperative to take strong action as soon as possible. As a result, MTD-strategies that give *maximum tolerated doses* with upfront dosing have become the standard of oncology for the initial phase of treatment, so-called *induction therapy*. Such procedures indeed are also the optimal solutions for mathematical models of cancer chemotherapy when only the cancerous cells are considered and the tumor consists of a homogeneous population of chemotherapeutically sensitive cells. Cell cycle specific models for cancer chemotherapy for this case have been formulated in the work of Swierniak et al. [42, 43]. Analyzing these models as optimal control problems with the objective to minimize the tumor volume while including the total amount of agents as a penalty term to measure the side effects of treatment gives optimal solutions that administer cytotoxic (killing) agents in one maximum dose therapy session with upfront dosing affirming an MTD-strategy (e.g., see [23, 28, 39, 44]) (Fig. 11.2).

Modern oncology, however, more realistically views a tumor as an agglomeration of possibly highly diverse subpopulations of cancerous cells with widely varying chemotherapeutic sensitivities. Cancer cells often are genetically highly unstable and, coupled with high proliferation rates, this leads to significantly higher mutation rates than in healthy cells [8]. As a result, a great variety of different cell types can exist within one tumor from the very beginning (*ab initio*) and it is possible that there exist small sub-populations of cells for which the activation mechanism of a certain drug (targeted or not) simply does not work (*intrinsic drug resistance*). In addition, growing

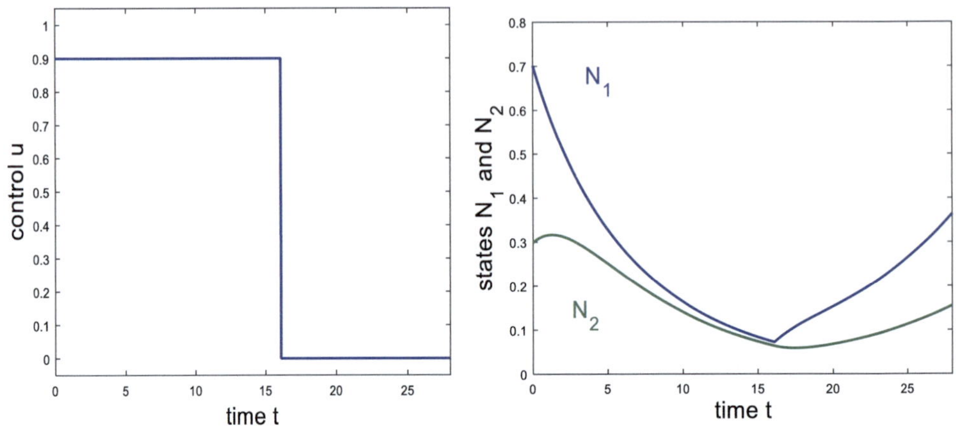

Fig. 11.2 Example of a typical solution for a 2-compartment model for cell-cycle specific cancer chemotherapy with one cytotoxic agent acting in the G_2/M phase of the cell cycle. The control is shown on the left and the corresponding response with N_1 denoting the number of cancer cells in the phases G_0, G_1 and S and N_2 cells in G_2/M are shown on the *right*

malignant tumors exhibit considerable evolutionary ability to enhance cell survival in an environment that is becoming hostile and this leads to the further development of strains of the cells that exhibit increased drug resistance (*acquired drug resistance*) [29, 30]. For example, the Norton-Simon hypothesis [33] postulates that tumors typically consist of faster growing cells that are sensitive to chemotherapy and slower growing populations of cells that have lower sensitivities or are resistant to the chemotherapeutic agent. The underlying rationale reflects the viewpoint that cells that duplicate fast will outperform those that replicate at slower rates, but at the same time these are also the cells that are more vulnerable to a cytotoxic attack since they have higher growth fractions in synthesis and mitosis where they are much more amenable to a chemotherapeutic attack. Once heterogeneity of the tumor cell population is taken into account, MTD-type solutions no longer need to be optimal [26]. Intuitively, in the presence of drug resistant strains, as the cytotoxic agent kills off the sensitive cells, the resistant cell population becomes increasingly more dominant and eventually more harm than good is done by an MTD-style treatment. Once tumor heterogeneity is taken into account, optimal solutions, after an initial period when full dose chemotherapy is given, either stop chemotherapy altogether or favor reduced dose rate administrations. In the medical literature such administration schedules have been called *chemo-switch* protocols [37]. Figure 11.3 shows an example of such an administration protocol for a mathematical model in which sensitive and resistant cancer cells are distinguished. Mathematically these optimal lower dose administration schedules are defined by what are called *singular controls* in optimal control. These correspond to a second class of natural candidates for optimality that do not take values in the boundary of the control set but are given by specific intermediate values that lie in the interior and are determined by necessary conditions for optimality. The other, and first class of candidates, are called *bang-bang* controls and they represent administration schedules that take values in the extreme points of the control set, $u = u_{max}$, and $u = 0$. Obviously, such controls correspond to sessions of full dose chemotherapy with rest periods.

While one can show mathematically that singular controls, and thus the administration of chemotherapy at intermediate values that they represent, are not optimal for cell cycle specific models for homogeneous tumor populations [23], this no longer holds for heterogeneous tumor populations [26] and, indeed, *as the degree of heterogeneity increases, singular controls become*

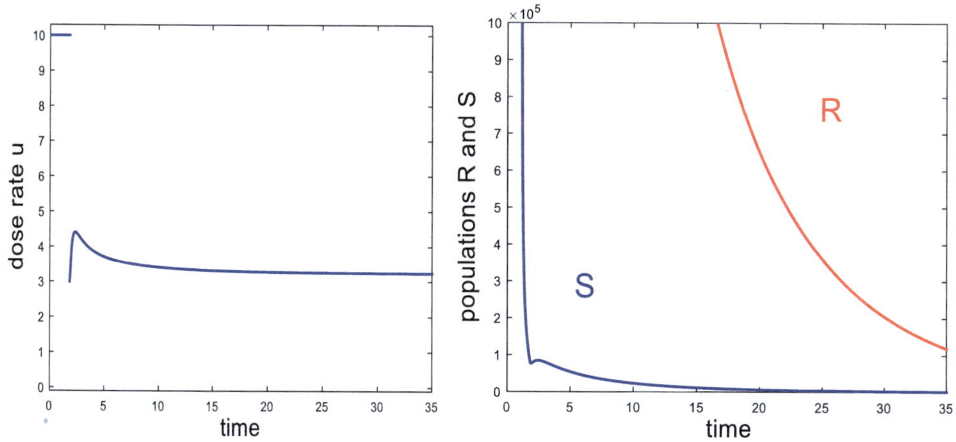

Fig. 11.3 Example of a chemo-switch type protocol for a mathematical model of chemotherapeutically 'sensitive' and 'resistant' cells with the terminology indicating that resistant cells have a much lower, but not necessarily zero resistance. The control shown on the left administers full dose chemotherapy for 1.8 [days] and then switches to a much lower dose for the remaining therapy horizon [35 days]. The initial tumor volume was 10^{10} cells with 94 % in the sensitive compartment. The initial high dose significantly reduces the sensitive cells (but not the more resistant group) and the lower subsequent dose still brings down the resistant compartment, but at a much slower pace

candidates for optimality from the very beginning [20]. There are also good biological reasons for such a behavior. Intuitively, if there exists an intrinsically resistant subpopulation of cancer cells, and if indeed this population is so to speak "outcompeted" by the sensitive population (and this would be in agreement with the Norton-Simon hypothesis), it could be argued that it may be beneficial to maintain a minimum level of the sensitive population (which at any rate can be killed off by the drug) to keep the more harmful resistant strain of cells in check. On the other hand, the typical MTD-style administration of chemotherapy annihilates the sensitive population entirely and in time thus promotes the development of the much more dangerous subpopulation, even if this may take years before it happens. Similar evolutionary ideas are the main rationale behind the concept of *adaptive therapy* by Gatenby et al. [6, 7] in the medical literature and they seem to find some resonance in mathematical models. This should be of particular interest since this is a fundamental issue that is related simply to the fact that tumors are heterogeneous. As such it remains a concern equally for traditional drugs that target all strongly proliferating cells as well as for modern strongly targeted therapeutic agents that have been developed in the past 10 years and are still the focus of so much pharmaceutical research. While these clearly limit the negative side effects of treatment and thus represent a tremendous advance in cancer treatments, nevertheless targeted therapies may be of no help with resolving the fundamental issue of heterogeneity. Simply put, if there exist subpopulations of cancer cells for which the specific activation mechanism of the drug does not work, then, as the sensitive populations are eliminated through therapy, eventually the resistant strain will become dominant. It is for this very reason that both traditional and targeted therapies eventually will fail in these cases. It is precisely here that the specific administration schedules of the drugs may make all the difference.

These observations reinforce the question whether it is possible to *optimize the overall effects of therapy by modulating the administration schedules of the therapeutic agents*. However, the answer to this question does not only depend on the cytotoxic effects of drugs on tumor cells, but also on an array of ancillary features that in various ways aid and abet the tumor, but also fight it and form the tumor microenvironment. The most important structure that sustains the tumor is its *vasculature* which provides the tumor with the oxygen and nutrients needed for further growth. The main example of an endogenous system that fights the tumor is the body's *immune system*. In modern oncology, therefore, the point of view of the tumor as a system of interacting components is becoming prevalent and modern treatments are multi-targeted therapies that not only aim to kill cancer cells, but include anti-angiogenic therapy that targets the vasculature, immunotherapy and other novel options such as cancer viruses.

11.3 Optimal Control of Anti-angiogenic Mono- and Combination Therapies

In the 1990s anti-angiogenic therapies were viewed as a new hope in anti-cancer therapies since they target the healthy and genetically stable endothelial cells that form the lining of blood vessels and capillaries and no developing drug resistance occurred [15, 16]. However, because of the indirect nature of the approach – treatment is only limiting the tumor's support mechanism without actually killing the cancer cells – anti-angiogenic therapy by itself only achieves a temporary, "pseudo-therapeutic" effect that goes away with time. Once treatment is halted the tumor grows back even more vigorously than before. While anti-angiogenic monotherapy thus is no longer considered a viable treatment option, it has become a staple of anti-cancer treatments in combination with both radio- and chemotherapy. The expectation is that *antiangiogenic therapy can enhance the efficacy of traditional approaches by normalizing a tumor's vasculature*. For example, Jain and Munn argue that a normalization of a tumor's irregular and dysfunctional vasculature [12, 13] through prior anti-angiogenic treatment enhances the delivery of chemotherapeutic agents and thus improves the effectiveness of chemotherapy.

Hahnfeldt et al. [9] have formulated a widely influential mathematical model that describes *tumor development under angiogenic signaling*. If we denote the tumor volume by p and the carrying capacity of the vasculature (measured in terms of the volume of the endothelial cells that provide the lining for the newly formed vessels and capillaries) by q, then based on an asymptotic expansion of the solutions for the underlying consumption-diffusion equation, the following dynamics is proposed to model the stimulatory and inhibitory interactions between the tumor and its vasculature:

$$p' = -\xi p \; ln\,(p/q) - \varphi_1 pv, \tag{11.1}$$

$$q' = bp - \left(\mu + dp^{2/3}\right) q - \lambda qu - \varphi_2 qv. \tag{11.2}$$

Equation (11.1) describes tumor growth using a Gompertzian model with coefficient ξ and Eq. (11.2) models the interplay between tumor derived stimulators and inhibitors. Stimulators act locally which is reflected in a fast clearing of these agents and this is modeled by the stimulation term bp with b a constant mnemonically labeled for 'birth'. Inhibitors, on the other hand, have a more systemic action and the inhibition term is taken in the form $dp^{2/3}q$ with d labeling a tumor stimulated 'death' term. The functional relation $p^{2/3}q$ reflects an interaction of the carrying capacity q with the tumor surface $p^{2/3}$ through which inhibitors need to be released. The constant μ denotes the natural rate of death for cells related to the carrying capacity and generally is small, often set to zero. The term $-\lambda qu$ represents the loss of vasculature due to anti-angiogenic treatment with u denoting the dose rate/concentration of the agent while the terms $-\varphi_1 pv$ and $-\varphi_2 qv$ multiplying v model the loss in the respective compartments under a cytotoxic agent v.

We only remark that in this chapter we identify the dose rate of the agents with their concentrations. While there clearly is a difference, this simplifies the mathematical aspects of our presentation. We have analyzed the effects that the inclusion of standard linear pharmacokinetic models has on optimal controls in various papers (e.g., see [24]). While there is a difference from the theoretical mathematical perspective, as far as the practical implications of the results are con-

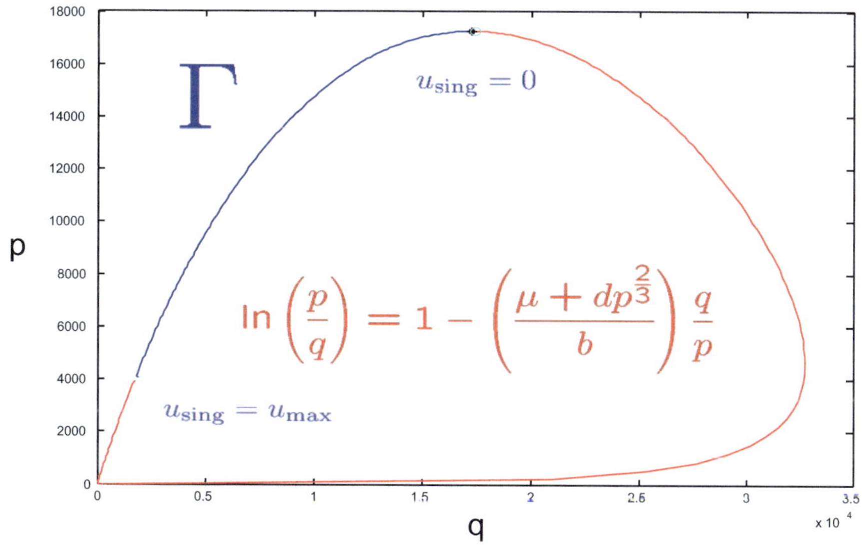

Fig. 11.4 The singular arc Γ defining the relation between the tumor volume p and the carrying capacity q along which the best tumor reductions can be achieved. The singular curve is the loop shown in the picture, but the dose rates are only admissible along the segment shown as a solid blue curve. This illustration is for the numerical values $\xi = 0.192$, $b = 5.85$, $d = 0.00873$, $\lambda = 0.015$, $\mu = 0$ that are based on data from Hahnfeldt et al. [9] and a sample maximum dose rate $u_{max} = 75$

cerned, the mathematical models considered here are sufficient to derive the qualitative structures of optimal protocols [39].

In the paper [25] we have given a complete solution for the optimal control problem that describes the monotherapy problem to *minimize the tumor volume with an a priori given amount of anti-angiogenic agents*. Interestingly, the mathematical optimal solutions point to a specific "path" Γ in (p, q)-space determined by an optimal singular arc that should be followed in order to obtain the best possible tumor reductions. In Fig. 11.4 this path is illustrated for a set of parameter values taken from [9]. Optimal controls typically (i.e., except for medically less realistic scenarios) employ the following strategy: give a brief and carefully measured maximum dose of anti-angiogenic inhibitors to reach the curve Γ and then reduce the dose rates to follow this specific path Γ until all inhibitors are exhausted. (A representative example is shown in Fig. 11.5.) Indeed, this strategy achieves the best possible minimum tumor volumes. For example, generally simple maximum dose rate strategies unnecessarily waste the limited amount of anti-angiogenic agents that can be utilized with higher efficacy using the lower dose rate singular controls. Maximum dose rates are only optimal if the value of the numerically computed singular control exceeds the maximum allowable dose rate. This occurs for low tumor volumes [25].

For the anti-angiogenic monotherapy problem singular controls form the core of the optimal solutions and this prevails when combinations with chemo- and radiotherapy are considered. Here we only briefly describe the case of combination with chemotherapy. In this case, optimal solutions implement the following strategy: again give a brief and carefully measured maximum dose of anti-angiogenic inhibitors to reach the curve Γ and then maintain this relation by judiciously choosing reduced dose rates for the anti-angiogenic agent (defined by the singular controls) until, at the best moment, chemotherapy is given in one full dose session. In the medical literature, the term *therapeutic window* [12] has been used to characterize this period. While both anti-angiogenic and chemotherapeutic agents are administered, the anti-angiogenic dose rate is adjusted to maintain the optimal relation between

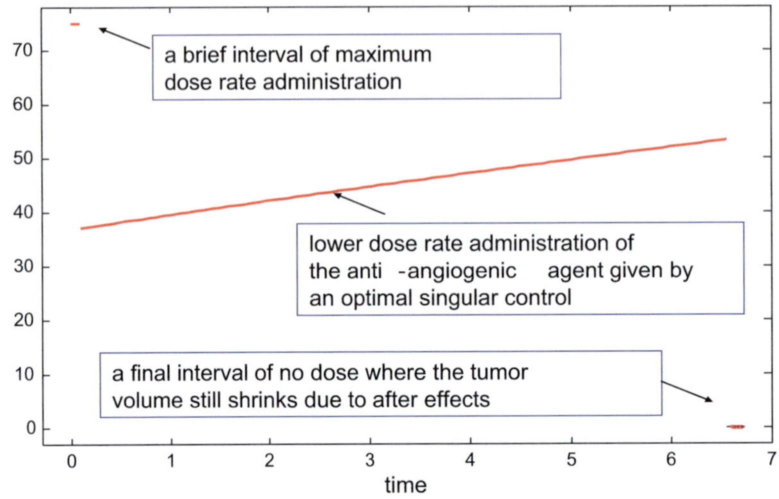

Fig. 11.5 Example of the time evolution of the optimal control (anti-angiogenic dose rate) for the initial values $p_0 = 12{,}000$ and $q_0 = 15{,}000$, system parameters $\xi = 0.192$, $b = 5.85$, $d = 0.00873$, $\lambda = 0.015$ and $\mu = 0$ and control data given by the maximum dose rate $u_{max} = 75$ and overall available amount of inhibitors given by 300. The example clearly illustrates the dominant portion in time when the optimal solution is given by a singular control with values in the interior of the control set $[0, u_{max}]$

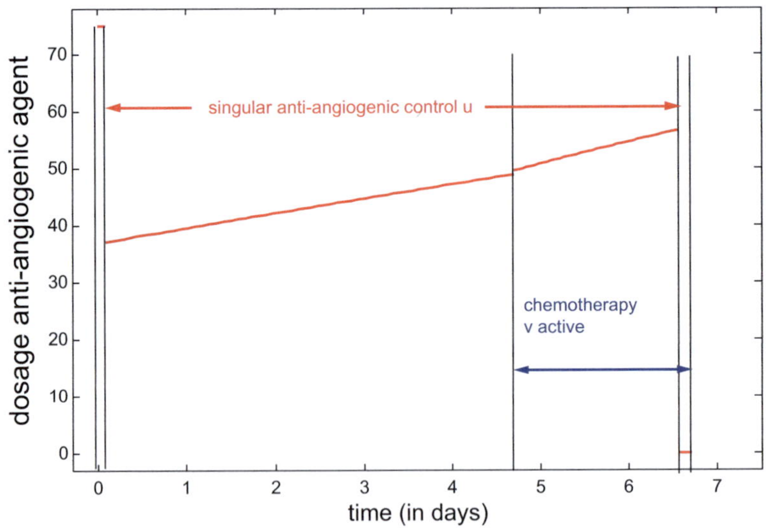

Fig. 11.6 Example of an optimal administration schedule for the anti-angiogenic agent *u* (shown in *red*) in combination with chemotherapy. The cytotoxic agent is active at maximum dose during the interval marked by the blue arrow (Adapted from Ledzewicz and Schättler [27])

tumor volume and carrying capacity [34]. This structure of the mathematically optimal solutions is in agreement with the medical hypothesis that the preliminary delivery of anti-angiogenic agents can regularize a tumor's vascular network with beneficial consequences for the successive delivery of cytotoxic chemotherapeutic agents [13] (Fig. 11.6).

11.4 Optimal Control with Tumor Immune System Interactions

Singular controls also come to the forefront of the structure of optimal therapy protocols when tumor immune system interactions are included in the mathematical model. The following equations are based on a classical model by Stepanova [41]:

$$p' = -\xi p \ln(p/q) - \theta pr - \varphi_1 pv, \qquad (11.3)$$

$$r' = \alpha\left(p - \beta p^2\right) r + \gamma - \delta r + \varphi_3 rw. \qquad (11.4)$$

Here, as before, p denotes the tumor volume and r is a non-dimensional, normalized, order of magnitude quantity related to the activities of various types of T-cells activated during the immune reaction. We summarily refer to it as the immunocompetent cell density. All other letters in these equations denote constant coefficients. Equation (11.4) models the tumor immune-system interactions. Various organs such as the spleen, thymus, lymph nodes and bone marrow, each contribute to the development of immune cells in the body and the parameter γ models a combined rate of influx of T-cells generated through these primary organs; δ is simply the rate of natural death of the T-cells. The first term in this equation models the proliferation of lymphocytes. For small tumors, it is stimulated by tumor antigen and this effect is taken to be proportional to the tumor volume p. Large tumors suppress the activity of the immune system. The reasons lie in an inadequate stimulation of the immune forces as well as a general suppression of immune lymphocytes by the tumor (see [41] and the references therein). This feature is expressed in the model through the inclusion of the term $-\beta p^2$. Thus $1/\beta$ corresponds to a threshold beyond which the immunological system becomes depressed by the growing tumor. The coefficients α and β are used to calibrate these interactions and collectively describe a state-dependent influence of the cancer cells on the stimulation of the immune system. As in the model for angiogenic signaling, Eq. (11.3) models tumor growth using a Gompertzian function with growth coefficient ξ. The additional term $-\theta pr$ in Eq. (11.3) models the beneficial effects of the immune system reaction on the cancer volume with θ denoting the rate at which cancer cells are eliminated through the activity of T-cells. As above, the term $-\varphi_1 pv$ describes the loss of tumor volume due to the administration of a cytotoxic agent at concentration v and $+\varphi_3 rw$ represents a rudimentary form of an immune boost.

Depending on the values of the parameters, the dynamical system (11.3) and (11.4) exhibits a wide range of behaviors that encompass a variety of medically realistic scenarios. These range from cases when tumor-immune system interactions are able to completely eradicate the tumor in the sense that all trajectories converge to the tumor free equilibrium point $(0, \gamma/\delta)$ (*immune surveillance*) to situations when *tumor dormancy* is induced (a unique, globally asymptotically stable, benign equilibrium point with small positive tumor volume exists) to multi-stable situations that have both persistent benign and malignant behaviors to situations when tumor growth simply is dominant and overcomes the immune system. Despite its simplicity, with just a few parameters this model rather accurately reflects the main qualitative aspects of tumor-immune interactions: the immune system can be effective in the control of small cancer volumes, but for large volumes the cancer dynamics suppresses the immune system and the two systems effectively become separated. For this reason, the underlying equations have been widely accepted as a basic model.

The most interesting and relevant scenario arises when the dynamics is characterized by bistable properties. Figure 11.7 shows the phase portrait for a typical such set of parameters. There exist two locally asymptotically stable equilibria, (p_b,r_b) and (p_m,r_m), marked with a green and red star, respectively, and a saddle point (p_s,r_s) marked with a black star. The *region of attraction* of a locally stable equilibrium point consists of all initial conditions from which the system converges to the equilibrium point. The tumor volume for the equilibrium point (p_m,r_m) is close to the carrying capacity of the system and it is by an order of magnitude larger than for (p_b,r_b). These values might be interpreted as a microscopic and a macroscopic locally asymptotically stable equilibrium point with the high value indicating that the patient will succumb to the disease. We call an equilibrium point *malignant* if the corresponding tumor volume is close to the carrying capacity of the system and *benign* if it is by an order of magnitude smaller. The corresponding regions of attractions are, respectively, the *malignant* and *benign* regions. In case of a microscopic benign

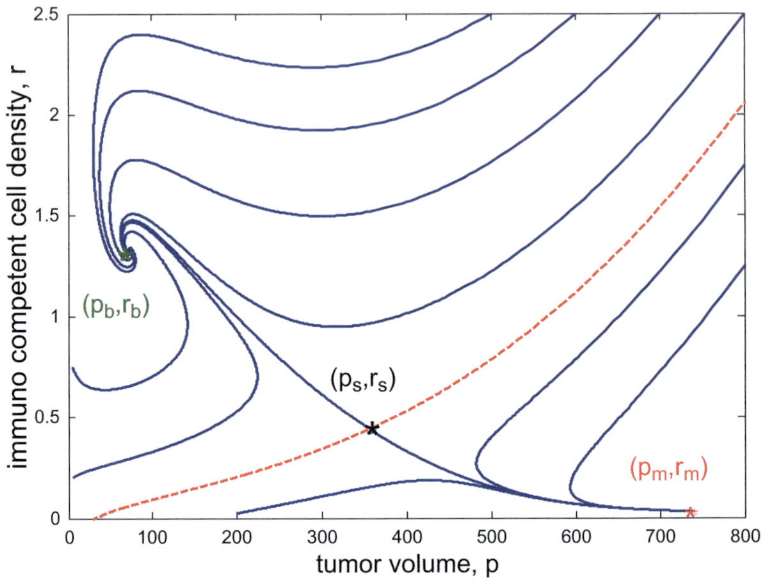

Fig. 11.7 Phaseportrait of the uncontrolled system (11.3) and (11.4) for parameter values $\xi = 0.5618$, $\theta = 1$, $\alpha = 0.00484$, $\beta = 0.00264$, $\gamma = 0.1181$ and $\delta = 0.3745$ that are based on data from Kuznetsov et al. [18]. The benign equilibrium point is at $(73, 1.33)$, the saddle point at $(355, 0.44)$ and the malignant equilibrium point is at $(737, 0.03)$. The boundary between the benign and malignant regions (separatrix) is formed by the stable manifold of the saddle

equilibrium, this region can be interpreted as the set of all states of the system where the immune system is able to control the cancer. This is one way of describing geometrically what medically has been called *tumor dormancy*. On the other hand, the region of attraction of the macroscopic equilibrium point corresponds to conditions when the system has escaped from this immune surveillance and the disease will become lethal. Obviously, the boundary between these two behaviors, called the separatrix in mathematics, is the critical object to study and it is formed by the stable manifold of the saddle point. This set consists of all initial data from which the system actually does converge to the saddle point as $t \to \infty$ and is shown as a dashed red curve in Fig. 11.7.

The dynamically only nontrivial – and at the same time medically most interesting case – arises in the bi-stable scenario. In this case, it is natural to consider the transfer of the state from the malignant into the benign region as an optimal control problem. In the papers [21, 22], we have formulated an objective functional which was designed to achieve this transfer by minimizing an appropriate penalty term that is based on the underlying geometry of the system, namely

$$J(u) = Ap(T) - Br(T) + \int_0^T (Mu(t) + Nv(t) + S)\, dt. \tag{11.5}$$

This objective function consists of three separate pieces. (i) The penalty term $Ap(T) - Br(T)$ at the final time is designed to induce the state of the system to move from the malignant into the benign region. This can be achieved by choosing the coefficients A and B as the coordinates of a properly oriented normal vector to the stable manifold at the saddle or as a properly oriented unstable eigenvector at the saddle point. In either case, both A and B are positive. (ii) The integral terms involving the cytotoxic agent u and the immuno therapy v measure the amounts of drugs given and are an indirect way of controlling their side effects. (iii) The penalty term ST on the final time is included to avoid solutions with infinitely long intervals and also makes the mathematical problem well-posed. All coefficients are positive. We emphasize that all these coefficients are variables of choice that should and need to be calibrated to fine-tune the response of the system. The choice of the weights aims at striking a balance between the benefit at the terminal time

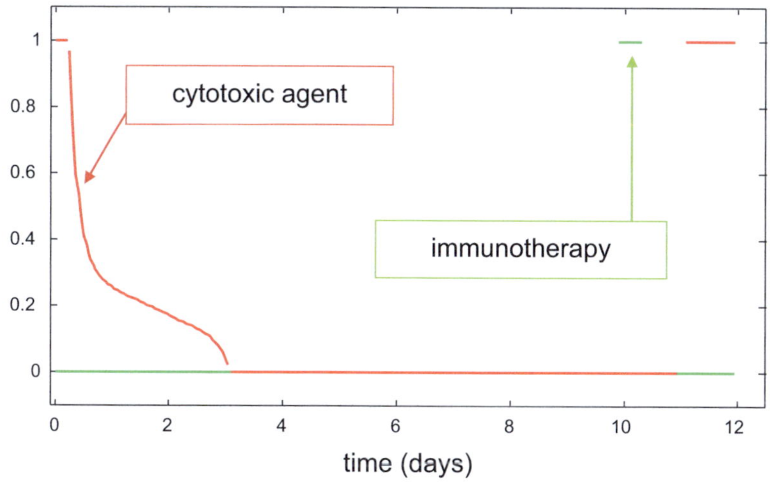

Fig. 11.8 Optimal controls for the transfer of the state from the malignant into the benign region are "chemo-switch" protocols. The parameter values used for this illustration are the same as in Fig. 11.7 for the dynamics, $\xi = 0.5618$, $\theta = 1$, $\alpha = 0.00484$, $\beta = 0.00264$, $\gamma = 0.1181$ and $\delta = 0.3745$, the pharmacodynamic coefficients are $\varphi_1 = 2$ and $\varphi_3 = 1$ and the initial condition is $(p_0, r_0) = (600, 0.1)$. The weights in the objective are given by $A = 0.00192$, $B = 1$, $M = 0.01$, $N = 0.025$ and $S = 0.001$ (Figure adapted from Ledzewicz and Schättler [27])

T, $Ap(T) - Br(T)$, and the overall side effects measured by the total amount of drugs given, while it guarantees the existence of an optimal solution by also penalizing the free terminal time T.

Figure 11.8 shows a typical example of an optimal control that achieves such a transition. The dose rates for the chemotherapeutic agent are shown in red while the immune boost is shown in green. After a brief initial period of maximum dose therapy, the dose rates for the cytotoxic agent are drastically reduced and follow an optimal singular arc from the malignant into the benign region. Essentially, as the beneficial aspects of the immune reaction increase, it simply is no longer necessary to give full dose chemotherapy. Since this dose is penalized in the objective (5), optimal controls reduce the dose rates and optimal protocols are of the *chemo-switch* type [37]. For a large range of parameter values, similar to the case shown here, after a prolonged rest period, optimal protocols often still administer a brief boost of maximum dose therapy at the end.

It clearly is the mitigating influence of the immune system which, for smaller tumor volumes, leads to the abandonment of the strict bang-bang scheme that was seen as optimal in the cell-cycle specific models. Intuitively, if the system is in a condition where it is able to control the cancer, why administer chemotherapy if this might destroy this innate ability of the organism? Despite the model's simplicity, optimal solutions show qualitative structures that are robust and conform to results in the medical research. Indeed, such *chemo-switch* protocols have shown effectiveness for certain types of cancer [37].

11.5 Metronomic Chemotherapy: A Mathematical Model for Its Effects on the Tumor Microenvironment

Another nontraditional way of administering chemotherapy that has shown itself effective precisely because of the effects it has on the tumor microenvironment are *metronomic* protocols that administer specific chemotherapeutic agents (such as cyclophosphamide) at significantly lower dose rates, almost continuously, with only short interruptions to increase the efficacy of the drugs (e.g., see [1, 10, 17, 35]). There exists mounting medical evidence that low-dose chemotherapy, while still having a moderate cytotoxic effect on cancerous cells, has both anti-angiogenic and immune stimulatory effects [11]. The rationale behind reducing dosage is that, in the absence of severe limiting toxic side effects, it will be possible to give chemotherapy over prolonged time intervals so that, because of the greatly extended time horizon, the overall effect may be improved when compared with repeated short MTD doses [2, 14, 45].

Because of its anti-angiogenic and immune stimulatory effects, a mathematical model for metronomic chemotherapy needs to take these fundamental aspects of the tumor microenvironment into account. Merging the mathematical model for angiogenic signaling from [4] defined by Eq. (11.2) with Stepanova's Eqs. (11.3) and (11.4) for tumor immune system interactions, we obtain the following minimally parameterized mathematical model for metronomic chemotherapy [19, 40]:

$$p' = -\xi p \, ln\,(p/q) - \theta pr - \varphi_1 pv, \tag{11.6}$$

$$q' = bp - \left(\mu + dp^{2/3}\right) q - \varphi_2 qv, \tag{11.7}$$

$$r' = \alpha\left(p - \beta p^2\right) r + \gamma - \delta r + \varphi_3 rv. \tag{11.8}$$

The variables and parameters are the same as described earlier with v denoting the concentration of some low-dose chemotherapeutic agent. For a number of cytotoxic drugs for which experimental data are available (e.g., cyclophosphamide), low dose metronomic chemotherapy has a strong anti-angiogenic effect while the cytotoxic and pro-immune effects are lower. Generally, however, these relations depend on the specific drug-tumor combination and are modeled by inequality relations between the pharmacodynamic parameters φ_i.

This model exhibits the same wide range of dynamical behaviors as Stepanova's model (11.3) and (11.4) [41]. These range from cases when low-dose metronomic chemotherapy is able to completely eradicate the tumor (in the sense that all trajectories converge to a tumor free equilibrium point) to situations when tumor dormancy is induced (a unique, globally asymptotically stable benign equilibrium point with small positive tumor volume exists) to multi-stable situations that have both persistent benign and malignant behaviors (the typical multi-stable scenario of mathematical models for tumor-immune system interactions) to situations when tumor growth simply is dominant and the disease cannot be cured by low-dose metronomic chemotherapy.

As before, the most important practical scenario arises when the system is bi-stable with both a benign and a malignant equilibrium point. Here the state space is 3-dimensional and the stable manifold of the saddle is a surface that separates the benign and malignant regions. Once more, we consider the problem to minimize an objective $J(u)$ that is designed to move an initial condition (p_0,q_0,r_0) that lies in the malignant region into the benign region. Analogously to (5), such a performance measure is constructed as

$$J(u) = Ap(T) + Bq(T) - Cr(T) + \int_0^T (Mu(t) + S)\, dt. \tag{11.9}$$

In this case, because of the dimension of the state space, singular controls become smooth functions of the state (p,q,r) and, in principle, always are a viable candidate for optimality. Figure 11.9 shows a slice of the singular control as a function of (p,q) for a fixed value r. However, numerical computations indicate that the actual values these controls would take are negative for high tumor volumes and carrying capacities. While the necessary conditions for optimality are satisfied in either case, for high tumor volumes the controls are inadmissible. The theoretical analysis of these models is still in progress, but these numerical computations again point to optimal controls that follow a chemo-switch strategy for initial conditions in the malignant region: start with a brief maximum dose rate chemotherapy and then, once the system moves into or close to the benign region, lower the dose rate to follow singular controls.

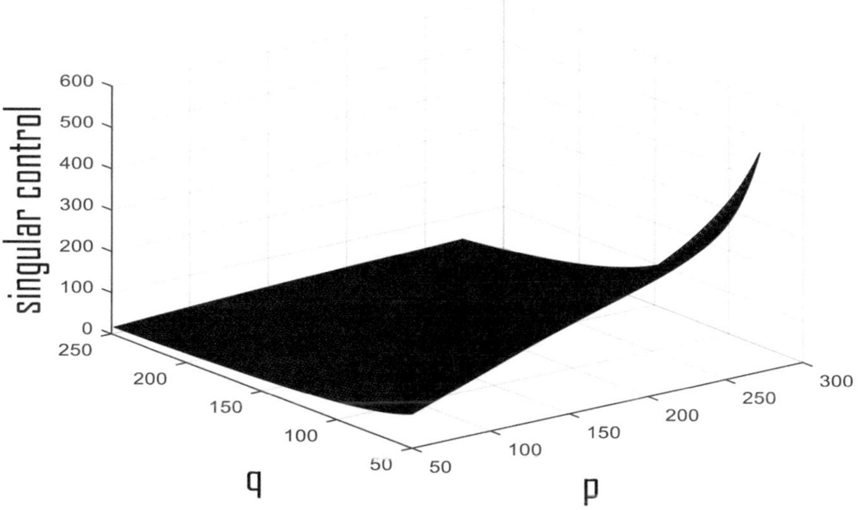

Fig. 11.9 Values of the singular control as function of *(p,q)* for the constant value $r = 0.2$. In the relevant region where *p* and *q* are comparable the values correspond to a low dose administration of agents

11.6 Conclusion

In this paper, we have summarized some results about the structure of optimal therapy protocols for chemotherapeutic agents that can be inferred from a mathematical analysis of minimally parameterized models. As important aspects of the tumor microenvironment are taken into account, such as the tumor vasculature and tumor immune system interactions, optimal solutions deviate from the customary MTD (maximum tolerable dose) approach still dominant in medical practice. While these type of protocols are well established and do make perfect sense under certain conditions, especially early in treatment and for homogeneous, chemotherapeutically sensitive tumor cell populations, once tumor heterogeneity and tumor-immune system interactions are considered as well, the rationale for an MTD type approach becomes blurry. For anti-angiogenic treatments clearly an ideal relationship between tumor volume and carrying capacity is sought that is realized with time-varying lower doses. Also, if the immune system can be recruited in the fight against cancer, then lower doses with less toxic side effects become viable as it is the case in the mathematical models described above. All these observations lead to the conclusion that alternative drug administration schedules that have been considered in medical practice such as chemo-switch protocols and metronomic chemotherapy at a minimum should be seriously considered. *In certain stages of the disease, mathematical models strongly support the hypothesis that "more is not necessarily better"* put forward in the medical literature [5, 27]. Clearly, there still remain many questions to be answered how exactly therapy protocols should be designed in order to optimize the effects of treatment in the sense of dose rates, frequency and also of sequencing if multiple drugs are involved. The analysis of minimally parameterized mathematical models like they were discussed in this paper allows to give qualitative insights that, with their rather robust conclusions, provide a theoretical basis for evaluation of classical and novel treatment strategies in the war on cancer.

Acknowledgment This material is based upon work supported by the National Science Foundation under collaborative research Grants Nos. DMS 1311729/1311733. Any opinions, findings, and conclusions or recommendations expressed in this material are those of the author(s) and do not necessarily reflect the views of the National Science Foundation.

References

1. André N, Padovani L, Pasquier E (2011) Metronomic scheduling of anticancer treatment: the next generation of multitarget therapy? Future Oncol 7(3):385–394
2. Benzekry S, Hahnfeldt P (2013) Maximum tolerated dose versus metronomic scheduling in the treatment of metastatic cancers. J Theor Biol 335:233–244
3. Billy F, Clairambault J, Fercoq O (2012) Optimisation of cancer drug treatments using cell population dynamics. In: Ledzewicz U, Schättler H, Friedman A, Kashdan E (eds) Mathematical methods and models in biomedicine. Springer, New York, pp 265–309
4. Friedman A (2012) Cancer as multifaceted disease. Math Model Nat Phenom 7:1–26
5. Friedman A, Kim Y (2011) Tumor cell proliferation and migration under the influence of their microenvironment. Math Biosci Engr – MBE 8(2):371–383
6. Gatenby RA (2009) A change of strategy in the war on cancer. Nature 459:508–509. doi:10.1038/459508a
7. Gatenby RA, Silva AS, Gillies RJ, Frieden BR (2009) Adaptive therapy. Cancer Res 69:4894–4903
8. Goldie JH, Coldman A (1998) Drug resistance in cancer. Cambridge University Press, Cambridge
9. Hahnfeldt P, Panigrahy D, Folkman J, Hlatky L (1999) Tumor development under angiogenic signaling: a dynamical theory of tumor growth, treatment response, and postvascular dormancy. Cancer Res 59:4770–4775
10. Hanahan D, Bergers G, Bergsland E (2000) Less is more, regularly: metronomic dosing of cytotoxic drugs can target tumor angiogenesis in mice. J Clin Invest 105(8):145–147
11. Hao YB, Yi SY, Ruan J, Zhao L, Nan KJ (2014) New insights into metronomic chemotherapy- induced immunoregulation. Cancer Lett 354(2):220–226
12. Jain RK (2001) Normalizing tumor vasculature with anti-angiogenic therapy: a new paradigm for combination therapy. Nat Med 7:987–989
13. Jain RK, Munn LL (2007) Vascular normalization as a rationale for combining chemotherapy with antiangiogenic agents. Princ Pract Oncol 21:1–7
14. Kamen B, Rubin E, Aisner J, Glatstein E (2000) High-time chemotherapy or high time for low dose? J Clin Oncol 18:2935–2937
15. Kerbel RS (1997) A cancer therapy resistant to resistance. Nature 390:335–336
16. Kerbel RS (2000) Tumor angiogenesis: past, present and near future. Carcinogensis 21:505–515
17. Klement G, Baruchel S, Rak J, Man S, Clark K, Hicklin DJ, Bohlen P, Kerbel RS (2000) Continuous low-dose therapy with vinblastine and VEGF receptor-2 antibody induces sustained tumor regression without overt toxicity. J Clin Invest 105(8):R15–R24
18. Kuznetsov VA, Makalkin IA, Taylor MA, Perelson AS (1994) Nonlinear dynamics of immunogenic tumors: parameter estimation and global bifurcation analysis. Bull Math Biol 56:295–321
19. Ledzewicz U, Amini B, Schättler H (2015) Dynamics and control of a mathematical model for metronomic chemotherapy. Math Biosci, MBE 12(6):1257–1275. doi:10.3934/mbe.2015.12.1257
20. Ledzewicz U, Bratton K, Schättler H (2014) A 3-compartment model for chemotherapy of heterogeneous tumor populations. Acta Appl Math 135(1):191–207. doi:10.1007/s10440-014-9952-6
21. Ledzewicz U, FarajiMosalman MS, Schättler H (2013) Optimal controls for a mathematical model of tumor-immune interactions under targeted chemotherapy with immune boost. Discr Cont Dyn Syst Ser B 18:1031–1051. doi:10.3934/dcdsb.2013.18.1031
22. Ledzewicz U, Naghnaeian M, Schättler H (2012) Optimal response to chemotherapy for a mathematical model of tumor-immune dynamics. J Math Biol 64:557–577. doi:10.1007/s00285-011-0424-6
23. Ledzewicz U, Schättler H (2002) Optimal bang-bang controls for a 2-compartment model in cancer chemotherapy. J Optim Theory Appl – JOTA 114:609–637
24. Ledzewicz U, Schättler H (2005) The influence of PK/PD on the structure of optimal control in cancer chemotherapy models. Math Biosci Eng (MBE) 2(3):561–578
25. Ledzewicz U, Schättler H (2007) Antiangiogenic therapy in cancer treatment as an optimal control problem. SIAM J Control Optim 46(3):1052–1079
26. Ledzewicz U, Schättler H (2014) On optimal chemotherapy for heterogeneous tumors. J Biol Syst 22(2):1–21
27. Ledzewicz U, Schättler H (2014) Tumor microenvironment and anticancer therapies: an optimal control approach. In: A d'Onofrio, A Gandolfi (eds) Mathematical Oncology, Springer
28. Ledzewicz U, Schättler H, Reisi Gahrooi M, Mahmoudian Dehkordi S (2013) On the MTD paradigm and optimal control for multi-drug cancer chemotherapy. Math Biosci Eng (MBE) 10(3):803–819. doi:10.3934/mbe.2013.10.803
29. Lorz A, Lorenzi T, Hochberg ME, Clairambault J, Berthame B (2013) Population adaptive evolution, chemotherapeutic resistance and multiple anti-cancer therapies. ESAIM: Math Model Numer Anal 47:377–399. doi:10.1051/m2an/2012031
30. Lorz A, Lorenzi T, Clairambault J, Escargueil A, Perthame B (2015) Effects of space structure and combination therapies on phenotypic heterogeneity and drug resistance in solid tumors. Bull Math Biol 77:1–22
31. Moore H, Li NK (2004) A mathematical model for chronic myelogeneous leukemia (CML) and T cell interaction. J Theor Biol 227:513–523
32. Nanda S, Moore H, Lenhart S (2007) Optimal control of treatment in a mathematical model of chronic myelogenous leukemia. Math Biosci 210:143–156
33. Norton L, Simon R (1986) The Norton-Simon hypothesis revisited. Cancer Treat Rep 70: 41–61

34. d'Onofrio A, Ledzewicz U, Maurer H, Schättler H (2009) On optimal delivery of combination therapy for tumors. Math Biosci 222:13–26. doi:10.1016/j.mbs.2009.08.004
35. Pasquier E, Kavallaris M, André N (2010) Metronomic chemotherapy: new rationale for new directions. Nat Rev Clin Oncol 7:455–465
36. Pasquier E, Ledzewicz U (2013) Perspective on "more is not necessarily better": metronomic chemotherapy. Newsl Soc Math Biol 26(2):9–10
37. Pietras K, Hanahan D (2005) A multi-targeted, metronomic and maximum tolerated dose 'chemo- switch' regimen is antiangiogenic, producing objective responses and survival benefit in a mouse model of cancer. J Clin Oncol 23:939–952
38. Schättler H, Ledzewicz U (2012) Geometric optimal control. Springer, New York
39. Schättler H, Ledzewicz U (2015) Optimal control for mathematical models of cancer therapies. Springer, New York
40. Schättler H, Ledzewicz U, Amini B (2016) Dynamical properties of a minimally parameterized mathematical model for metronomic chemotherapy. J Math Biol 72:1255–1280. doi:10.1007s/00285-015-0907y
41. Stepanova NV (1980) Course of the immune reaction during the development of a malignant tumour. Biophysics 24:917–923
42. Swierniak A (1988) Optimal treatment protocols in leukemia – modelling the proliferation cycle, Proc. 12th IMACS World Congress, Paris, vol 4, pp 170–172
43. Swierniak A (1995) Cell cycle as an object of control. J Biol Syst 3:41–54
44. Swierniak A, Ledzewicz U, Schättler H (2003) Optimal control for a class of compartmental models in cancer chemotherapy. Int J Appl Math Comp Sci 13:357–368
45. Weitman SD, Glatstein E, Kamen BA (1993) Back to the basics: the importance of 'concentration x time' in oncology. J Clin Oncol 11:820–821

Progress Towards Computational 3-D Multicellular Systems Biology

12

Paul Macklin, Hermann B. Frieboes, Jessica L. Sparks, Ahmadreza Ghaffarizadeh, Samuel H. Friedman, Edwin F. Juarez, Edmond Jonckheere, and Shannon M. Mumenthaler

Abstract

Tumors cannot be understood in isolation from their microenvironment. Tumor and stromal cells change phenotype based upon biochemical and biophysical inputs from their surroundings, even as they interact with and remodel the microenvironment. Cancer should be investigated as an adaptive, multicellular system in a dynamical microenvironment. Computational modeling offers the potential to detangle this complex system, but the modeling platform must ideally account for tumor heterogeneity, substrate and signaling factor biotransport, cell and tissue biophysics, tissue and vascular remodeling, microvascular and interstitial flow, and links between all these sub-systems. Such a platform should leverage high-throughput experimental data, while using open data standards for reproducibility. In this chapter, we review advances by our groups in these key areas, particularly in advanced models of tissue mechanics and interstitial flow, open source simulation software, high-throughput phenotypic screening, and multicellular data standards. In the future, we expect a transformation of computational cancer biology from individual groups modeling isolated parts of cancer, to coalitions of groups combining compatible tools to simulate the 3-D multicellular systems biology of cancer tissues.

P. Macklin (✉) • A. Ghaffarizadeh • S.H. Friedman
S.M. Mumenthaler
Lawrence J. Ellison Institute for Transformative Medicine, University of Southern California, Los Angeles, CA, USA
e-mail: Paul.Macklin@MathCancer.org

H.B. Frieboes
Department of Bioengineering, University of Louisville, Louisville, KY, USA

J.L. Sparks
Department of Chemical, Paper, and Biomedical Engineering, Miami University, Oxford, OH, USA

E.F. Juarez
Lawrence J. Ellison Institute for Transformative Medicine, University of Southern California, Los Angeles, CA, USA

Department of Electrical Engineering, University of Southern California, Los Angeles, CA, USA

E. Jonckheere
Department of Electrical Engineering, University of Southern California, Los Angeles, CA, USA

K.A. Rejniak (eds.), *Systems Biology of Tumor Microenvironment*, Advances in Experimental Medicine and Biology 936, DOI 10.1007/978-3-319-42023-3_12

Keywords
Multicellular systems biology • Computational modeling • Tissue engineering • Cancer microenvironment

12.1 Introduction

Tumors cannot be understood without the context of their microenvironments. (See [1–3] and references therein.) Tumor cells depend upon growth substrates like oxygen, glucose, and growth factors for survival and proliferation. They release signaling factors that influence the behavior of other tumor cells and "normal" cells in the surrounding tissue (the stroma). Fibroblasts may respond to tumor-released signals by increasing motility and remodeling the extracellular matrix (ECM: a complex scaffolding that supports a tissue and its cells). Large populations of tumor cells directly alter the biochemical landscape through their uptake and depletion of growth substrates, creating gradients of these substrates that change the overall spatiotemporal distribution of substrates. When tumor cells experience low levels of oxygen (hypoxia), they may release factors (e.g., VEGF-A_{165}) that promote angiogenesis: endothelial cells detach from the existing blood vasculature, migrate, and proliferate to form new blood vessels [4]. Necrotic tumor cells (those that undergo uncontrolled death due to energy depletion—see our recent review [5]) and even viable tumor cells may release signaling factors that promote inflammatory responses, including infiltration by white blood cells and further matrix remodeling by stromal cells. All these cells cross-communicate in myriad, poorly understood ways.

The complexity of the tumor-tissue system is not merely biochemical. The ECM is a mixture of elastic fibers (e.g., collagen IV) and matrix-bound signaling factors [6]. Tissue remodeling (e.g., by fibroblasts, or by migrating tumor and endothelial cells) can release these matrix-bound factors, provoking new tumor and stromal cell responses. The mechanics of the ECM also plays a role: stiffer ECM can promote increased migration and proliferation, whereas softer matrices can down-regulate proliferation and motility [7]. Moreover, the stiffness and density of the ECM affect the speed of tumor cell migration. Even the geometry of the ECM matters: tumor cells use completely different migratory mechanisms on 2-D surfaces (e.g., basement membranes) and within 3-D matrix scaffolds [8]. Tumor cells can change their phenotype (current behavioral characteristics) based upon adhesive contact with 2-D or 3-D matrix and other cells [9].

Hence, a tumor is in constant, dynamical communication with the microenvironment through biochemical and biophysical processes. The microenvironment shapes tumor cell behavior, even while tumor cells reshape the microenvironment directly (e.g., by matrix remodeling) and indirectly (e.g., by secreted signals). Tumor growth cannot be understood in isolation—it must be studied as a 3-D multicellular *system*, in the presence of a dynamical biochemical and biophysical environment. In spite of recent advances in biomimetic materials, bioengineered tissues, and animal models, the complex tumor-tissue system is difficult to study solely through experiments.

Computational modeling, however, can provide a platform to ask questions and test new hypotheses on this complex system. To study cancer, a 3-D multicellular simulation platform should:

- simulate the birth, death, and motion of tumor cells,
- simulate biochemical microenvironments with multiple diffusing substrates,
- simulate the biomechanics of cells and the extracellular matrix,
- simulate the evolving blood vasculature,
- simulate interstitial and microvascular flow,
- integrate the above models, along with molecular-scale models to drive cell phenotype,

- integrate high-throughput experimental data to calibrate and validate models, and
- do so reproducibly, using interoperable data formats.

In this chapter, we shall discuss progress by our groups on these problems, and outline key steps for advancing from models of individual tumor and microenvironment subsystems, to true 3-D multicellular simulation systems that are adequate for attacking the complexity of cancer.

12.2 Progress Towards 3-D Multicellular Systems Biology

12.2.1 Simulating Tumor Growth in a Heterogeneous Microenvironment

Our earliest work with John Lowengrub and collaborators [10–14] focused on describing the evolving morphology of tumors, as driven by gradients of growth substrate. Building upon work by Cristini, Lowengrub and Nie [15], we developed a sharp interface model, where a well-defined tumor boundary Σ is represented with a level set function [16, 17] φ, satisfying $\varphi<0$ inside the tumor, $\varphi=0$ on the boundary, and $\varphi>0$ outside the tumor. Tumor tissue was assumed incompressible with constant cell density. $\nabla \cdot \mathbf{u}$ gives the net rate of tissue creation, where $\mathbf{u}$ is the tissue velocity field. The tumor boundary moved with normal velocity $V = \mathbf{u}\cdot\mathbf{n}$. Cell proliferation was proportional to available growth substrate σ, which diffused from the tumor boundary. (See Sect. 12.2.2 below.) Wherever the growth substrate concentration was below a threshold σ_N, tumor cells became necrotic, giving one of the first detailed models of necrosis [11, 12].

We modeled tumor tissue mechanics with a tissue pressure p that obeyed Darcy's law (porous flow), simulating tumor tissue as a fluid moving through the ECM. The Darcy coefficient μ modeled several biophysical effects, including cell-matrix adhesion and matrix stiffness. Combining Darcy's law with the incompressibility assumption gave an elliptic partial differential equation (PDE) for the pressure; a curvature boundary condition (surface tension) modeled cell-cell adhesion. These level set models took the general form:

$$\nabla \cdot \mathbf{u} = \begin{cases} G(\sigma - A) & \text{inside the viable tumor} \\ -G_N & \text{in the necrotic core} \\ 0 & \text{elsewhere} \end{cases}$$

$$\mathbf{u} = -\mu \nabla p + \text{chemotactic terms (where needed)}$$

$$V = \mathbf{u}\cdot\mathbf{n} \text{ on the tumor boundary, and}$$

$$\frac{\partial \phi}{\partial t} = -\tilde{V}\,|\nabla \phi|\,,$$

where $\tilde{V}$ is an extension of V off the tumor boundary Σ. In the work above, G is a "tumor aggressiveness" parameter that combines the effects of cell-cell adhesion and cell proliferation, A is the relative rate of tissue loss due to apoptosis, and G_N is the rate of tissue loss in necrotic regions [13]. Constitutive relations between the model parameters and microenvironmental factors could be used to model molecular-scale biological effects. See Sect. 12.2.3 for further discussion.

A typical simulation result can be found in Fig. 12.1a [15]. Our later work [10–14] improved the biological accuracy (by separately tracking the position of the tumor's necrotic core [13] and allowing the substrate diffusivity and Darcy coefficient to vary spatially [14]) and numerical accuracy (a more robust curvature discretization [12], a more accurate jump boundary condition discretization, and a faster numerical solver for steady-state diffusion [14]). These improvements allowed simulation of tumor growth in complex tissues, such as brain tissues with white and grey matter, cerebrospinal fluid, and cranium [10, 14]. See Fig. 12.1b.

For improved modeling of tumor tissue mechanics, Wise, Lowengrub, Frieboes, Cristini, and others developed "phase field" or "mixture" models: each mesh site was modeled as a mixture of one or more cell types, matrix, and water [21, 22]. Each phase of this mixture was

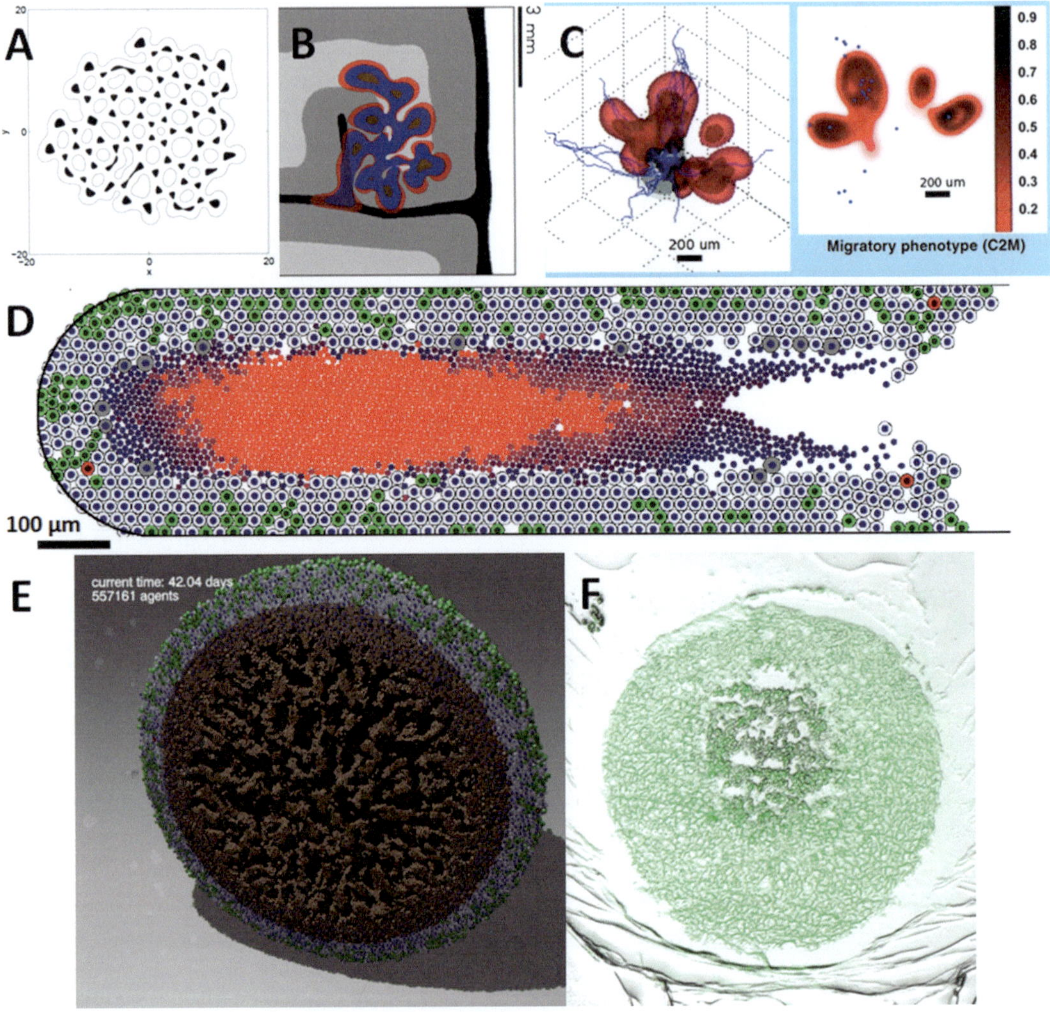

Fig. 12.1 (**a**) A level set simulation of a tumor with viable regions (*white*) and necrotic tissue (*black*). The tumor shape can undergo complex topological changes, based upon the balance of growth and mechanics parameters (Adapted with permission from [11]). (**b**) Numerical refinements allowed simulation of growth in heterogeneous tissues, such as this simulated brain tumor. Red regions are proliferating, blue regions are hypoxic, and brown are necrotic. The brain tissue has white matter (*light grey*), grey matter (*dark grey*), cerebrospinal fluid (*black*), and cranium (*white*) (Adapted with permission from [52]). (**c**) A phase field simulation of a highly-motile subclone (*red*) emerging due to hypoxic signaling from a glioblastoma (*grey*) [7] (Adapted with permission from [7]). (**d**) Agent-based models—like this patient-calibrated simulation of ductal carcinoma in situ (DCIS) [19]—can simulate small-scale tissue mechanics, with more direct calibration to experimental and clinical data (Adapted with permission from [19]). (**e**) The agent-based model has been extended to 3D [29]. Here, we plot a cut-away view of a necrotic tumor spheroid. Green cells are proliferating, gray cells are quiescent, red cells are apoptotic, and brown cells are necrotic. Note the "crackly" structure in the necrotic core. (**f**) A hanging tumor drop spheroid (HCC827 non-small cell lung carcinoma) showing a similar structure in the necrotic center. Image courtesy Mumenthaler lab, Lawrence J. Ellison Institute for Transformative Medicine, University of Southern California

governed by conservation laws for mass and momentum; energy laws were used to govern mixing between the phases. The approach led to the introduction of Cahn-Hilliard equations of the form:

$$\frac{\partial \rho_i}{\partial t} + \nabla \cdot (\mathbf{u}_i \rho_i + \mathbf{J}_i) = S_i, \quad i \in \{V, D, H\}.$$

The rate of change in the density ρ_i of cell species i (V: viable tumor; D: dead tumor; H: host) is determined by balancing net creation (S_i: proliferation minus cell death) with cell advection ($\nabla \cdot (\mathbf{u}_i \rho_i)$) by its velocity field $\mathbf{u}_i$, and cell-cell and cell-ECM mechanical interactions ($\nabla \cdot \mathbf{J}_i$), where the flux $\mathbf{J}_i$ generalized Fick's law to include adhesion, cell incompressibility, chemotaxis, haptotaxis, and other biomechanical effects [21, 22].

The viable tumor cell density ρ_V increased through proliferation and decreased through apoptosis and necrosis. We assumed that normal host cells (ρ_H) do not proliferate but can apoptose (A) or necrose (N); the total dead cell density is ρ_D. These primarily affect tumor mass through water transport in the tissue; their solid fraction is neglected for simplicity [22]. Proliferation was assumed to increase with nutrient substrate σ above a threshold level σ_{N} [22], resulting in the creation of cells by removing the equivalent water volume from the interstitium. Cells experiencing a substrate level below σ_{N} were considered quiescent (e.g., due to hypoxia). Apoptosis transferred cells from the viable tumor and host cell species to the dead cell species, where cells degraded and released their water content. Necrosis occurred when the nutrient substrate concentration falls below the threshold σ_N and ultimately releases cell's water content. The resulting model is

$$\begin{aligned} S_V &= (\lambda_{M,V}(\sigma - \sigma_N) - \lambda_{A,V}) H[\sigma - \sigma_N] \rho_V \\ &\quad - \lambda_{N,V} H[\sigma_N - \sigma] \rho_V \\ S_D &= H[\sigma - \sigma_N](\lambda_{A,V}\rho_V + \lambda_{A,H}\rho_H) \\ &\quad + H[\sigma_N - \sigma](\lambda_{N,V}\rho_V + \lambda_{N,H}\rho_H) - \lambda_D \rho_D, \end{aligned}$$

where $\lambda_{\mathrm{M},i}$, $\lambda_{\mathrm{A},i}$, and $\lambda_{\mathrm{N},i}$ are mitosis, apoptosis, and necrosis rates, λ_{D} is the cell degradation rate (with different values in apoptotic and necrotic tissue), and H is the Heaviside "switch" function.

Each cell species moves under the balance of proliferation-generated oncotic pressure, cell-cell and cell-ECM adhesion, chemotaxis (due to substrate gradients), and haptotaxis (due to gradients in the ECM density). The motion of cells and interstitial fluid through the ECM is modeled as flow of a viscous, inertialess fluid in a porous medium. We made no distinction between interstitial fluid hydrostatic pressure and mechanical pressure due to cell-cell interactions. Cell velocity is a function of cell mobility μ_i and tissue oncotic (solid) pressure (Darcy's law); cell-cell adhesion is modeled with an energy approach from continuum thermodynamics [22]. For simplicity, the interstitial fluid is modeled as moving freely through the ECM at a faster time scale than the cells. These assumptions yield a constitutive relation for the tumor tissue velocity field $\mathbf{u}_i$:

$$\begin{aligned} \rho_i \mathbf{u}_i + \mathbf{J}_i = &- \mu_i(\rho_i, f)\left(\nabla p - \sum_j \frac{\gamma_j \delta E}{\delta \rho_j} \nabla \rho_j\right) \\ &+ \chi_\sigma(\rho_i, f, \sigma) \nabla \sigma \\ &+ \chi_h(\rho_i, f) \nabla f, \quad i \in \{V, D, H\}. \end{aligned}$$

The variational derivative $\delta E/\delta \rho_i$, combined with the remaining contributions to the flux $\mathbf{J}$ (due to pressure, haptotaxis, and chemotaxis; see [22]), yields a generalized Darcy-type constitutive law for the cell velocity $\mathbf{u}_i$ of a cell species i, determined by the balance of proliferation-generated oncotic pressure p, cell-cell and cell-ECM adhesion, chemotaxis (due to gradients in the cell substrates σ), and haptotaxis (due to gradients in the ECM density f). The Darcy coefficient μ_i is cellular mobility, reflecting the response to pressure gradients and cell-cell interactions by breaking integrin-ECM bonds and deforming the host tissue. γ_j is the cell adhesion parameter, and χ_σ and χ_h are the chemotaxis and haptotaxis coefficients, respectively.

Solving this system required sophisticated numerical techniques [23], but the work was worthwhile: it allowed modeling new tissue biomechanics (see Sect. 12.2.3 below) to address drawbacks in the level set approach [11–14]. In particular, the phase field model could simulate mixed populations of tumor sub-clones without sharp boundaries between them. In [18], we simulated the 3-D growth of glioblastoma multiforme using the new phase field model. In the work, hypoxic tumor cells could mutate into a more motile subclone, modeled as a new phase in the phase field model. See Fig. 12.1c. In Sect. 12.2.3, we give another example of this model for simulating lymphoma [24]. Similar models were also developed to account for cell type and mechanical response heterogeneity of the solid and liquid tumor phases. See the review [25] for further discussion.

One difficulty for continuum models is that individual cell phenotypes (particularly for heterogeneous cell populations) cannot be fully resolved at the continuum scale. Allowing model parameters to vary at cell-scale resolution (e.g., 20 μm) and solving for the cell densities at cell-scale resolution rather than tissue-scale resolution (e.g., 100–200 μm) can result in small protrusions and other features that, while numerically accurate, are in violation of the models' continuum hypotheses; these cannot be regarded as meaningful scientific results. Another difficulty for continuum models is matching to experimental and clinical data. Parameters such as μ in the models above incorporate multiple biophysical and biological effects, so calibration may require iteratively testing the model parameters until shape and other metrics match data at multiple time points. Such matching risks overfitting an underconstrained model, bringing scientific conclusions into doubt. "Bottom-up" calibration from direct cell-scale measurements can help overcome these problems, but direct mappings of such cell-scale measurements onto multiple-effect parameters are unclear. Both these difficulties can be addressed with cell-scale (discrete) models.

In [19], we developed an agent-based model of cancer, with application to ductal carcinoma in situ of the breast (DCIS). In this work, tumor cells in a duct (represented as a level set function) can be quiescent (Q), cycling (P), apoptotic (A), or necrotic (N) in regions of insufficient oxygen. Tumor cells obeyed conservation of momentum, with cell motion determined by a balance of adhesive and "repulsive" forces exchanged with other cells and the duct wall and fluid drag. The probability of changing state (Q, P, A, and N) depended upon the microenvironmental conditions. Cycling cells divided and regrew volume, apoptotic cells shrunk, and necrotic cells shrunk and calcified; this was the first model of calcifications in DCIS with comedonecrosis (a centralized core of necrotic material, which may be partially calcified) [5]. See Fig. 12.1d for a typical DCIS simulation. Individual cell phenotypes can be clearly observed and tracked (green cycling cells, gray quiescent cells, red apoptotic cells, and a central core of necrotic cells in varying states of degradation and calcification). A movie of this simulation can be found at [26].

We recently extended the model to 3D and increased its simulation capacity from thousands of cells to 10^5 to 10^6 cells on desktop workstations. The extended code (PhysiCell: physics-based cell simulator) is being prepared for a 2016 open source release [20]. See Fig. 12.1e for a PhysiCell simulation, showing a cross-section of a 3-D hanging drop spheroid with a necrotic core. The competing effects of the 3-D multicellular geometry, necrotic cell contraction (from water loss), and necrotic cell adhesion result in a fractured necrotic core structure, which can be observed in experimental data (Fig. 12.1f). For PhysiCell project updates, see http://PhysiCell.MathCancer.org. Several other open source model frameworks can simulate many cells in 3D, including Chaste [27], CompuCell3D [28], Morpheus [29], and iDynoMiCS [30]. In Sect. 12.2.5, we shall revisit the agent-based model in the context of direct calibration to experimental and pathology data.

12.2.2 Simulating the Chemical Microenvironment with Many Substrates

Substrate transport is critical to representing the tumor microenvironment. Growth substrates are released by the vasculature, transported within the tissue, and consumed by cells, which subsequently release metabolic waste products. Cells exchange diffusible signaling factors that alter phenotype. Apoptotic and necrotic cells may also release diffusing substrates that affect phenotype [5]. In our earliest work [11–14], we modeled substrate transport as quasi-steady, requiring a steady state solution after evolving the tumor morphology. Thus, we solved PDEs of the form

$$0 = \nabla\cdot(D\nabla\sigma) - \lambda_{\text{background}}\sigma - \lambda(\text{cells}) + \text{Sources}$$

with appropriate boundary conditions. In the first level set models [10–14, 31], we imposed a Dirichlet boundary condition on the tumor boundary to simulate growth into a well-vascularized tissue. Later, we embedded the evolving tumors into a larger domain with substrate diffusion and Dirichlet conditions as a far-field condition, to model growth into a locally-affected region [13]. We used a jump boundary condition to enforce continuity of substrate flux across the tumor boundary, using the ghost fluid method [11]. In 2D, the method yielded a banded linear system, which we solved with the stabilized biconjugate gradient method. (See the references in [11].) However, this technique was slow, not terribly robust, and difficult to extend to more accurate discretizations of the jump boundary conditions. In [14], we introduced a pseudotime τ:

$$\frac{\partial\sigma}{\partial\tau} - \nabla\cdot(D\nabla\sigma) \quad \lambda_{\text{background}}\sigma - \lambda(\text{cells}) + \text{Sources}$$

and solved to steady state using a semi-implicit finite difference scheme. The method was stable and second-order accurate in the steady-state solutions, but it could not (and was not designed to) capture the dynamics of the evolving substrate distribution. Many models (e.g., [32]) solve the time-dependent problem using explicit finite difference methods. While these methods are straightforward to implement and accurate, their stringent time step restrictions make them unfeasible for 3-D simulations or for simulating many substrates. Implicit methods like ADI (alternating directions implicit, method; see [33] for a model using this method) can remove the time step restrictions, but they often require linking to linear algebra libraries that can complicate cross-platform compatibility. (See the discussion in [34].) In [21, 22], Wise, Lowengrub, Frieboes, and co-workers took an alternative approach by using a fully adaptive, nonlinear multigrid/finite difference method to efficiently solve the equations [23].

One drawback of these approaches is that they do not scale very well to larger numbers of substrates, particularly in large 3-D domains. Each evolving substrate distribution requires solving a PDE. Most of these codes—including those above—solve the PDEs sequentially, and so simulating ten diffusing substrates requires ten times the computational effort of simulating one. This does not scale well as the number of substrates and the domain size are increased.

We recently addressed these problems by creating BioFVM (finite volume method for biological problems), an open source 3-D diffusion solver that was designed for both standalone simulations and for integration with existing simulation packages [34]. See http://BioFVM.MathCancer.org. BioFVM solves for a family of diffusing substrates vectorially:

$$\frac{\partial \mathbf{p}}{\partial t} = \mathbf{D} \circ \nabla^2 \mathbf{p} - \boldsymbol{\lambda} \circ \mathbf{p} + \mathbf{S} \circ (\mathbf{p}^* - \mathbf{p}) - \mathbf{U} \circ \mathbf{p} + \sum_{\text{cells } i} (\mathbf{S}_i \circ (\mathbf{p}_i^* - \mathbf{p}) - \mathbf{U}_i \circ \mathbf{p})\, \delta_i V_i$$

where $\mathbf{p}$ is a vector of diffusing substrates, $\mathbf{D}$ and $\boldsymbol{\lambda}$ are vectors of diffusion and decay constants (respectively), $\mathbf{S}$ and $\mathbf{U}$ terms are vectors of source and uptake rates (respectively), $\mathbf{p}^*$ terms are vectors of saturation densities, and all products ($\circ$) are component-wise. Here, V_i is the volume of the ith cell in a simulation environment, and δ_i is a Dirac delta function centered at its position.

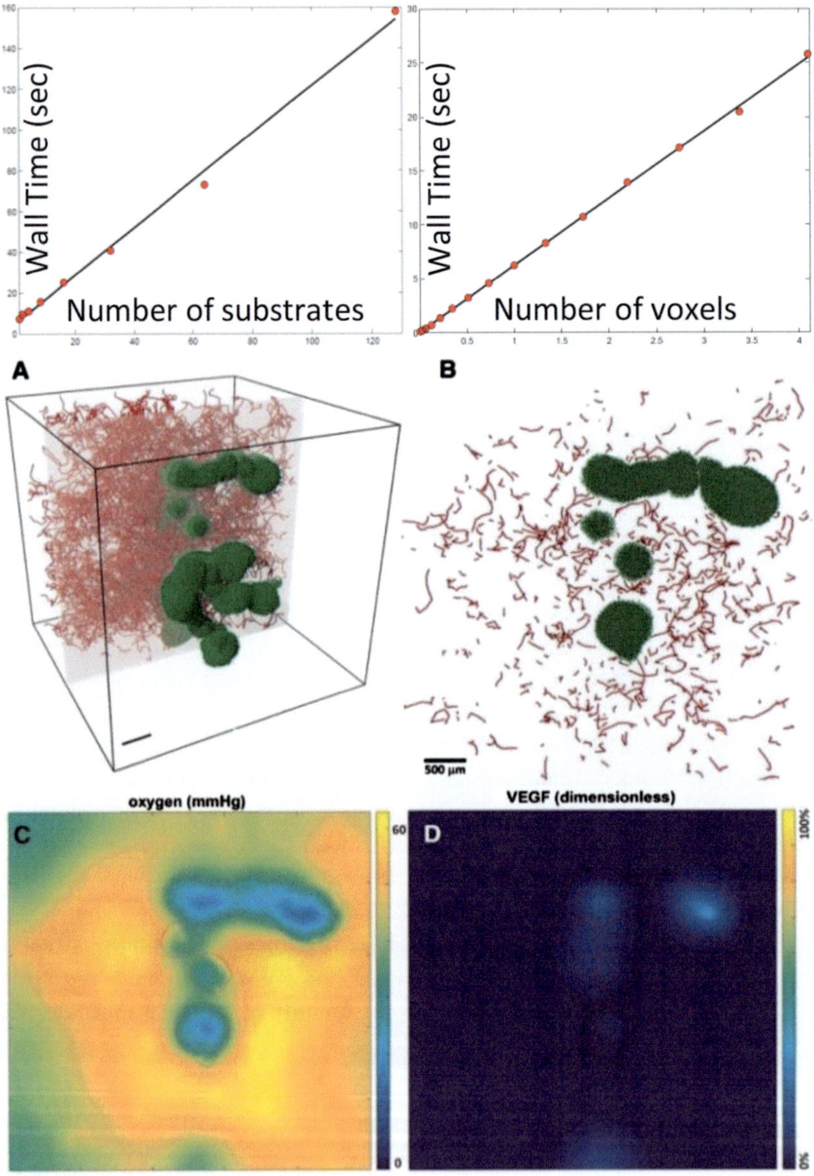

Fig. 12.2 *Top*: BioFVM shows roughly linear cost scaling with the number of substrates and voxels. (See the supplementary material in [34]). *Bottom*: Simulation of O_2 release by blood vessels (*red line sources* in **a**, **b**), diffusion, and consumption by tumor cells (*green*), and VEGF release by hypoxic tumor cells. (**c**, **d**) plot the pO_2 and VEGF distributions in the gray plane in (**a**) (Adapted with permission from [34])

By using a combination of operator splitting, vectorized operations, tailored ODE and Thomas solvers, specialized overloaded operators to reduce the memory-bound costs, and OpenMP parallelization, BioFVM can simulate diffusion of ten or more substrates on 3-D domains with a million or more voxels (enough to simulate diffusion in a 8 mm^3 tissue at 20 μm resolution), using just a desktop computer. Simulating ten substrates requires only 2.6 times more effort than simulating one, and we have simulated up to 128 substrates. See Fig. 12.2 (top). The code is first-order accurate in time, second-order accurate in space, and numerically stable. Computational time scales linearly with the number of substrates, number of voxels, and the number of cell-based source/sink terms. In Fig. 12.2 (bottom), we simulate oxygen and VEGF transport in a 125 mm^3 tissue (15,625,000 voxels) with over 2.8 million discrete source/sink terms, using a quad-core desktop workstation. Larger simulations are be feasible on individual HPC compute nodes, which often have 16 computational cores (32 virtual CPUs when including hyperthreading) and 32 GB of memory.

12.2.3 Simulating the Physical Microenvironment

In [13], we performed our first investigations of the impact of the microenvironment on tumor shape. We enclosed the tumor in a larger circle of tissue where growth substrate σ diffused with coefficient D_{tissue} (the diffusion coefficient was set at D_{tumor} inside the tumor), and the mechanical pressure p dissipated with Darcy coefficient μ_{tissue} outside the tumor and μ_{tumor} inside the tumor. We assumed that the substrate was continuous across the tumor boundary Σ with no jump in flux ($D_{\text{tumor}} \nabla\sigma \cdot \mathbf{n} - D_{\text{tissue}} \nabla\sigma \cdot \mathbf{n} = 0$ on Σ). We modeled the mechanical pressure as solving $p_{\text{tumor}} - p_{\text{tissue}} = k$ (an extension of the curvature-based cell-cell adhesion in [11, 12]), with no jump in the flux of the pressure across the tumor boundary ($\mu_{\text{tumor}} \nabla p \cdot \mathbf{n} - \mu_{\text{tissue}} \nabla p \cdot \mathbf{n} = 0$ on Σ). We varied by the microenvironmental conditions through the ratios

$$D = \frac{D_{\text{tissue}}}{D_{\text{tumor}}} \text{ and } \mu = \frac{\mu_{\text{tissue}}}{\mu_{\text{tumor}}}.$$

Large values of D represented growth into a well-oxygenated tissue; large values of μ modeled growth into a soft, mechanically-compliant tissue. We found that the (D,μ) parameter space divided into three phases: tumors broke into small fragments when growing in poorly-oxygenated tissues (low D), whereas they developed invasive fingers in mechanically stiff, vascularized tissues (high D, low μ). In well-oxygenated, soft tissues (high D, high μ), relatively compact tumor shapes emerged. See Fig. 12.3a. These results were similar to cellular automaton model findings by Anderson and co-workers [32, 37], which examined the impact of oxygen heterogeneity and cell-cell adhesion upon tumor shape stability, but without modeling tissue mechanics.

In [10, 14, 31], we extended the level set framework to let μ vary continuously across the tissue, as a model of mechanical variations. In our first such studies [10, 14] (tumors growing in brain tissue), the Darcy coefficient μ was set to a higher value ($\mu = 1.5$) in relatively pliant white matter, a lower value ($\mu = 0.5$) in stiffer grey matter, a high value ($\mu = 10$) in cerebrospinal fluid, and a very low value ($\mu = 0.0001$) in non-deformable cranium. In the simulations, tumor grew preferentially in regions of higher μ, and mechanical pressure built up between the growing tumor and the cranium. See Fig. 12.1b. In later work [31], we studied tissue remodeling by simulating the release of a diffusing matrix-degrading enzyme that removed ECM (E). Moreover, we introduced a constitutive relation between the ECM density and the Darcy coefficient:

$$\mu = \frac{\mu_0}{a + bE},$$

so that increasing ECM density decreased the Darcy coefficient, thus increasing the overall tissue stiffness and reducing its permeability to cells. Interestingly, this constitutive relation can be obtained mechanistically by considering the individual force terms in agent-based models discussed above, under the inertialess assumption that dissipative forces like fluid and ECM drag balance quickly with adhesive, motile, and other forces. (See the supplementary materials in [19].) A movie of an example simulation including the effects of tissue degradation and ECM-dependent cell mobility can be found at [38]. We note that later, higher-resolution studies of

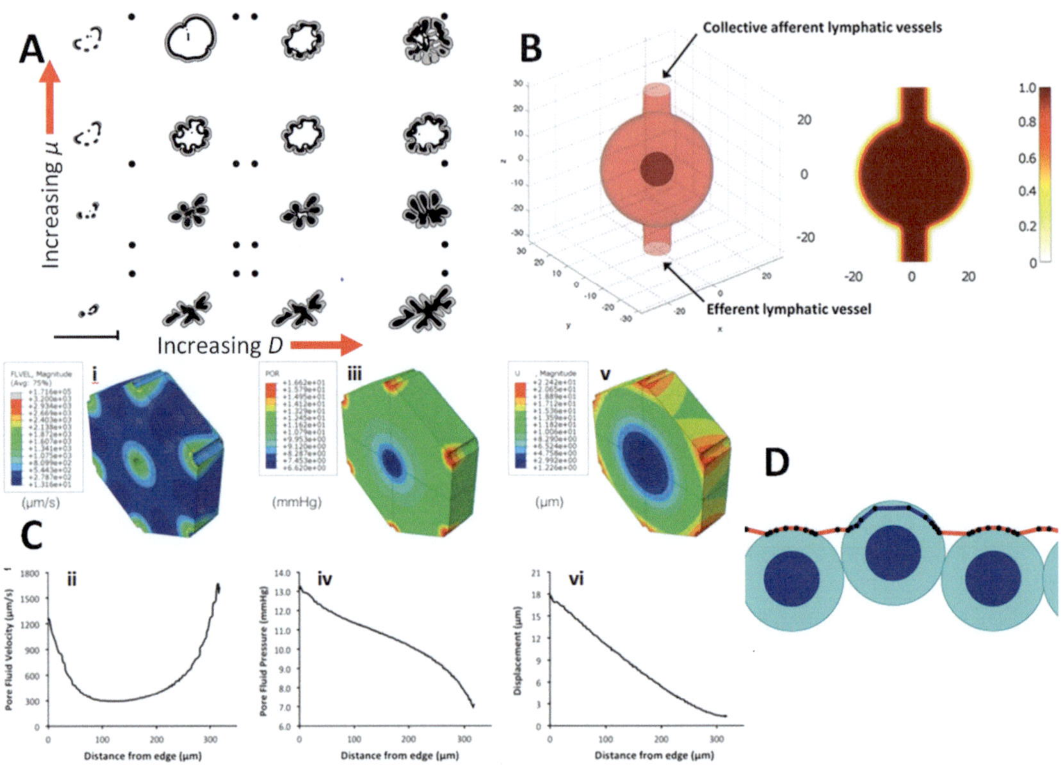

Fig. 12.3 (**a**) A phase diagram of tumor shape (*grey regions* are viable; *black regions* are necrotic) of the (D,μ) parameter space [13]. Increasing D models increasing vascularization in the tissue surrounding the tumors; increasing μ models decreasing stiffness in the surrounding tissue. Three characteristic morphologies are seen: fragmenting (low D), fingering/invasive (high D, low μ), and compact growth (high D, high μ) (Adapted with permission from [13]). (**b**) Simulation of lymphoma in a lymph node [25], using the phase field method to represent the live and dead tumor cell populations, fluid, and the lymphatic vessel outer wall (Adapted with permission from [25]). (**c**) A poroviscoelastic (PVE) simulation of perfused liver tissue [66]. **i,ii**: pore fluid velocity. **iii,iv**: pore fluid pressure. **v,vi**: solid matrix deformation (Adapted with permission from [66]). (**d**) Discrete-scale models [13] can investigate smaller-scale tissue biophysics, such as this simulation of passive cell invasion through a weakened section of basement membrane (Adapted with permission from [13])

matrix degradation found that common reaction-diffusion type models cannot adequately model tissue remodeling without including the impact of activation factors and inhibitors (otherwise, they over-predict tissue degradation), and that the more common membrane-bound matrix metalloproteinases should not be modeled as diffusing species, even with very small diffusion coefficients [39].

While the level set models were able to simulate changing tumor shapes in heterogeneous, evolving microenvironments, they had several shortcomings. The tissue biomechanics were constrained by the constant cell density assumption. Cell-to-cell adhesion could not vary throughout the cell tissues (since it was modeled as a surface tension). Individual cell effects were difficult to include, particularly on smaller scales. While the sharp interface assumption was well-suited to strongly-adherent epithelial tumors, it could not model low-adhesion, highly motile cell species such as glioblastoma cells [40] and stromal cells [41]. Hence, the sharp interface tumor model could not readily apply to other cell types of interest in cancer biology, making it difficult to study the biomechanical dynamics of tissue remodeling and the impact on tumor growth. Even in cases where the sharp interface assumptions were valid, new cell populations could only be modeled by introducing new level set functions, with a large increase in computational cost.

Phase field/mixture models (see Sect. 12.2.1) can simulate sophisticated tissue biomechanics in more generalized cases where different cell types are mixed without sharp boundaries, or where the extracellular matrix itself must be evolved. For example, in [24], the lymph node was modeled as a surface Γ that is stretched by the growing tumor, with membrane normal velocity proportional to the proliferation-generated pressure gradient. It was assumed that the tissue surrounding the organ can be deformed sufficiently to accommodate expansion [42]. The geometry is described by a phase field variable ψ governed by a modified Cahn-Hilliard equation:

$$\frac{\partial \psi}{\partial t} + \mathbf{v} \cdot \nabla \psi = \nabla \cdot \left(B(\psi) \nabla \mu_{\psi} \right), \text{ where } \psi(\mathbf{x}, t) = \begin{cases} 1 \text{ inside the lymph node} \\ 0 \text{ outside the lymph node} \end{cases}$$

specifies the position of the interface Γ through the narrow transition region characterized by a thickness parameter ε in the Cahn-Hilliard potential μ_{ψ}. $B(\psi)=36\,\psi^2(1-\psi)^2$ specifies the interfacial region [43, 44] where the Cahn-Hilliard potential takes effect. The surface Γ is advected by the mechanical pressure (P) generated by tumor proliferation and the tissue surface tension [22]:

$$\mathbf{v} = -\mu \left(\nabla P - \frac{\gamma}{\epsilon} \mu_{\psi} \nabla \psi \right),$$

where γ specifies the strength of the surface tension, and μ is the tissue mobility in response to these exerted forces. In Fig. 12.3b, we show an example of simulating non-Hodgkin's lymphoma in a lymph node from [24]. The phase ψ was used to represent the outer wall of the lymph node as discussed above, allowing later simulation of lymph node swelling—a common feature in lymphoma and metastatic carcinoma [45, 46].

The phase field mixture approach allows generalized modeling of a wide range of tissue mechanics, with separate constitutive relations and parameters for each phase of the simulated cell-ECM-fluid mixture. However, water was modeled as flowing freely through the simulated domain, and hence decoupled from the evolving ECM and cell phases. In some tissues, advective interstitial and microvascular flow couple significantly with the solid components of the tissue, causing deformation. Tissues in this flow regime can demonstrate viscoelastic properties [47].

Continuum models of flow through deformable porous media [48] are computationally efficient and do not require precise spatial and geometric information about every fiber or cell in a tissue [49]; hence they are well-suited for modeling perfusion in porous materials [50]. Similarly to the phase field model, poroviscoelastic (PVE) models use a continuum approximation to simulate both tissue mechanics and pore fluid behavior in tissue parenchyma. PVE is an extension of poroelastic or biphasic theory, to model material as a porous fluid-saturated linear elastic solid in which the fluid flows relative to the deforming solid [35, 48]. A poroviscoelastic model incorporates time-dependent (viscous) effects from two different sources: pore fluid movement through the matrix and intrinsic viscoelasticity of the solid matrix itself [51]. Thus, PVE models are well-suited for materials which exhibit significant viscoelastic behavior, such as liver [47], brain [52], or cartilage [51]. Since PVE models predict pore fluid pressure and velocity in addition to solid matrix stress, this modeling strategy is attractive for examining perfused tissue in both native and decellularized states [53, 54]. For example, we recently characterized the biomechanical response of perfused native and decellularized liver on both macroscopic and microscopic length scales via spherical indentation tests, then used PVE finite element models to extract the fluid and solid mechanical properties from the experimental data [53]. In another recent study, we used PVE modeling to predict lobule-scale stresses and deformations associated with experimental perfusion rates for native and decellularized livers [35]; the work was able to effectively predict flow rates and mechanical deformation in both decellularized and native liver tissues. See Fig. 12.3c. On the whole, PVE theory offers an effective technique for determining the tissue-scale spatial distribution of key microenvironmental variables related to ECM mechanics and interstitial flow.

Discrete models can simulate tissue mechanics while incorporating localized, micron-scale biology, particularly for thin basement membranes that cannot be accurately modeled with tissue-scale continuum models [39]. In [36], we developed a 2-D discrete plasto-elastic model of basement membrane. The membrane was written as a linked list of basement membrane agents centered at $\mathbf{x}_k$, each of which was linked to two neighboring agents at $\mathbf{x}_{k-1}$ and $\mathbf{x}_{k+1}$. The total force $\mathbf{F}^k$ acting on the portion of basement membrane at $\mathbf{x}_k$ was modeled as:

$$\mathbf{F}^k = \sum\nolimits_{\text{cells } i} \left(\mathbf{F}^{ik}_{\text{cba}} + \mathbf{F}^{ik}_{\text{cbr}}\right) + \mathbf{F}^k_{\text{BM}} + \mathbf{F}^k_{\text{ECM}} - \nu \mathbf{v}_k,$$

where $\mathbf{F}_{\text{cba}}{}^{ik}$ and $\mathbf{F}_{\text{cbr}}{}^{ik}$ are the cell-BM adhesive and repulsive forces, respectively (modeled with potential functions as in [19]), $\mathbf{F}_{\text{BM}}{}^{k}$ is the force exchanged with neighboring portions of basement membrane, $\mathbf{F}_{\text{ECM}}{}^{k}$ is the elastic force between the portion of basement membrane and the nearby stroma, and $\upsilon \mathbf{v}_k$ is the dissipative, drag-like force acting against the basement membrane's velocity $\mathbf{v}_k$. We modeled $\mathbf{F}_{\text{BM}}{}^{k}$ as elastic:

$$\mathbf{F}^k_{\text{BM}} = -\frac{K_{k+1}\left(\ell_{k+1} - \ell_{k+1,0}\right)}{\ell_{k+1}}\left(\mathbf{x}_k - \mathbf{x}_{k+1}\right) - \frac{K_k\left(\ell_k - \ell_{k,0}\right)}{\ell_k}\left(\mathbf{x}_k - \mathbf{x}_{k-1}\right)$$

where K_k is the elastic modulus of the basement membrane at $\mathbf{x}_k$, and ℓ_k and $\ell_{k,0}$ are the current and resting lengths of the basement segment centered at $\mathbf{x}_k$, respectively. We introduced additional constitutive relations to relate the elastic modulus K_k to the amount of material present in the basement membrane section at $\mathbf{x}_k$ and to the thickness of the basement membrane at that section; see [36] for further details. While the basement membrane is elastic over relatively short time scales, it can undergo plastic rearrangement over longer time scales as elastic fiber cross-links break and reform. We modeled this as an evolution of the resting length $\ell_{k,0}$:

$$\frac{d\ell_{k,0}}{dt} = r\ell_k \max\left(0, \ell_k - \ell_{k,0} - \tau_{\text{act}}\right)$$

where τ_{act} is a threshold stress level above which ECM cross-links begin to break [55].

An example of this model can be seen in Fig. 12.3d. We were able to model small-scale interactions between basement membrane and cells. In particular, we found that if a small section of basement membrane is weakened (by reducing the amount of matrix material in the kth BM agent, and then reducing its elastic modulus via the constitutive relation), then passive elastic forces alone can result in epithelial cell protrusion into the stroma (Fig. 12.3d) [36]. In [39], we used the framework to investigate the time scales of basement membrane degradation by matrix metalloproteinases, finding that realistic, 100 nm thick basement membranes can readily be penetrated in just 10–15 min. However, these models have proven difficult to implement efficiently in 3D. Moreover, solving basement membrane mechanics with micron-scale resolution is not scientifically meaningful when cell morphologies are not resolved, as in our present agent-based models. (These resolve cell position and volume, but not morphology.)

Other groups have addressed this problem by modeling extracellular matrix at the multicellular scale using modified agents. For example, to emulate the invasion of breast cancer from a duct to the surrounding stroma, Bani Baker and coworkers used small agents with different properties to model both basement membrane (BM) and extracellular matrix (ECM) [56]. Modeling ECM as a matrix of small particles allows changing the ECM structure by varying the number and type of interactions between matrix particles. For example, to model the ECM stiffening as the result of lysyl oxidase presence [57], one can increase the crosslinking between the particles.

12.2.4 Simulating the Evolving Microvasculature and Interstitial Flow

In [58–60], Anderson, Chaplain, and McDougall developed a sophisticated 2-D cellular automaton model of tumor-driven sprouting angiogenesis. In the model, each lattice site on a regular Cartesian mesh could contain ECM, a tumor cell, an endothelial cell, or a migrating sprout tip. Blood vessels released oxygen, which diffused through the tissue and was consumed by tumor cells. Hypoxic tumor cells released VEGF, which diffused through the tissue and could activate endothelial cells and "convert" them to migrating sprout tips. The sprout tips followed a random walk up gradients of VEGF (chemotaxis) and ECM (haptotaxis) to form new vessels by leaving a trail of endothelial cells behind them. They solved for blood flow in the connected network of endothelial cells, including the nonlinear effects of the plasma and solid hematocrit phases of the blood. Sheer stresses drove network remodeling.

In [31], we coupled the level set tumor growth model (see Sect. 12.2.1) to this angiogenesis model. As before, the vasculature released oxygen, but with an improved source function: oxygen release was proportional to hematocrit (as an indicator of flow) and the difference between the vascular and tumor pressure. Hypoxic tumor regions released VEGF, tumor tissue could remodel the ECM by secretion and degradation processes, and notably, regions of high tumor pressure could collapse vessels, thereby interrupting flow in the vascular network and creating new regions of hypoxia and renewed angiogenesis. See [31, 61] for a simulation movie. In [62], we extended the work to include interstitial fluid flow. See Fig. 12.4 left. This work provided key insights on coupling the biomechanics and biochemistry of the microenvironment, tumor growth, and angiogenesis. However, it had several drawbacks, most notably being restricted to 2D, and its reliance upon a non-physical Cartesian arrangement of blood vessels.

In [21], we built a 3-D lattice-free model of angiogenesis, building upon this earlier cellular automaton work [58–60] and refining off-lattice models by Plank and Sleeman [63, 64]. The model generates a vascular network dependent on tumor angiogenic factors [65] (e.g., VEGF), implemented via a single continuum variable c reflecting the net balance of pro- and anti-angiogenic regulators. Hypoxic tumor cells released c, which caused endothelial cells to proliferate and grow vessels towards the tumor by haptotaxis and chemotaxis [66]. Migrating endothelial sprout tips were assigned a fixed probability of branching at each time step of the simulation. Vessels were required to form loops (anastomose) before delivering growth substrates [67]; the vessels could connect if a leading endothelial cell crossed the trailing path of another vessel. Tumor proliferation-induced solid tissue pressure could cause vessels to spontaneously shut and regress [68]. We did not model microvascular blood flow or flow-induced changes in the vasculature (e.g., shear stress-induced branching). Instead, we focused on assessing the effect of local heterogeneity of growth substrates on the tumor species. A typical simulation can be seen in Fig. 12.4 right. We have used this model extensively in other investigations of 3-D tumor growth [18, 24].

The flow component in the PVE model (Sect. 12.2.3) is also well-suited to modeling interstitial flow in conjunction with these angiogenesis models. In [35] we used PVE models to create tissue-scale predictions of the distribution of interstitial fluid pressures and velocities across a decellularized hepatic lobule. Models were coupled to varying experimental perfusion flow rates [69]; results for trials at the 9 ml/min flow rate are highlighted here. Terminal and pre-terminal portal vein branches, located at the periphery of the lobular hexagonal prism (Fig. 12.3c) were used as the fluid inlets, and the central vein served as the fluid outlet. Prescribed pressures were applied to inlet and outlet vessel surfaces based on generation-

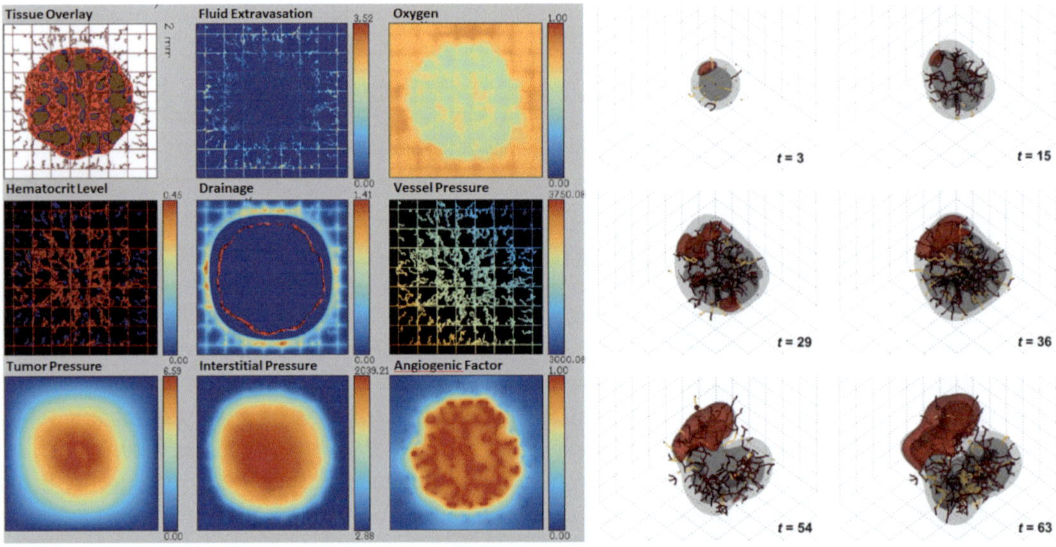

Fig. 12.4 *Left*: Simulation of vascularized tumor growth and interstitial flow from [95], in which we extended our prior coupling [31] of a discrete angiogenesis model [4, 56, 57] with a level set tumor growth model [11–14] (Adapted from [95] with permission). *Right*: We later developed off-lattice, 3-D models of vascular growth [23] that included vascular pruning due to tumor-generated mechanical pressure (Adapted with permission from [23])

specific pressures predicted by a previously developed electrical analog model of liver hemodynamics [35], for the appropriate flow rate. Poroviscoelastic material properties of perfused decellularized liver found in our previous work [53] were used as reported, while hydraulic conductivity was calculated from fluid properties (density, viscosity), vessel geometry, and void ratio data obtained from microscopy images of decellularized liver. Distributions of pore fluid pressure, pore fluid velocity, and ECM strain were then calculated from the finite element simulation. Average pore fluid pressure across the decellularized lobule (1.90 mmHg) agreed well with experimental interstitial fluid pressure for decellularized liver perfused at 9 mL/min (1.95 $\pm$ 1.16 mmHg, [69]). Fluid velocities ranged from 300 to 1700 μm/s over the decellularized lobule, with an average value of 618 μm/s. This result falls at the high end of the 100–750 μm/s plasma velocity range reported for in vivo native liver [70–73], as expected since vascular resistance is lower in decellularized liver compared to native [69]. It was also found that our native model pressure and velocity results for physiological flow rates were consistent with literature values [70–74].

12.2.5 Calibration to and Validation Against Clinical and Experimental Data

If multicellular systems modeling is to have an impact in explaining biological phenomena and predicting tumor growth dynamics, models must be calibrated to and validated against experimental and clinical data. If data are fitted by iteratively refining parameter values to obtain a match, then independent measurements must be used for model validation. In the level set and phase field models presented in Sect. 12.2.1, parameters such as G and μ incorporate multiple biophysical and biological effects. Early calibration efforts (e.g, [75]) calibrated a level set tumor model by (1) estimating the apoptosis and cell division time scales in the apoptosis parameter A, (2) fitting experimental tumor steady radius measurements to theoretical shape stability analyses [15] to constrain the tumor aggressiveness parameter G, and (3) matching growth curves. While the fitted models gave meaningful insights on tumor growth dynamics, we were motivated to move towards more direct calibration techniques. Using the agent-based model [19] described in Sect. 12.2.1, we took the approach of estimating or calibrating a larger set of simpler, biophysically meaningful parameters, most of which could be directly calibrated to cell-scale measurements.

In [19], we developed the first patient-specific calibration technique (for DCIS) that could fully constrain an agent-based model to pathology data from a single time point. After estimating cell cycle, apoptosis, and necrosis time scales, we coarse-grained the model to derive a system of differential equations for the fractions of apoptotic (AI) and proliferative (PI) cells in the viable rim:

$$\frac{d\,[\mathrm{PI}]}{dt} = \langle \alpha_P \rangle \left(1 - [\mathrm{AI}] - [\mathrm{PI}]\right) - \frac{1}{\tau_P}\left([\mathrm{PI}] + [\mathrm{PI}]^2\right) + \frac{1}{\tau_A}[\mathrm{AI}]\,[\mathrm{PI}]$$

$$\frac{d\,[\mathrm{AI}]}{dt} = \langle \alpha_A \rangle \left(1 - [\mathrm{AI}] - [\mathrm{PI}]\right) - \frac{1}{\tau_A}\left([\mathrm{AI}] - [\mathrm{AI}]^2\right) + \frac{1}{\tau_P}[\mathrm{AI}]\,[\mathrm{PI}]$$

where $\langle \alpha_P \rangle$ and $\langle \alpha_A \rangle$ are the mean transition rates for quiescent cells to the proliferative and apoptotic states, respectively, and τ_P and τ_A are previously-estimated durations of the proliferative and apoptotic states. Assuming that the relative fractions of proliferative, apoptotic, and quiescent cells reach a balance after several days in a steady microenvironment [13], we solved the system above to steady state with patient measurements of PI (via Ki-67, a standard nuclear immunohistochemical marker for cell proliferation [76]) and AI (via cleaved Caspase-3, a cytoplasmic marker for apoptosis [77]) to obtain patient-specific estimates of $\langle \alpha_P \rangle$ and $\langle \alpha_A \rangle$. We used a similar approach to calibrate cell-cell mechanical interactions: published experimental data on cell mechanical relaxation were used to estimate the overall cell mechanics timescale, and we used patient-specific measurements of cell density to calibrate the mean cell volume and equilibrium spacing between cell centers. We set the cell-cell adhesive force parameter by setting cell adhesion and repulsion in equilibrium at the mean cell-cell spacing. We estimated oxygen boundary conditions by matching steady-state, radially-symmetric solutions of the oxygen transport equations to the patient's (mean) ductal geometry. We later refined this calibration protocol to better account for cell confluence, and the fact that Ki-67 stains positive not only in cells preparing to divide, but also those in G_1 state immediately after division [78].

We seeded a 2-D section of a patient's breast duct with calibrated DCIS cells and simulated 45 days of growth along a 1.5 mm length of the duct. After several days, a viable rim and necrotic core emerged (See Fig. 12.1d) with sizes consistent with our calibration data. We tracked the leading viable and calcified cells: these positions advanced linearly in time, due to the balance between substrate-limited growth in the viable rim and tissue volume loss in the necrotic center. Moreover, the simulated growth curve predicted that DCIS grows along the ducts at approximately 1 cm/year, similar to prior clinical measurements [79, 80]. The model also predicted a linear mammography-pathology size correlation, and after extrapolating this relationship over two orders of magnitude, the mammography-pathology correlation was consistent with an earlier clinical study [81]. Hence, a "bottom-up" calibration to cell-scale data can yield meaningful tissue-scale predictions. In later work [82], we matched the equations in the coarse-grained agent-based model to the level set model of tumor growth in [15], allowing us to directly calibrate A to pathology scale data without fitting. We then used prior theoretical results [15] to compute the steady-state DCIS resection area, with successful predictions (as validated against post-mastectomy pathology size measurements not used in the calibration process) in 14 of 17 patients.

In [24], we extended the calibration technique to simulation studies of non-Hodgkin's lymphoma. The work aimed to attain a deeper understanding of lymphoma growth in the inguinal lymph node and transport barriers to effective treatment. Cell-scale data were obtained by fine sectioning across whole tumors within lymph nodes, yielding 3-D cell-scale information. After calibrating the phase field model (Sects. 12.2.1

and 12.2.3) to these cell-scale data for two different lymphoma cell lines, the simulations correctly predicted tissue-scale, in vivo observations of growth dynamics and tumor size, without fitting to the data. This work also gave new insights on the physical causes for drug therapy failure in resistant Eμ-myc Arf-/- lines: it found that the cells were more densely packed in the lymph nodes than sensitive lines, thus exacerbating drug delivery gradients. This is a critical consideration when attempting to quantify and predict the treatment response.

These examples demonstrate that computational models can be successfully calibrated to experimental and clinical data, and used to make scientifically and clinically useful predictions. However, most of this work required substantial manual coding to make use of the measurements, which themselves required custom image analysis routines. For multicellular modeling to be both useful and widespread, it must transition from single-use, custom-built prototypes to a generalized workflow that can automatically extract model parameters from high-throughput data.

We have developed experimental platforms to characterize cell phenotype (cell birth and death rates, motility, and other parameters) in controlled microenvironmental conditions. Using the Operetta high content screening (HCS) platform and Harmony image analysis software (PerkinElmer), we can convert cell-based images into detailed quantitative phenotypic information across different timescales, environmental contexts, and in high-throughput. Our automated image analysis protocol can rapidly generate single-cell data for millions of cells. Cells can be identified and segmented at the nuclear level to determine live and dead cell counts over time using specific nuclear and dead cell stains. Filter criteria, including nuclear size and nuclei clustering, are used to identify individual cells, and must be optimized for each cell type. The dead cell stain intensity (e.g. PI, TO-PRO-3, or DRAQ7) is subsequently calculated for each cell, and a threshold is defined to identify cells with stain intensities indicative of cell death. Thus, for any set of microenvironmental conditions, we can obtain live and dead cell counts at several time points, which can then be used to determine context-dependent birth and death rates [83, 84].

At the population level, it is important to not only characterize mean phenotype, but also phenotypic heterogeneity across the population [33]. We can address this by tracking individual cell nuclei with nuclear fusion proteins (e.g., histone-2B-GFP) across time. Readouts from these experiments include changes in total cell count and motility parameters (e.g., speed, direction, displacement). Using this high-throughput imaging platform, we can assess the impact of a heterogeneous tumor microenvironment on cellular dynamics and treatment response in real-time. Such a comprehensive view of cellular behavior under the unique control of individual and co-occurring gradients of environmental factors is a considerable improvement over current models based on qualitative *in vitro* experiments/assays.

Lastly, cell count and other high-throughput measurements must be analyzed to obtain biophysical parameters. We recently have developed CellPD (cell phenotype digitizer), which gives a user-friendly interface to input cell count data (as an Excel spreadsheet), obtains best-fit parameters and uncertainty estimates for several "canonical" mathematical forms (e.g., exponential and logistic growth), ranks the fits, and summarizes the results (parameter values and publication-quality plots) as user-friendly HMTL pages [85, 86]. CellPD will be open sourced in 2016.

12.2.6 Data Standards and Reproducibility

High-throughput screening platforms can generate many cell phenotype parameters under a variety of microenvironmental conditions, and for many cell types. This can yield a vast collection of phenotype parameters, but they cannot be used by mathematical models without systematic recording. Similarly, mathematical models output quantitative data on cell positions, phenotypes, and substrate distributions, but many papers ultimately discard these outputs in favor of simplified analytics (e.g., tumor size vs. time) and visualizations. Even when the data are stored and disseminated, each model tends to use a customized data format. This vastly complicates replication studies and new analyses of prior works. Moreover, using different formats for simulation and experimental data hinders efforts to directly compare simulation and validation datasets.

We are working to overcome these difficulties. While good standardizations exist for subcellular data (e.g., the Gene Ontology (GO) [87] is used for annotating genomics data), few exist for multicellular data. In [19], we introduced MultiCellXML (multicellular extensible markup language) to describe the model outputs of our agent-based model. The key data elements for MultiCellXML described biophysical cell agent parameters that are common to many discrete models: cell position, volume, phenotypic state (e.g., cycling, apoptotic, or quiescent), and elapsed time in the state. Indeed, many of these key data elements were incorporated into the Cell Behavior Ontology (e.g., the Boolean data element IsApoptotic) [88]. In 2014, we expanded this effort to form the MultiCellDS (multicellular data standard) Project. After assembling a multidisciplinary panel of biologists, clinicians, mathematicians, and computer scientists, we set out to form a data standard that was complementary to most ontology efforts: a method to systematically record microenvironment-dependent phenotype data (digital cell lines), and a method to consistently report continuum and discrete simulation data (digital snapshots). To accelerate the project, we are incorporating data elements from existing ontologies (e.g., Cell Behavior Ontology (CBO) [88] and Chemical Entities of Biological Interest (ChEBI) [89]) when they are available. Much of this work focuses on giving a logical, hierarchical structure to the diverse set of phenotype and biochemical descriptors in use today. By focusing on data interchange, we hope to improve the cross-model compatibility, encourage data sharing, and ease the creation of configuration, analysis, and visualization software. We note that the same standard can be used for segmented experimental and clinical data. Visit http://MultiCellDS.org for up-to-date project information.

A standardized, model-independent recording of simulation output data is key to reproducibility and open science. To reproduce a modeling result, both the same computational model and independent models or implementations should simulate the same physical system, and their outputs should be directly compared, either voxel-by-voxel (for continuum models) or on a statistical basis (for stochastic, agent-based models). However, even if model outputs are made openly available as part of a publication (i.e., as open data), this step is either complicated or impossible if the simulation inputs are model-dependent. Hence, today it is difficult to use a lattice-free model (e.g., Chaste [27]) to reproduce a cellular Potts result (e.g., from CompuCell3D [28]).

Lastly, we note that for work to be truly reproducible and open, the underlying computational code should be distributed as open source. Otherwise, version-dependent bugs are difficult to eliminate when replicating simulation results. Moreover, non-open licensing can prevent scientists from fully stating their method. For example, BioCellion [90] can simulate millions to billions of cells on supercomputers, but its non-OSI [91] academic license [92] is very restrictive: it stipulates internal non-commercial use only, with no distribution of source code (e.g., as a method section) or sublicensing without written permission; this hampers reproducibility [93].

12.3 Next Steps and Closing Thoughts

We have seen great strides in building simulation platforms to understand 3-D multicellular systems in complex, dynamical microenvironments. Tumor growth models can simulate millions of cells with individual cell effects, or large masses of mixed cell populations. Simulations can include detailed tissue biomechanics, coupled to the fluid mechanics of interstitial and microvascular flow. Simulated tumors can alter the mechanical and vascular landscape, with feedback loops affecting tumor cell phenotype. We have seen that models can be calibrated to experimental and clinical data to give meaningful insights, which can be shared and replicated with open data and open source codes. We have seen an emerging consensus on the need to document and share data and models, and new ontologies and data standards are emerging to accomplish the task.

But key ingredients are missing. The work we described focused on the biophysics of the tumor and the microenvironment; however, molecular-scale effects (cell signaling, mutation networks, metabolic/energy models, etc.) still need to be integrated. This can be achieved by integrating fast ordinary differential equation solvers for systems biology, such as libRoadrunner [94]. For better efficiency, tumor growth models should combine both discrete cell models and continuum models, with mass and force exchange between the discrete and continuum cells [25].

Perhaps more notably, the models presented in this paper (and most in the field) only couple two or three key components, often by manually combining the codes. No one simulator brings to bear all these aspects in a single platform. It is impractical to expect any single simulator to model all biochemical and biophysical aspects of cancer and the microenvironment well, from subcellular effects to tissue-scale dynamics and coupled vascular networks. Even if there were such a monolithic platform, it would be scientifically risky: a single bug could undermine scores of papers built upon it. And investment (and "sunk costs") in a single simulator may well discourage development of "competing" simulation engines for independent replication studies.

Instead, we need modular software infrastructures for combining open source models. Multiple groups could contribute tumor growth, vascularization, diffusion, and other modules, which read and modify shared data structures through standardized protocols (similarly to message passing in MPI, or TCIP/IP packets in networks). To encourage the widest possible participation, such a platform needs to support many programming languages (e.g., C++, Python, Java, Julia), rather than require compiler-level binary linking. With standardized data structures (e.g., MultiCellDS) now emerging, such a development effort is feasible.

The community will require better shared data resources. As sophisticated multiphysics models emerge, we will need correspondingly sophisticated validation datasets, including information on cell phenotype and distribution, substrate distribution, tissue mechanics, and interstitial flow, with sufficient time resolution to validate model dynamics as well as steady-state behavior. We will need further advances in novel bioengineered and biomimetic models (such as organoids grown in bioreactors and "organ-on-a-chip" systems [95]) that can drive development and validation of tumor-vasculature interaction models. These data sets must be released openly with standardized formats, so that all modelers may test their models and contribute to the community. As open data become more widespread, centralized, searchable repositories will be needed to help make data discoverable. To ensure research quality, these repositories will need to be curated, based upon (as yet undetermined) community standards for assessing quality and deciding when a newly-submitted measurement should replace an existing measurement.

No group can do this alone. Individual models of tumor growth and the microenvironment are growing to such sophistication that no one group could hope to develop a model of everything. Experiments, too, are requiring large efforts that are better realized through teams. We expect that in the future, computational systems biology will make the leap from isolated, single-lab efforts to coalitions of scientists working with open source codes and open, standardized data, allowing us to take the best of each and grow beyond the sum of the parts.

Acknowledgements This research was supported by University of Southern California (USC) Center for Applied Molecular Medicine (CAMM), the Breast Cancer Research Foundation, the NIH (5U54CA143907, 1R01CA180149), and the USC James H. Zumberge Research and Innovation Fund. We thank Nathan Choi for his 3-D hanging spheroid work in Fig. 12.1.

We thank Alexander Anderson (Moffitt Cancer Center), Mark Chaplain (University of St. Andrews), Vittorio Cristini (University of Texas Health Science Center-Houston), Jasmine Foo (University of Minnesota-Twin Cities), John Lowengrub (University of California-Irvine),

Steve McDougall (Heriot-Watt University), Greg Reese (Miami University), Shay Soker (Wake Forest University), and Steven Wise (University of Tennessee-Knoxville), for past and present collaborations. This work would not be where it is today without such valuable collaborations.

References

1. Deisboeck TS, Wang Z, Macklin P, Cristini V (2011) Multiscale cancer modeling. Annu Rev Biomed Eng 13(1):127–155. doi:10.1146/annurev-bioeng-071910-124729
2. Macklin P (2010) Biological background. In: Cristini V, Lowengrub JS (eds) Multiscale modeling of cancer: an integrated experimental and mathematical modeling approach. Cambridge University Press, Cambridge, pp 8–23
3. Weinberg RA (2013) Biology of Cancer, 2nd edn. Garland Science, New York
4. Weis SM, Cheresh DA (2011) Tumor angiogenesis: molecular pathways and therapeutic targets. Nat Med 17(11):1359–1370. doi:10.1038/nm.2537
5. Macklin P, Mumenthaler S, Lowengrub J (2013) Modeling multiscale necrotic and calcified tissue biomechanics in cancer patients: application to ductal carcinoma in situ (DCIS). In: Gefen A (ed) Multiscale computer modeling in biomechanics and biomedical engineering, vol 14, Studies in mechanobiology, tissue engineering and biomaterials. Springer, Berlin/Heidelberg, pp 349–380. doi:10.1007/8415_2012_150
6. Butcher DT, Alliston T, Weaver VM (2009) A tense situation: forcing tumour progression. Nat Rev Cancer 9(2):108–122. doi:10.1038/nrc2544
7. Nagelkerke A, Bussink J, Rowan AE, Span PN (2015) The mechanical microenvironment in cancer: how physics affects tumours. Semin Cancer Biol 35:62–70. doi:10.1016/j.semcancer.2015.09.001
8. Fraley SI, Feng Y, Krishnamurthy R, Kim DH, Celedon A, Longmore GD, Wirtz D (2010) A distinctive role for focal adhesion proteins in three-dimensional cell motility. Nat Cell Biol 12(6):598–604. doi:10.1038/ncb2062
9. Huang S, Ingber DE (2005) Cell tension, matrix mechanics, and cancer development. Cancer Cell 8(3):175–176. doi:10.1016/j.ccr.2005.08.009
10. Frieboes HB, Lowengrub JS, Wise S, Zheng X, Macklin P, Bearer EL, Cristini V (2007) Computer simulation of glioma growth and morphology. NeuroImage 37(Suppl 1):S59–S70. doi:10.1016/j.neuroimage.2007.03.008
11. Macklin P, Lowengrub J (2005) Evolving interfaces via gradients of geometry-dependent interior Poisson problems: application to tumor growth. J Comput Phys 203(1):191–220. doi:10.1016/j.jcp.2004.08.010
12. Macklin P, Lowengrub J (2006) An improved geometry-aware curvature discretization for level set methods: application to tumor growth. J Comput Phys 215(2):392–401. doi:10.1016/j.jcp.2005.11.016
13. Macklin P, Lowengrub J (2007) Nonlinear simulation of the effect of microenvironment on tumor growth. J Theor Biol 245(4):677–704. doi:10.1016/j.jtbi.2006.12.004
14. Macklin P, Lowengrub JS (2008) A New ghost cell/level set method for moving boundary problems: application to tumor growth. J Sci Comput 35(2-3):266–299. doi:10.1007/s10915-008-9190-z
15. Cristini V, Lowengrub J, Nie Q (2003) Nonlinear simulation of tumor growth. J Math Biol 46(3):191–224. doi:10.1007/s00258-002-0174-6
16. Osher S, Fedkiw RP (2003) Level set methods and dynamic implicit surfaces. Applied mathematical sciences, vol 153. Springer, New York
17. Sethian JA (1999) Level set methods and fast marching methods : evolving interfaces in computational geometry, fluid mechanics, computer vision, and materials science. Cambridge monographs on applied and computational mathematics, vol 3, 2nd edn. Cambridge University Press, Cambridge/New York
18. Bearer EL, Lowengrub JS, Frieboes HB, Chuang YL, Jin F, Wise SM, Ferrari M, Agus DB, Cristini V (2009) Multiparameter computational modeling of tumor invasion. Cancer Res 69(10):4493–4501. doi:10.1158/0008-5472.CAN-08-3834
19. Macklin P, Edgerton ME, Thompson AM, Cristini V (2012) Patient-calibrated agent-based modelling of ductal carcinoma in situ (DCIS): from microscopic measurements to macroscopic predictions of clinical progression. J Theor Biol 301:122–140. doi:10.1016/j.jtbi.2012.02.002
20. Ghaffarizadeh A, Friedman SH, Macklin P (2015) Agent-based simulation of large tumors in 3-D microenvironments. bioRxiv. doi:10.1101/035733
21. Frieboes HB, Jin F, Chuang YL, Wise SM, Lowengrub JS, Cristini V (2010) Three-dimensional multispecies nonlinear tumor growth-II: tumor invasion and angiogenesis. J Theor Biol 264(4):1254–1278. doi:10.1016/j.jtbi.2010.02.036
22. Wise SM, Lowengrub JS, Frieboes HB, Cristini V (2008) Three-dimensional multispecies nonlinear tumor growth – I Model and numerical method. J Theor Biol 253(3):524–543. doi:10.1016/j.jtbi.2008.03.027
23. Wise SM, Lowengrub JS, Cristini V (2011) An adaptive multigrid algorithm for simulating solid tumor growth using mixture models. Math Comput Model 53(1-2):1–20. doi:10.1016/j.mcm.2010.07.007
24. Frieboes HB, Smith BR, Chuang YL, Ito K, Roettgers AM, Gambhir SS, Cristini V (2013) An integrated computational/experimental model of lymphoma growth. PLoS Comput Biol 9(3), e1003008. doi:10.1371/journal.pcbi.1003008
25. Lowengrub JS, Frieboes HB, Jin F, Chuang YL, Li X, Macklin P, Wise SM, Cristini V (2010) Nonlinear modelling of cancer: bridging the gap between cells and tumours. Nonlinearity 23(1):R1–R91. doi:10.1088/0951-7715/23/1/R01

26. Macklin P (2010) Ductal carcinoma in situ (DCIS) simulation – extended results. https://www.youtube.com/watch?v=b_GVnZWVhgk
27. Mirams GR, Arthurs CJ, Bernabeu MO, Bordas R, Cooper J, Corrias A, Davit Y, Dunn SJ, Fletcher AG, Harvey DG, Marsh ME, Osborne JM, Pathmanathan P, Pitt-Francis J, Southern J, Zemzemi N, Gavaghan DJ (2013) Chaste: an open source C plus plus library for computational physiology and biology. PLoS Comput Biol 9(3). doi:ARTN e1002970 10.1371/journal.pcbi.1002970
28. Swat MH, Thomas GL, Belmonte JM, Shirinifard A, Hmeljak D, Glazier JA (2012) Multi-scale modeling of tissues using CompuCell3D. Methods Cell Biol 110:325–366. doi:10.1016/B978-0-12-388403-9.00013-8
29. Starruss J, de Back W, Brusch L, Deutsch A (2014) Morpheus: a user-friendly modeling environment for multiscale and multicellular systems biology. Bioinformatics 30(9):1331–1332. doi:10.1093/bioinformatics/btt772
30. Lardon LA, Merkey BV, Martins S, Dotsch A, Picioreanu C, Kreft JU, Smets BF (2011) iDynoMiCS: next-generation individual-based modelling of biofilms. Environ Microbiol 13(9):2416–2434. doi:10.1111/j.1462-2920.2011.02414.x
31. Macklin P, McDougall S, Anderson ARA, Chaplain MAJ, Cristini V, Lowengrub J (2009) Multiscale modelling and nonlinear simulation of vascular tumour growth. J Math Biol 58(4-5):765–798. doi:10.1007/s00285-008-0216-9
32. Anderson AR (2005) A hybrid mathematical model of solid tumour invasion: the importance of cell adhesion. Math Med Biol: J IMA 22(2):163–186. doi:10.1093/imammb/dqi005
33. Robertson-Tessi M, Gillies RJ, Gatenby RA, Anderson AR (2015) Impact of metabolic heterogeneity on tumor growth, invasion, and treatment outcomes. Cancer Res 75(8):1567–1579. doi:10.1158/0008-5472.CAN-14-1428
34. Ghaffarizadeh A, Friedman SH, Macklin P (2015) BioFVM: an efficient, parallelized diffusive transport solver for 3-D biological simulations. Bioinformatics. doi:10.1093/bioinformatics/btv730
35. Nishii K, Reese G, Moran EC, Sparks JL (2016) Multiscale computational model of fluid flow and matrix deformation in decellularized liver. J Mech Behav Biomed Mater 57:201–214. doi:10.1016/j.jmbbm.2015.11.033
36. D'Antonio G, Macklin P, Preziosi L (2013) An agent-based model for elasto-plastic mechanical interactions between cells, basement membrane and extracellular matrix. Math Biosci Eng: MBE 10(1):75–101. doi:10.3934/mbe.2013.10.75
37. Anderson AR, Weaver AM, Cummings PT, Quaranta V (2006) Tumor morphology and phenotypic evolution driven by selective pressure from the microenvironment. Cell 127(5):905–915. doi:10.1016/j.cell.2006.09.042
38. Macklin P (2010) Tumour growth in a complex tissue structure (brain tumor). https://www.youtube.com/watch?v=cxsAgP_l1uw
39. Mumenthaler SM, D'Antonio G, Preziosi L, Macklin P (2013) The need for integrative computational oncology: an illustrated example through MMP-mediated tissue degradation. Front Oncol 3:194. doi:10.3389/fonc.2013.00194
40. Raman F, Scribner E, Saut O, Wenger C, Colin T, Fathallah-Shaykh HM (2016) Computational trials: unraveling motility phenotypes, progression patterns, and treatment options for glioblastoma multiforme. PLoS One 11(1), e0146617. doi:10.1371/journal.pone.0146617
41. Karagiannis GS, Poutahidis T, Erdman SE, Kirsch R, Riddell RH, Diamandis EP (2012) Cancer-associated fibroblasts drive the progression of metastasis through both paracrine and mechanical pressure on cancer tissue. Mol Cancer Res: MCR 10(11):1403–1418. doi:10.1158/1541-7786.MCR-12-0307
42. Peckham MJ (1973) Quantitative cytology and cytochemistry of Hodgkin's tissue labelled in vivo with tritiated thymidine. Br J Cancer 28(4):332–339
43. Li X, Lowengrub J, Ratz A, Voigt A (2009) Solving PDEs in complex geometries: a diffuse domain approach. Commun Math Sci 7(1): 81–107
44. Ratz A, Voigt A (2005) Edge diffusion in phase-field models for epitaxial growth. Int Ser Numer Math 149:115–125
45. Filly R, Bland N, Castellino RA (1976) Radiographic distribution of intrathoracic disease in previously untreated patients with Hodgkin's disease and non-Hodgkin's lymphoma. Radiology 120(2):277–281. doi:10.1148/120.2.277
46. Swartz MA (2001) The physiology of the lymphatic system. Adv Drug Deliv Rev 50(1-2):3–20
47. Raghunathan S, Evans D, Sparks JL (2010) Poroviscoelastic modeling of liver biomechanical response in unconfined compression. Ann Biomed Eng 38(5):1789–1800. doi:10.1007/s10439-010-9957-x
48. Simon BR (1992) Multiphase poroelastic finite element models for soft tissue structures. Appl Mech Rev 45(6):191–218. doi:10.1115/1.3121397
49. Whittaker RJ, Booth R, Dyson R, Bailey C, Chini LP, Naire S, Payvandi S, Rong ZM, Woollard H, Cummings LJ, Waters SL, Mawasse L, Chaudhuri JB, Ellis MJ, Michael V, Kuiper NJ, Cartmell S (2009) Mathematical modelling of fibre-enhanced perfusion inside a tissue-engineering bioreactor. J Theor Biol 256(4):533–546. doi:10.1016/j.jtbi.2008.10.013
50. Swartz MA, Fleury ME (2007) Interstitial flow and its effects in soft tissues. Annu Rev Biomed Eng 9:229–256. doi:10.1146/annure/bioeng.9.060906.151850
51. Mak AF (1986) The apparent viscoelastic behavior of articular-cartilage – the contributions from the intrinsic matrix viscoelasticity and interstitial fluid-flows. J Biomech Eng-T ASME 108(2): 123–130

52. Cheng S, Bilston LE (2007) Unconfined compression of white matter. J Biomech 40(1):117–124. doi:10.1016/j.jbiomech.2005.11.004
53. Evans DW, Moran EC, Baptista PM, Soker S, Sparks JL (2013) Scale-dependent mechanical properties of native and decellularized liver tissue. Biomech Model Mechan 12(3):569–580. doi:10.1007/s10237-012-0426-3
54. Moran EC, Raghunathan S, Evans DW, Vavalle NA, Sparks JL, LeRoith T, Smith TL (2012) Porohyperviscoelastic model simultaneously predicts parenchymal fluid pressure and reaction force in perfused liver. J Biomech Eng 134(9):091002. doi:10.1115/1.4007175
55. Preziosi L, Ambrosi D, Verdier C (2010) An elasto-visco-plastic model of cell aggregates. J Theor Biol 262(1):35–47. doi:10.1016/j.jtbi.2009.08.023
56. Bani Baker Q (2015) Computational modeling to study disease development: applications to breast cancer and an *in vitro* model of macular degeneration. All graduate theses and dissertations, Paper 4409. http://digitalcommons.usu.edu/etd/4409
57. Elbjeirami WM, Yonter EO, Starcher BC, West JL (2003) Enhancing mechanical properties of tissue-engineered constructs via lysyl oxidase crosslinking activity. J Biome Mater Res Part A 66(3):513–521. doi:10.1002/jbm.a.10021
58. Anderson ARA, Chaplain MAJ (1998) Continuous and discrete mathematical models of tumor-induced angiogenesis. Bull Math Biol 60(5):857–899. doi:10.1006/bulm.1998.0042
59. McDougall SR, Anderson AR, Chaplain MA (2006) Mathematical modelling of dynamic adaptive tumour-induced angiogenesis: clinical implications and therapeutic targeting strategies. J Theor Biol 241(3):564–589. doi:10.1016/j.jtbi.2005.12.022
60. McDougall SR, Anderson ARA, Chaplain MAJ, Sherratt JA (2002) Mathematical modelling of flow through vascular networks: implications for tumour-induced angiogenesis and chemotherapy strategies. B Math Biol 64(4):673–702. doi:10.1006/bulm.2002.0293
61. Macklin P (2010) Simulation of tumour growth and angiogenesis. https://www.youtube.com/watch?v=hkZJt27AgUY
62. Wu M, Frieboes HB, McDougall SR, Chaplain MAJ, Cristini V, Lowengrub J (2013) The effect of interstitial pressure on tumor growth: coupling with the blood and lymphatic vascular systems. J Theor Biol 320:131–151. doi:10.1016/j.jtbi.2012.11.031
63. Plank MJ, Sleeman BD (2003) A reinforced random walk model of tumour angiogenesis and anti-angiogenic strategies. Math Med Biol J IMA 20(2):135–181. doi:10.1093/imammb/20.2.135
64. Plank MJ, Sleeman BD (2004) Lattice and non-lattice models of tumour angiogenesis. Bull Math Biol 66(6):1785–1819. doi:10.1016/j.bulm.2004.04.001
65. Takano S, Yoshii Y, Kondo S, Suzuki H, Maruno T, Shirai S, Nose T (1996) Concentration of vascular endothelial growth factor in the serum and tumor tissue of brain tumor patients. Cancer Res 56(9):2185–2190
66. Jain RK (2003) Molecular regulation of vessel maturation. Nat Med 9(6):685–693. doi:10.1038/Nm0603-685
67. Augustin HG (2001) Tubes, branches, and pillars – the many ways of forming a new vasculature. Circ Res 89(8):645–647
68. Holash J, Maisonpierre PC, Compton D, Boland P, Alexander CR, Zagzag D, Yancopoulos GD, Wiegand SJ (1999) Vessel cooption, regression, and growth in tumors mediated by angiopoietins and VEGF. Science 284(5422):1994–1998. doi:10.1126/science.284.5422.1994
69. Moran EC, Baptista PM, Evans DW, Soker S, Sparks JL (2012) Evaluation of parenchymal fluid pressure in native and decellularized liver tissue. Biomed Sci Instrum 48:303–309
70. Aird WC (2007) Phenotypic heterogeneity of the endothelium: II. Representative vascular beds. Circ Res 100(2):174–190. doi:10.1161/01.RES.0000255690.03436.ae
71. Koo A, Liang IY (1979) Microvascular filling pattern in rat liver sinusoids during vagal stimulation. J Physiol 295:191–199
72. Müller M, Keimling R, Lang S, Pauli J, Dahmen U, Dirsch O (2009) Estimating blood flow velocity in liver vessels. In: Meinzer H-P, Deserno TM, Handels H, Tolxdorff T (eds) Bildverarbeitung für die Medizin 2009: Algorithmen—Systeme—Anwendungen Proceedings des Workshops vom 22. bis 25. März 2009 in Heidelberg. Springer Berlin Heidelberg, Berlin, Heidelberg, pp 36–40. doi:10.1007/978-3-540-93860-6_8
73. Oda M, Yokomori H, Han JY (2003) Regulatory mechanisms of hepatic microcirculation. Clin Hemorheol Microcirc 29(3-4):167–182
74. Grisham JW (2009) Organizational principles of the liver. In: The liver. John Wiley & Sons, Ltd, pp 1–15. doi:10.1002/9780470747919.ch1
75. Frieboes HB, Zheng X, Sun CH, Tromberg B, Gatenby R, Cristini V (2006) An integrated computational/experimental model of tumor invasion. Cancer Res 66(3):1597–1604. doi:10.1158/0008-5472.CAN 05 3166
76. Gerdes J, Lemke H, Baisch H, Wacker HH, Schwab U, Stein H (1984) Cell cycle analysis of a cell proliferation-associated human nuclear antigen defined by the monoclonal antibody Ki-67. J Immunol 133(4):1710–1715
77. Duan WR, Garner DS, Williams SD, Funckes-Shippy CL, Spath IS, Blomme EA (2003) Comparison of immunohistochemistry for activated caspase-3 and cleaved cytokeratin 18 with the TUNEL method for quantification of apoptosis in histological sections of PC-3 subcutaneous xenografts. J Pathol 199(2):221–228. doi:10.1002/path.1289
78. Hyun AZ, Macklin P (2013) Improved patient-specific calibration for agent-based cancer modeling. J Theor Biol 317:422–424. doi:10.1016/j.jtbi.2012.10.017

79. Carlson KL, Helvie MA, Roubidoux MA, Kleer CG, Oberman HA, Wilson TE, Pollak EW, Rochester AB (1999) Relationship between mammographic screening intervals and size and histology of ductal carcinoma in situ. Am J Roentgenol 172(2):313–317
80. Thomson JZ, Evans AJ, Pinder SE, Burrell HC, Wilson ARM, Ellis IO (2001) Growth pattern of ductal carcinoma in situ (DCIS): a retrospective analysis based on mammographic findings. Br J Cancer 85(2):225–227. doi:10.1054/bjoc.2001.1877
81. De Roos MA, Pijnappel RM, Post WJ, De Vries J, Baas PC, Groote LD (2004) Correlation between imaging and pathology in ductal carcinoma in situ of the breast. World J Surg Oncol 2:4. doi:10.1186/1477-7819-2-4
82. Edgerton ME, Chuang YL, Macklin P, Yang W, Bearer EL, Cristini V (2011) A novel, patient-specific mathematical pathology approach for assessment of surgical volume: application to ductal carcinoma in situ of the breast. Anal Cell Pathol 34(5):247–263. doi:10.3233/Acp-2011-0019
83. Mumenthaler SM, Foo J, Choi NC, Heise N, Leder K, Agus DB, Pao W, Michor F, Mallick P (2015) The impact of microenvironmental heterogeneity on the evolution of drug resistance in cancer cells. Cancer Informat 14(Suppl 4):19–31. doi:10.4137/CIN.S19338
84. Mumenthaler SM, Foo J, Leder K, Choi NC, Agus DB, Pao W, Mallick P, Michor F (2011) Evolutionary modeling of combination treatment strategies to overcome resistance to tyrosine kinase inhibitors in non-small cell lung cancer. Mol Pharm 8(6):2069–2079. doi:10.1021/mp200270v
85. Juarez EF, Lau R, Friedman SH, Ghaffarizadeh A, Jonckheere E, Agus DB, Mumenthaler SM, Macklin P (2016 (in review)) Quantifying Differences in Cell Line Population Dynamics Using CellPD
86. Juarez Rosales EF, Ghaffarizadeh A, Friedman SH, Jonckheere E, Macklin P (2015) Estimating cell cycle model parameters using systems identification. bioRxiv. doi:10.1101/035766
87. Gene Ontology Consortium: going forward (2015) Nucleic acids research 43 (Database issue):D1049–1056. doi:10.1093/nar/gku1179
88. Sluka JP, Shirinifard A, Swat M, Cosmanescu A, Heiland RW, Glazier JA (2014) The cell behavior ontology: describing the intrinsic biological behaviors of real and model cells seen as active agents. Bioinformatics 30(16):2367–2374. doi:10.1093/bioinformatics/btu210
89. Degtyarenko K, de Matos P, Ennis M, Hastings J, Zbinden M, McNaught A, Alcantara R, Darsow M, Guedj M, Ashburner M (2008) ChEBI: a database and ontology for chemical entities of biological interest. Nucleic Acids Res 36(Database issue):D344–D350. doi:10.1093/nar/gkm791
90. Kang S, Kahan S, McDermott J, Flann N, Shmulevich I (2014) Biocellion: accelerating computer simulation of multicellular biological system models. Bioinformatics 30(21):3101–3108. doi:10.1093/bioinformatics/btu498
91. OSI (2008–present) OSI-Approved Open Source Licenses (Listed by name). Open Source Initiative. https://opensource.org/licenses/alphabetical
92. Biocellion (2014–present) Biocellion 1.1 academic individual end user license agreement. http://biocellion.com/download.1.1/biocellion1.1.academic.eula.html
93. Prlic A, Lapp H (2012) The PLOS computational biology software section. PLoS Comput Biol 8(11). doi:ARTN e1002799 10.1371/journal.pcbi.1002799
94. Somogyi ET, Bouteiller JM, Glazier JA, Konig M, Medley JK, Swat MH, Sauro HM (2015) libRoadRunner: a high performance SBML simulation and analysis library. Bioinformatics 31(20):3315–3321. doi:10.1093/bioinformatics/btv363
95. Villasante A, Vunjak-Novakovic G (2015) Tissue-engineered models of human tumors for cancer research. Expert Opin Drug Discovery 10(3):257–268. doi:10.1517/17460441.2015.1009442

Index

K.A. Rejniak (eds.), *Systems Biology of Tumor Microenvironment*, Advances in Experimental Medicine and Biology 936, DOI 10.1007/978-3-319-42023-3

MIX
Papier aus verantwortungsvollen Quellen
Paper from responsible sources
FSC® C105338

If you have any concerns about our products,
you can contact us on
ProductSafety@springernature.com

In case Publisher is established outside the EU,
the EU authorized representative is:
Springer Nature Customer Service Center GmbH
Europaplatz 3, 69115 Heidelberg, Germany

Printed by Libri Plureos GmbH
in Hamburg, Germany